Trace Elements
in
Soils and Plants

Authors

Alina Kabata-Pendias, Ph.D., D.Sc.
Professor
Trace Elements Laboratory
Institute of Soil Science
and Cultivation of Plants
Pulawy, Poland
and
Department of Petrography
and Mineralogy
Geological Institute
Warsaw, Poland

Henryk Pendias, Ph.D.
Associate Professor
Department of Petrography
and Mineralogy
Geological Institute
Warsaw, Poland

CRC Press, Inc.
Boca Raton, Florida

Library of Congress Cataloging in Publication Data

Kabata-Pendias, Alina.
　　Trace elements in soils and plants.

　　Bibliography: p.
　　Includes index.
　　1. Soils—Trace element content. 2. Plants,
Effect of trace elements on. 3. Plants—Chemical
analysis. 4. Trace elements. 5. Plant-soil relation-
ships. I. Pendias, Henryk. II. Title.
S592.6.T7K3 1984　　　631.4′1　　　83-15083
ISBN 0-8493-6639-9

　　Direct all inquiries to CRC Press, Inc., 2000 Corporate Blvd., N.W., Boca Raton, Florida, 33431.

© 1984 by CRC Press, Inc.
Second Printing, 1985
Third Printing, 1985
Fourth Printing, 1986

International Standard Book Number 0-8493-6639-9

Library of Congress Card Number 83-15083
Printed in the United States

FOREWORD

Investigations of the chemical properties and environmental interactions of trace elements have increased in number and scope at an astonishing rate in the past 2 decades. This increase was fostered by the almost simultaneous convergence of factors whose importance could not be ignored by scientific establishments, governments, and the population at large. The rapid industrialization of preceding decades, which led to population shifts from rural locations to urban centers, continued unabated worldwide. The concentrations of population led to unnatural concentrations of metals and other trace elements as the energy requirements and abundance of consumer goods became highly localized. These element concentrations however, do not always remain localized, but often are redistributed, even across national or continental boundaries. In a biological world that is adapted to element concentrations that are generally related to crustal abundance, these concentrations were manifest as one aspect of environmental contamination or pollution. At the same time, studies of the essentiality of certain trace elements in plant and animal nutrition emphasized the fine distinction between adequate and toxic concentrations. The awareness of these problems and their effects on the health and well-being of organisms, including man, coincided with the development of the technical means to investigate geochemical properties and relationships more thoroughly and economically in respect to industrial emissions, agricultural practices, and general environmental conditions. This also led to the development of an extensive literature on the abundance and behavior of trace elements in the biosphere.

It is appropriate at this stage of progress in the understanding of trace elements in soils and plants to pause and see where we now stand. Trace element research has been worldwide in scope; a summary of this work must, therefore include research results from many countries. The present authors are eminently qualified for making such a synthesis because of their scientific competence in the field of geochemistry and their familiarity with worldwide literature on the subject, especially the literature not readily available to many English-speaking investigators.

The trace element data presented in this book follow the sequence of chemical periodicity. These data cover the general geochemical and biochemical properties of trace elements and provide new insights on chemical relationships which influence their environmental behavior.

An understanding of what has been accomplished in geochemical research should indicate the direction for future investigations of the intricate relationships of trace elements in soil and plants, and their effects on specific ecosystems and on the biosphere as a whole.

Hansford T. Shacklette
U.S. Geological Survey
Denver, Colorado

PREFACE

This book aims to review the current knowledge of some geochemical and biochemical processes which directly or indirectly influence the distribution of trace elements in soils and plants, but does not attempt to give an encyclopedic coverage of the subject. Therefore, a number of original papers and textbooks have not been cited although they may be as substantial as those that are listed. A real effort has been made, however, to gather information from different countries, and to relate the available data to the complex interrelationships that govern the natural and man-influenced abundance of the trace elements in soils and plants.

Soils and plants are significant components of the biological environment that may be broadly defined as the enormously complex and constantly changing part of the biosphere. The abundance of trace elements in soils governs the appropriate supply of these elements to living organisms and therefore is of great concern in environmental and health studies.

The term *trace elements* is not precisely defined since it refers to those elements whose terrestrial abundance is low (most often below 0.1%) as well as to those which occur in living matter only in trace amounts. Trace elements are also called *microelements* and, when they are essential to vital processes, *micronutrients*.

The first five chapters of this book are devoted to general processes in soils and plants that govern natural cycling of trace elements in the soil-plant system. The following eight chapters deal with the occurrence, forms, and fates of the trace elements in rocks, soils, and plants. The trace elements are discussed in the sequence of chemical periodicity, which allows a clear understanding of the geochemistry and biochemistry that influence their behavioral properties.

Trace element determinations differ in precision, accuracy, and reliability, and are known to bear both systematic and nonsystematic analytical errors. Therefore it becomes difficult to answer questions of the extent to which differences in concentrations recorded for various soils and plants represent actual variations. However, most of the results collected in this book show that, on the average, the values given by different authors for various areas of the world are of reasonable quality, and may provide reliable data for trace element abundance in soils and plants.

We will be delighted if scientists, experts, and students of various disciplines such as agronomy, biology, ecology, and environmental sciences find useful information in this book.

Alina Kabata-Pendias
Henryk Pendias

THE AUTHORS

Alina Kabata-Pendias, Ph.D., D.Sc., Professor of Soil Chemistry is a head of the Trace Elements Laboratory of the Institute of Soil Science and Cultivation of Plants in Pulawy, Poland, where she has worked for over 30 years. She is an author of more than a hundred publications on the occurrence of trace elements in natural and contaminated soil-plant environments.

Professor Kabata-Pendias also works in the Geological Institute in Warsaw, on the mobility of trace elements, and on the alteration of minerals in weathered zones of various geological formations. She has been involved in analytical and methodical studies on trace elements and clay minerals.

She was a visiting research scientist at various scientific centers in Poland, Chechoslovakia, Great Britain, and the U.S. She was a Fellow of the Rockefeller Foundation in 1958, and joined the staff of the U.S. Soil Plant and Nutrition Laboratory in Ithaca, N.Y.

Dr. Kabata-Pendias is a member of international scientific societies and belongs to many scientific societies and committees in Poland. She is involved in counsel activities of several societies.

Henryk Pendias, Ph.D., Associate Professor, is a head of the Department of Petrography and Mineralogy of the Geological Institute in Warsaw, Poland. He has been a member of the staff of this institute since 1950. His geochemical studies have resulted in about 50 publications.

The chemistry of various geological formations, the distribution of trace elements in rocks and minerals, and the impact of biogeochemical processes on the mobility of elements in geological environments are the main subjects of his research. He initiated research on the distribution of trace elements in coals of Lower Silesia. He has also devoted much time and attention to methodical and analytical studies in geochemistry.

Dr. Pendias has attended many national and international scientific meetings. He is a founding member of the Mineralogical Society of Poland, and a chairman of its Warsaw branch. For several years he has been a member of the Geological Society of Poland.

LIST OF UNITS, SYMBOLS, AND ABBREVIATIONS

Basic units of the International System (SI) have been guidelines for the units used in this book. Some notable exceptions have been made in order to maintain the necessary links with practical usage.

Unless otherwise specified, the concentration of a trace element in soils and plants is based on the total content by weight of the element in air-dried or oven-dried material.

Units

kg	— kilogram; $kg = 10^3$ g
g	— gram $= 10^{-3}$ kg
mg	— milligram; $mg = 10^{-3}$ g
μg	— microgram; $\mu g = 10^{-3}$ mg
ng	— nanogram; $ng = 10^{-3}$ μg
pg	— picogram; $pg = 10^{-3}$ ng
°C	— temperature in degrees Celsius
Bq	— Becquerel
Ci	— Curie
mCi	— milliCurie; μCi, microCurie; nCi, nanoCurie; pCi, picoCurie
ha	— hectare; $ha = 10,000$ meters^2
ℓ	— liter; $\ell = 1$ dm^3; $m\ell = 10^{-3}$ ℓ
$m\ell$	— milliliter; 10^{-3} ℓ
m^3	— cubic meter; $m^3 = dm^3 \times 10^3$

Symbols

ppm	— $\mu g\ g^{-1} = mg\ kg^{-1} = g\ t^{-1}$
ppb	— 10^{-3} ppm $= ng\ g^{-1} = \mu g\ kg^{-1} = mg\ t^{-1}$
M	— concentration in terms of moles per liter
mM	— millimole; μM, micromole; nM, nanomole
N	— concentration in terms of milliequivalents per liter
hr	— hour

Constants

Eh	— electrical potential (volts)
pH	— minus logarithm, base 10, of H^+ concentration
p	— minus logarithm, base 10, of an ion activity
G	— free energy (enthalpy) (kcal)
ΔG_f°	— standard free energy of a reaction (kcal mole^{-1})
K	— equilibrium constant of a reaction

Abbreviations

AW	— ash weight basis of samples
DW	— air dried or oven dried (up to 70° C) weight basis of samples
FW	— fresh or wet weight basis of samples (used mainly in relation to plant material)
EDTA	— ethylenediaminetetraacetic acid
EDDHA	— ethylenediamine -di(o-hydroxyphenylacetic acid)
DTPA	— diethylenetriaminepentaacetic acid
DNA	— deoxyribonucleic acid
RNA	— ribonucleic acid
ADP	— adenosine 5′-(trihydrogen diphosphate)
ATP	— adenosine 5′-(tetrahydrogen triphosphate)

FA   — fulvic acid
HA   — humic acid
B.P.   — before the present time (used in expressing geologic time)
vs.   — versus

THE PERIODIC TABLE OF ELEMENTS

The elements are arranged in order of atomic number, within the groups and subgroups. The symbols given in thin letters indicate the elements not occurring naturally in the environment.

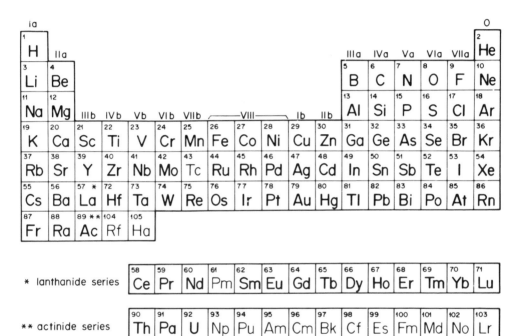

* lanthanide series

** actinide series

ACKNOWLEDGMENTS

Many people helped, supported, and encouraged the authors during the completion of this book. The help and support given the authors in many different matters was very important in enabling them to devote all their time, after fulfilling their scientific obligations, to writing this manuscript.

Many colleagues, too numerous to mention, have assisted in other ways in the preparation of this book. Our particular thanks must go to Dr. A. Čumakov (Czechoslovakia), Dr. B. E. Davies (U.K.), Dr. J. N. Firth (U.K.), Prof. Dr. E. Schlichting (W. Germany), and Dr. K. G. Tiller (Australia). Great encouragement and unstinted help have also been received from Dr. A. W. Taylor (U.S.).

Dr. H. T. Shacklette (U.S.) helped more than can be said in brief by editing our poor English text, typing the manuscript, and acting as the best helpmate, without whom the completion of the book during this period would have been impossible.

The truly devoted assistance of our colleagues of the Institute of Soil Science and Cultivation of Plants in Pulawy, in particular of the Trace Elements Laboratory, and of the Geological Institute in Warsaw was also indispensable.

Our acknowledgments would not be complete without mention of our mother, Helena Kabata, who was always ready to help, carrying much of the home burden.

The Authors

TABLE OF CONTENTS

Chapter 1

THE BIOSPHERE

The biosphere, also called the ecosphere, is the natural environment of living things and is the complex biological epidermis of the earth whose dimensions are not precisely defined. It consists of the surficial part of the lithosphere, a lower part of the atmosphere, and the hydrosphere. Chemical and physical characteristics of the biosphere are determined by these other spheres, which have created relatively constant environments for the existence of living matter in a given ecosystem. The ecosystem is a fundamental environmental system consisting of the community of all living organisms in a given area and having a balanced cycling of elements and energy flow.

In general, the biosphere consists of three main ecosystems — the land environment, the fresh-water environment, and the marine environment. These fundamental ecosystems include several smaller systems of variable dimensions and conditions. A significant part of the ecosystems has already been considerably modified by man, and such modification will continue.

The energy of life is derived from the radiant energy of the sun, which drives the chemical reaction of photosynthesis. The other sources of energy, such as geothermal energy, gravitation, and electric potentials, are of negligible importance in the total energy flow, however, they may determine some conditions of ecosystems.

Biochemical processes of producing organic matter in the earth's environment are dated from the Early Precambrian Period, i.e., about 3×10^9 years B.P. (before the present time). The entire process of photosynthesis has been calculated to have developed before about 1.5×10^9 years B.P. Since that period several million kinds of living organisms have developed evolutionarily and have adjusted to their natural environments.

Most mineral nutrients for all life on land are supplied mainly from the soil overlying the surficial lithosphere. The atmosphere is a source of only some of the essential nutrients (N_2, O_2, and CO_2), and the hydrosphere is the main source of water — a basic constituent of all life.

The bulk mass of living matter (above 90%) is composed mainly of organic compounds and water. Organo-mineral compounds and mineral compounds form a relatively small portion of living matter. The bulk of living matter is formed from the chemical elements C, O, H, and N. Such elements as K, P, Ca, Mg, S, Na, and Cl are present in living organisms in smaller and variable amounts. All these elements are readily mobile in the biosphere and are known to form either volatile or easily soluble compounds that are involved in major environmental cycles.

Many elements occur in trace amounts in living matter. Some of these elements are essential for the growth, development, and health of organisms. Usually the quantitative differences between essential amounts and biological excesses of these elements are very small. Some trace elements seem to be essential to vital processes but their biochemical functions are not yet understood. The essentiality of other trace elements may be discovered in the future.

The chemical composition of living matter has developed and adjusted to the chemistry of environments over long periods of geologic time. However, all organisms, in order to survive in the complex geochemical composition of their surroundings, had to develop mechanisms of active selection of elements involved in vital processes and of rejection of toxic excesses of other elements. These processes are fundamental for homeostasis that is requisite to the existence of each organism.

Although all living organisms, plants in particular, show a natural ability for the selection of chemical elements, they are also highly dependent on the geochemistry of their surroundings. Any environmental factor which has an adverse effect on plants may cause either evolutionary or drastic changes even over short periods of time, involving only a few generations in the life of a plant species. These phenomena have been easily observed in evolving tolerance of populations, especially of microorganisms, to high concentrations of trace elements in either natural geochemical provinces or under man-induced conditions.

Even though biological selection of the chemical elements allows plants to control their chemical composition to a certain extent, these controls are somewhat limited in respect to trace elements. Therefore, concentrations of trace elements in plants often are positively correlated with the abundance of these elements in soils and even with the abundances in underlying rocks. This correlation creates several problems for both plants and animals that may be associated either with a deficiency or an excess of trace elements. The questions surrounding how, and how much of, a trace element is taken up by plants from their environments have been very active topics of research in biological circles in recent years. At a time when food production and environmental quality are of major concern to man, a better understanding of the behavior of the trace elements in the soil-plant system seems to be particularly significant.

Chapter 2

THE ANTHROPOSPHERE

I. INTRODUCTION

The role of man in the biosphere has been so important recently that it has become necessary to distinguish the anthroposphere — the sphere of man's settlement and activity. The anthroposphere, however, does not represent a separate sphere, but can be applied to any part of the biosphere that has been changed under an influence of technical civilization.

Man's impact on the biosphere has been very broad and complex, and most often has led to irreversible changes. While geological and biological alterations of the earth's surface have been very slow, changes introduced and/or stimulated by man have accumulated extremely quickly in recent years. All man-made changes disturb the natural balance of each ecosystem which has been formed evolutionarily during a long period of time. Thus, these changes lead most often to a degradation of the natural human environment. Since the development of agricultural activities several ecosystems have been altered into artificial agroecosystems. Although man's impact on the biosphere has been dated from the Neolithic Period, the problems of the deterioration of ecosystems due to pollution have become increasingly acute during the latter decades of the 20th century.

Environmental pollution, especially by chemicals, is one of the most effective factors in the destruction of the biosphere components. Among all chemical contaminants, trace elements are believed to be of a specific ecological, biological, or health significance. Many textbooks have been published recently on trace elements as pollutants in the biosphere or in particular ecosystems, and all have pointed out the triangular relationships between contents of inorganic trace pollutants in air, soil, and plants.

Energy and mineral consumption by man is the main cause of trace element pollution in the biosphere. An estimation of the global release of trace elements as contaminants into the environment may be based on the established world mineral and energy consumption and demand (Tables 1 and 2). Bowen[94] has suggested that when the rate of mining of a given element exceeds the natural rate of its cycling by a factor of ten or more, the element must be considered a potential pollutant. Thus, the potentially most hazardous trace metals to the biosphere may be Ag, Au, Cd, Cr, Hg, Mn, Pb, Sb, Sn, Te, W, and Zn. This list does not correspond closely to the list of elements considered to be of great risk to environmental health — Be, Cd, Cr, Cu, Hg, Ni, Pb, Se, V, and Zn.

Trace elements released from anthropogenic sources have entered the environment and have followed normal biogeochemical cycles. The transport, residence time, and fate of the pollutants in the particular ecosystem have been of special environmental concern. The behavior of trace elements in each ecosystem is very complex and therefore has been usually studied separately for air, water, soil, and biota.

II. AIR POLLUTION

Most air pollution has arisen from the burning of coal and other fossil fuels and from smelting of iron and nonferrous metals. The steady global increase of trace element concentrations in the atmosphere is illustrated in Table 3. Some trace pollutants, most likely Se, Au, Pb, Sn, Cd, Br, and Te, can exceed 1000 times their normal concentration in air. In general, elements which form volatile compounds, or are present at a lower particle radius, may be readily released into the atmosphere from the burning of coal and other industrial processes. Materials released by man's activities are not the only contribution to global air

Table 1
WORLD MINING OF THE ELEMENTS
AND FORECAST FOR ELEMENT
DEMAND IN THE YEAR 2000

Element	Consumption		Forecast for the year 2000 (tonnes)
	Year	Tonnes	
Iron	1974	$507,000 \cdot 10^3$	$1,000,000 \cdot 10^3$
Aluminum	1974	$13,900 \cdot 10^3$	$60,000 \cdot 10^3$
Manganese	1974	$9,500 \cdot 10^3$	$18,000 \cdot 10^3$
Copper	1975	$7,470 \cdot 10^3$	$12,000 \cdot 10^3$
Zinc	1975	$6,190 \cdot 10^3$	$11,000 \cdot 10^3$
Lead	1975	$3,590 \cdot 10^3$	$5,000 \cdot 10^3$
Barium	1974	$3,025 \cdot 10^3$	$5,000 \cdot 10^3$
Fluorine	1975	$2,350 \cdot 10^3$	$3,500 \cdot 10^3$
Chromium	1974	$2,250 \cdot 10^3$	$3,750 \cdot 10^3$
Titanium	1974	$1,250 \cdot 10^3$	$1,800 \cdot 10^3$
Boron	1974	$1,000 \cdot 10^3$	—
Nickel	1975	$740 \cdot 10^3$	$1,500 \cdot 10^3$
Zirconium	1974	$250 \cdot 10^3$	$500 \cdot 10^3$
Tin	1975	$205 \cdot 10^3$	$300 \cdot 10^3$
Molybdenum	1974	88,940	130,000
Strontium	1973	85,280	120,000
Antimony	1975	73,400	100,000
Arsenic	1974	45,000	51,000
Tungsten	1974	36,420	47,000
Bromine	1974	22,700	35,000
Cobalt	1975	21,100	30,000
Vandium	1974	20,180	35,000
Uranium	1974	18,900	250,000
Cadmium	1975	15,550	20,000
Tantalum	1974	10,150	32,000
Silver	1973	9,430	12,000
Mercury	1974	9,240	14,000
Iodine	1974	7,000	—
Niobium	1969	5,175	—
Lithium	1974	4,430	33,000
Bismuth	1974	4,400	6,000
Beryllium	1974	3,610	15,000
Gold	1974	1,467	2,000
Selenium	1974	1,200	2,000
Thorium	1974	700[a]	—
Tellurium	1974	192	250
Germanium	1974	75	200
Indium	1975	60	—
Thallium	1974	30	—
Gallium	1974	15	—
Rhenium	1974	7.1	14
Hafnium	1974	1.3	—
Rubidium	1975	1	3
Cesium	—	—	30

[a]   Data only for capitalistic countries.

Table 2
WORLD DEMAND FOR RAW ENERGY
MATERIALS

Raw product	Mining in 1974 ($\times 10^9$)	Forecast for the year 2000 ($\times 10^9$)
Natural gas (m³)	1255.250	2500.0
Crude oil (ton)	2.792	4.5
Coal (ton)	2.227	4.5
Lignite coal (ton)	0.842	2.0

pollution; some natural sources such as eolitic dusts, volcanic eruptions, evaporation from water surfaces, and others should also be taken into account.

The atmospheric deposition of trace elements, mainly the heavy metals, contributes to contamination of all other components of the biosphere, e.g., waters, soils, and vegetation. This deposition has been widely reviewed, especially by Rühling and Tyler,[669] Folkeson,[238] and Thomas,[780] and it has been established that mosses and lichens are the most sensitive organisms to atmospheric pollution by trace metals, although sensitivity varies decidedly among species. Tops of all plants are collectors for all air pollutants, and their chemical composition may be a good indicator for contaminated areas when it is assessed against background values obtained for unpolluted vegetation (Table 4).

The environmental characteristics of inorganic trace pollutants in air are the following:

1. Wide dispersion and long-distance transport
2. Bioaccumulation, most often affecting the chemical composition of plants without causing easily visible injury
3. Reaction in living tissues by disturbing the metabolic processes and by causing the reduction of sunlight entering plant tissues
4. Resistance to metabolic detoxification, therefore entering the food chain

Indirect effects of air pollutants through the soil are of real importance because of the large-scale sustained exposure of soil to both wet and dry deposition of trace elements. These environmental effects should also be given greater attention.

III. WATER POLLUTION

Trace elements are present in natural waters (ground and surface), and their sources are associated with either natural processes or man's activities. The basic natural processes contributing trace elements to waters are chemical weathering of rocks and soil leaching. Both processes also may be largely controlled by biological and microbiological factors. The anthropogenic sources of trace elements in waters are associated mainly with mining of coal and mineral ores and with manufacturing and municipal waste waters. Water pollution by trace elements is an important factor in both geochemical cycling of these elements and in environmental health.

Most trace elements, especially heavy metals, do not exist in soluble forms for a long time in waters. They are present mainly as suspended colloids or are fixed by organic and mineral substances. Thus, their concentration in bottom sediments or in plankton is most often an adequate indication of water pollution by trace elements. As Dossis and Warren[191] concluded, sediments may be regarded as the ultimate sink for heavy metals that are discharged into the aquatic environment. On the other hand, easily volatile elements such as Br and I can reach higher concentrations in surface waters, from which they can also readily

Table 3
TRACE ELEMENTS IN AIR FROM DIFFERENT LOCATIONS

			Europe				America		
Element	South Pole	Greenland	Shetland Islands	Norway	West Germany	Japan	North	Central	South
ng m^{-3}									
Al	0.32—0.81	240—380	60	32	160—2,900	40—10,600	600—2,330	760—880	460—15,000
As	0.007	—	0.6	1.9	1.5—53	0.3—120	2—40	—	—
Be	—	—	—	—	0.9—4	5—100	0.1—0.3	—	—
Br	0.38—1.41	14—20	15	4.4	30.5—2,500	1.6—150	6—1,200	65—460	2—200
Cd	0.015	0.003—0.63	0.8	—	0.5—620	0.5—43	1—41	—	—
Cr	0.003—0.01	0.6—0.8	0.7	0.7	1—140	1.3—167	5—1,100	1—2	1—8
Cu	0.03—0.06	—	20	2.5	8—4,900	11—200	3—153	70—100	30—180
Fe	0.51—1.19	166—171	90	48	130—5,900	47—14,000	829—2,000	554—1,174	312—9,225
I	0.08	—	4	0.6	3—15	6	40—6,000	—	—
Mn	0.004—0.02	2.8—4.5	3	3	9—210	5.3—680	60—900	14—16	4—330
Mo	—	—	0.2	—	0.2—3.2	—	1—10	—	—
Ni	—	—	4	1.2	4—120	1—150	120	—	—
Pb	0.19—1.2	15—22	21	—	120—5,000	19—1,810	45—13,000	0.2—317	11—344
Sb	0.001—0.003	0.9—45	0.4	0.3	2—51	0.13—63	1—55	0.8—15	1—24
Sn	—	—	—	—	1.5—800	—	10—70	—	—
Ti	—	0.8—1.4	10	2.6	22—210	5—690	10—230	7—200	—
V	0.0006—0.002	—	3	1.9	5—92	1.5—180	4—174	—	7—91
Zn	0.002—0.051	18—41	15	10	550—16,000	14—6,800	88—741	60—182	25—1,358
pg m^{-3}									
Ce	0.8—4.9	—	100	60	360—14,000	100—18,000	20—13,000	—	—
Co	0.1—1.2	70—150	60	60	390—6,790	44—6,000	130—2,200	250—650	120—360
Cs	—	—	40	20	60—1,500	16—1,500	70—300	—	—
Eu	0.004—0.02	—	4	—	5—80	7.3—27	10—1,700	—	—

Hf	—	40—60	—	—	300	18—590	0.5—290	60—70	40—760
Hg	—	40—80	40	10	170—11,200	1,600	70—3,800	70—120	70—690
In	0.05	—	20	—	30—360	1,200	20—140	—	—
La	0.2—1.4	50—110	200	30	610—3,420	53—3,000	490—9,100	440	290—3,400
Sc	0.06—0.21	30—40	15	5	30—700	5—1,300	80—3,000	150—220	60—3,000
Se	4.2—8.2	170—360	500	260	150—11,000	160—21,000	60—30,000	280—1,210	50—1,530
Sm	0.03—0.09	10—12	10	3	240—420	9.8—320	70—1,000	30—80	30—630
Ta	—	10—30	—	—	—	6—100	50—280	20—50	20—150
Tb	—	1—5	—	—	10	—	19—34	8—27	20—120
Th	0.02—0.08	20—40	20	11	30—1,000	16—1,300	50—1,300	10—20	30—1,050
U	—	—	—	—	20[a]	—	<500	—	—

[a] Median values for Europe.[94,381,395]

**Table 4
ENRICHMENT
FACTORS (EF) FOR
TRACE POLLUTANTS
IN THE TOPS OF
VEGETATION GROWN
IN INDUSTRIAL AREAS**

Pollutant	Range of EF
Cd	4.5—450
Cu	10—68
F	7—34
Pb	2.4—134
Zn	1.4—164
^{226}Ra	1—5.5
U	1.4—2.4
Th	1—23

Note: EF = ratio against background value.

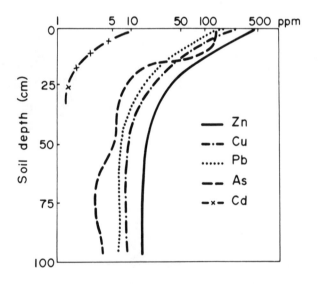

FIGURE 1.   Vertical distribution of Zn, Cu, Pb, and Cd in a profile
of the soil sludged during a period of 15 years and the distribution
of As in light soil polluted by metalliferous mine.[297,395]

vaporize under favorable climatic conditions. Microbial alkylation of the group of metals
including Hg, Se, Te, As, and Sn that occurs mainly in sediments and on suspended particles
in waters is also of great importance in their mobility.[359]

Both phytoplankton and vascular water plants are known to selectively concentrate trace
elements. As a result of this selectivity, concentrations of some trace elements in waters
may decrease in some seasons, while other elements may become soluble during the decay
of plants.

Waste water used on farmland is generally a source of several trace elements. Therefore,
the possibility of contamination should limit this method of waste water disposal because
of the accumulation of hazardous amounts of heavy metals in surface soil (Figure 1).

IV. SOIL

A. Soil Contamination

Soil is a very specific component of the biosphere because it is not only a geochemical sink for contaminants, but also acts as a natural buffer controlling the transport of chemical elements and substances to the atmosphere, hydrosphere, and biota. Trace elements originating from various sources may finally reach the surface soil, and their further fate depends on soil chemical and physical properties. Although the chemistry of soil contaminants recently has been the subject of many studies, our knowledge of the behavior of polluting trace elements is far from complete. The persistence of contaminants in soil is much longer than in other components of the biosphere, and contamination of soil, especially by heavy metals, appears to be virtually permanent. Metals accumulated in soils are depleted slowly by leaching, plant uptake, erosion, or deflation. The first half-life of heavy metals, as calculated by Iimura et al.[336] for soils in lysimetric conditions, varies greatly — for Zn, 70 to 510 years; for Cd, 13 to 1100 years; for Cu, 310 to 1500 years; and for Pb, 740 to 5900 years.

The input-output balance of metals in soils discussed in Chapter 3, Section III. B. shows that trace metals concentrations in surface soil are likely to increase, on a global scale, with growing industrial and agricultural activities. There are several indications that the composition of surface soil may be influenced by both local contamination and long-range transport of pollutants. Purves[634] concluded that the extent of soil contamination in the urban environment is now so great that it is possible to identify most soils as urban or rural on the basis of their content of a few trace metals that are known to be general urban contaminants. The annual increment of heavy metals caused by dust fallout in Tokyo is estimated for Cd to be 0.05 ppm and for Pb and Mn to be about 0.5 ppm.[395]

The regional contamination of soils, as reported most commonly, occurs mainly in industrial regions and within centers of large settlements where factories, motor vehicles, and municipal wastes are the most important sources of trace metals. However, due to the long distance aerial transport of trace pollutants, especially those which form volatile compounds (e.g., As, Se, Sb, and Hg), it has become difficult to estimate the natural background values for some trace elements in soils.

In addition to aerial sources of trace pollutants, fertilizers, pesticides, and all sewage-derived materials have added to the trace element pool in soils. The mobilization of heavy metals from smelter and mine spoil by transport with seepage waters or by windblown dust may also be an important source of soil contamination in some industrial regions. The variability of trace element concentrations in materials used in agriculture is presented in Table 5. Goodroad,[273] Piotrowska and Wiacek,[620] and Stenström and Vahter[755a] reported that long-term use of inorganic phosphate fertilizers adds substantially to the natural levels of Cd and F in soils, while other elements such as As, Cr, Pb, and V do not increase significantly. Effects of sewage sludge applications on soil composition are especially of great environmental concern and have been the subject of many studies and much legislation. Advisory standards and guidelines for safe addition of trace elements in sewage sludge to land is still in the stage of experiment and negotiation; however, several authors have given threshold values for the maximum addition of trace elements in one dose and over a period of time (Table 6). In spite of some diversity of opinion, there is general agreement, especially regarding the maximum concentrations of heavy metals in soils. Maximum allowable limits set up for paddy soils in Japan are somewhat different.[395] Cu content was established at 125 ppm (0.1 N HCl soluble) and As was established at 15 ppm (1 N HCl soluble) as critical for rice growth. The hazardous concentration in soils of Cd is limited by allowable Cd in rice, which should not exceed 1 ppm. It should be emphasized, however, that all the allowable limits need to be related not only to the given plant-soil system, but also to ratios between single elements as well as to their total burden in soil.

In several soils the threshold levels already have been exceeded either in gardens and

Table 5
AGRICULTURAL SOURCES OF TRACE ELEMENT CONTAMINATION IN SOILS (PPM DW)

Element	Sewage sludges[a]	Phosphate fertilizers[b]	Limestones[c]	Nitrogen fertilizers[d,l]	Manure[c]	Pesticides[e] (%)
As	2—26	2—1,200	0.1—24.0	2.2—120	3—25	22—60
B	15—1,000	5—115	10	—	0.3—0.6	—
Ba	150—4,000	200	120—250	—	270	—
Be	4—13	—	1	—	—	—
Br	20—165	3—5	—	185—716[f]	16—41	20—85
Cd	2—1,500	0.1—170	0.04—0.1	0.05—8.5	0.3—0.8	—
Ce	20	20	12	—	—	—
Co	2—260	1—12[g]	0.4—3.0	5.4—12	0.3—24[g]	—
Cr	20—40,600	66—245	10—15	3.2—19	5.2—55	—
Cu	50—3,300	1—300	2—125	<1—15	2—60	12—50
F	2—740	8,500—38,000[h]	300	—	7	18—45
Ge	1—10	—	0.2	—	19	—
Hg	0.1—55	0.01—1.2[d]	0.05	0.3—2.9	0.09—0.2	0.8—42
In	—	—	—	—	1.4	—
Mn	60—3,900	40—2,000	40—1,200	—	30—550	—
Mo	1—40	0.1—60	0.1—15	1—7	0.05—3	—
Ni	16—5,300	7—38[d]	10—20	7—34	7.8—30	—
Pb	50—3,000	7—225[i]	20—1,250	2—27	6.6—15	60
Rb	4—95	5	3	—	0.06	—
Sc	0.5—7	7—36	1	—	5	—
Se	2—9	0.5—25[j]	0.08—0.1	—	2.4	—
Sn	40—700	3—19[d]	0.5—4.0	1.4—16.0	3.8	—
Sr	40—360	25—500	610	—	80	—
Te	—	20—23	—	—	0.2	—
U	—	30—300[k]	—	—	—	—
V	20—400	2—1,600[d]	20	—	—	45
Zn	700—49,000	50—1,450	10—450	1—42	15—250	1.3—25
Zr	5—90	50	20	—	5.5	—

[a]   Refs. 70, 249, 593.
[b]   Refs. 94, 381, 399.
[c]   Refs. 20, 25, 249, 532.
[d]   Ref. 701.
[e]   Ref. 510.
[f]   Ref. 875.
[g]   Ref. 744.
[h]   Refs. 55, 620.
[i]   Ref. 755a.
[j]   Ref. 809.
[k]   Ref. 306.
[l]   Mainly ammonium sulfate.

orchards or in other locations by contamination from industrial emissions or heavy and repeated applications of sewage sludges. A high heavy metal content of sludges is the most important hindrance to their use in agriculture. Although Purves[634] reported that in practice the concern with using sludges commonly is only their phytotoxicity due to excesses of Zn, Cu, and Ni, their content of Cd in particular, as well as of Pb and Hg, should be of concern as serious health risks. As Andersson and Nilsson[25] have observed, long-term use of sewage sludge increased the soil levels of Zn, Cu, Ni, Cr, Pb, Cd, and Hg. Of these elements, however, only Zn, Cu, Ni, and Cd were increased in cereal grains, and Zn, Cu, Cr, and Pb were increased in cereal straw. Chaney[127] and Sikora et al.[726] recommended higher doses of sewage sludges because of the relatively low availability of heavy metals to plants. Beckett

Table 6
TOTAL CONCENTRATIONS OF TRACE ELEMENTS CONSIDERED AS PHYTOTOXICALLY EXCESSIVE LEVELS IN SURFACE SOILS (PPM DW)

Element	Concentrations as given by various authors					
	a	b	c	d	e	f
Ag	—	—	2	—	—	—
As	—	50	25	30	20	15
B	30	100	—	100	25	—
Be	—	10	—	10	10	—
Br	—	—	—	20	10	—
Cd	—	5	8	5	3	—
Co	30	50	25	50	50	50
Cr	—	100	75	100	100	—
Cu	60	100	100	100	100	125
F	—	500	—	1000	200	—
Hg	—	5	0.3	5	2	—
Mn	3000	—	1500	—	—	—
Mo	4	10	2	10	5	—
Ni	—	100	100	100	100	100
Pb	—	100	200	100	100	400
Sb	—	—	—	10	5	—
Se	—	10	5	10	10	—
Sn	—	—	—	50	50	—
Tl	—	—	—	—	1	—
V	—	—	60	100	50	—
Zn	70	300	400	300	300	250

Note: Sources are the following: a, 419; b, 206; c, 479; d, 376a; e, 398; and f, 395.

et al.[59] concluded that in addition to the commonly monitored levels of Cu, Ni, Zn, Cd, Cr, and Pb during the disposal of sewage sludge on farm land, it may be necessary to monitor levels of Ag, Ba, Co, Sn, As, and Hg and also possibly Mo, Bi, Mn, and Sb, until their likely accumulations in surface soil can be shown to be harmless.

Soil contaminated with heavy metals can produce apparently normal crops that may be unsafe for human or animal consumption. Kloke[398] calculated that if the content of Hg, Cd, and Pb in the soil is not higher than the threshold values (Table 6), it can be expected that the contents of these metals in human diets will not exceed weekly tolerable intakes established by FAO/WHO.[398] Therefore, safe use of sewage sludge must be assessed on the basis of a safe addition of trace metals into soils.

Permissible levels of trace elements, particularly heavy metals, used on farmland can be calculated based on several factors. It is most important, however, to evaluate acceptable application rates in relation to:

1. Initial trace element content of soil
2. Total amount added of one element and of all heavy metals
3. Cumulative total load of heavy metals
4. Heavy metal dose limitation
5. Equivalency of trace element toxicity to plants
6. Threshold values of trace element concentrations in soils
7. Relative ratios between interacting elements

8.   Soil characteristics, e.g., pH, free carbonates, organic matter, clay content, and moisture
9.   Input-output balance
10. Plant sensitivity

Lewin and Beckett[469] widely reviewed monitoring of heavy metal accumulation in agricultural soils treated with sewage sludge and pointed out that it will be unreasonable to assume, without checks, whether heavy metals in soil will become immobilized with time or not.

Different soil types, plant species, and growing conditions contribute to the divergent influences of soil contamination on trace element status in plants. Some authors use a term "soil resistance to heavy metal contamination", which is related to the critical levels of metallic pollutants that exhibit toxic effects on plants and environments. This term is largely related to the cation exchange capacity (CEC) of soils (see Chapter 3, Section III.D.). Usually the resistance of a nonacid heavy soil with a higher content of organic matter exceeds several times the resistance of a light sandy acid soil. Loamy neutral soils may accumulate a higher amount of trace elements with much less environmental risk. However, a general chemical imbalance of such soils usually results in decreased biological activity, decreased or increased pH, and, as a further consequence, in degradation of organic and mineral sorption complexes.

Contamination of agricultural soils has already become relatively common and is likely to continue. Noticeable, also, is the fact that most often soils become contaminated by several metallic pollutants that are accompanied quite frequently by acid rains (mainly SO_2 and HF). Such an association of pollutants in soil greatly complicates their impact on the environment.

B. Soil Reclamation

The improvement of soils damaged and contaminated by pollutants has recently become a great practical problem. Reclamation of the particular soil requires, in so far as possible, a full understanding of soil properties and of the deteriorating factors. Soil contamination with heavy metals is usually quite permanent, as has been reported by Davies,[166] Johnson et al.,[365] Purves,[634] and Kitagishi and Yamane.[395] Therefore, it is necessary to emphasize that a soil heavily contaminated, especially by trace metals, is likely to be the sink of these contaminants, resulting in degradation of biological and chemical properties of the soil.

Several specific techniques for amelioration of various industrial wastes and for their revegetation have been described by Gemmell.[260] For soils contaminated by trace elements, the practices advised to prevent plant pollution are based on two main reactions — the leaching of easily soluble elements and the immobilization of microcations in soils. Heavily contaminated soil may need some special treatment, as was done by Kobayashi et al.[405] who removed an excess of soil Cd by repeated treatment with EDTA solution and lime (the Cd content of the surface soil decreased from 27.9 to 14.4 ppm). Mixing polluted topsoils with unpolluted soil material, as well as covering over the polluted soils, or replacement of the polluted topsoils, as reported by Kitagishi and Yamane[395] have been used for arable soils in Japan.

Reclamation of soils contaminated by heavy metals is usually based on the application of lime and phosphates and the addition of organic matter. The addition of lime, resulting in increased soil pH, however, does not always bring the expected results in the immobilization of some trace metals. The metals that are most likely to occur in soil as organic chelates in larger particulates may become soluble quite easily after heavy liming, as has been reported mainly for Cu, Zn, and Cr.[152,260,618] In most cases, however, lime and phosphate are quite effective in lowering heavy metal concentrations in plants, especially those growing on acid sandy soils. This response is an effect of both chemical and physical reactions in soil materials and cation interactions physiologically characteristic of a plant.

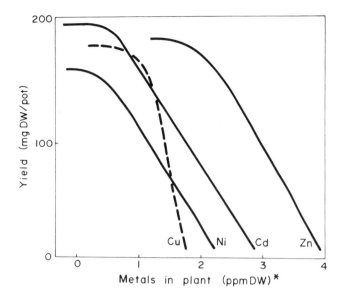

FIGURE 2.   Response of young barley plants to heavy metal concentrations in their tissues. Asterisk indicates concentration of metals is given in powers of ten.[59]

A review of the literature does not reveal any generally adequate method for rapid reclamation of soils heavily contaminated by trace metals. The effects of each treatment will depend upon soil properties, mainly on CEC and on plant response. Therefore, the reclamation or improvement of arable land polluted with trace elements needs to be designed for a specific plant-soil system.

V. PLANTS

The significant role of plants in both cycling of trace elements and contaminating the food chain has been well illustrated for various ecosystems and published in numerous papers. Plants can accumulate trace elements, especially heavy metals, in or on their tissues due to their great ability to adapt to variable chemical properties of the environment, thus plants are intermediate reservoirs through which trace elements from soils, and partly from waters and air, move to man and animals. As Tiffin[789] has concluded, plants may be passive receptors of trace elements (fallout interception or root adsorption), but they also exert control over uptake or rejection of some elements by appropriate physiological reactions.

One of the basic environmental problems relates to the quantities of accumulated metals in plant parts used as food. Special attention also should be given to the forms of metals distributed within plant tissues, for the metal forms in plants seem to have a decisive role in metal transfer to other organisms.

Several authors have observed that the yield of various crops can be decreased due to metallic pollution (Figure 2). The generalized effects of metal concentrations in nutrient solution on yield and metal content of plants are shown in Figure 3. Most important, however, are the biological and health effects on man and animals caused by metallic pollution in plants. This subject has been reviewed in detail by many textbooks on environmental health.

Each case of plant pollution is unique and should be studied for a specific environment. There is an increasing awareness that results of studies based on simulation-type systems cannot be related to those in a natural system. This fact is supported by de Vries and Tiller,[831] who reported a much lower absorption of heavy metals by lettuce and onions grown in a market garden soil than by those grown under greenhouse and miniplot conditions.

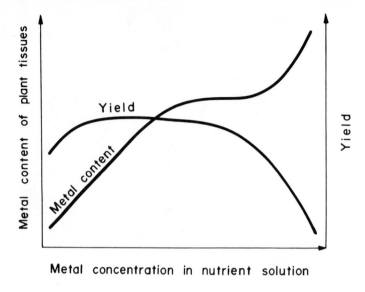

FIGURE 3.   Generalized effects of trace metal concentrations in nutrient
solution on yield and metal content of plants.

Trace pollutants entering plant tissues are active in metabolic processes, but also can be stored as inactive compounds in cells and on the membranes. In each case, however, they may affect the chemical composition of plants without causing easily visible injury. The most common symptoms of phytotoxicity of several trace elements are rather nonspecific and are described in detail in Table 29.

Chapter 3

SOILS AND SOIL PROCESSES

I. INTRODUCTION

A valid concept of the nature of soil must avoid the common error that soil is simply a mixture of unconsolidated material resulting from the weathering processes of underlying rocks. Soil is a natural body, having both mineral and organic components as well as physical, chemical, and biological properties. Soil properties, therefore, cannot be a simple reflection of the combined properties of all soil components.

Any classification of soils suffers from the disadvantage that it is impossible to relate it to the great complexities of soil genesis and properties. The terms used in defining the soils in different systems seldom are exactly equivalent. The definitions of soil units used in this book are adopted from FAO/UNESCO.[226] Short descriptions of the soil units are presented in Table 7. The names of soils described in this book were taken from the original publications, translated, and the associated soil characteristics were fitted, insofar as possible, into the soil units of the FAO/UNESCO system.

The composition of soils is extremely diverse and, although governed by many different factors, climatic conditions and parent material predominate most commonly. An approximation of soil composition is shown in Figures 4 and 5. Soil is composed of three phases — solid (mineral and organic), liquid, and gaseous — and exhibits properties resulting from the physical and chemical equilibriums of these phases. Moreover, not only the chemical composition of the solid components of soil, but also its mineral structure and the state of dispersion are important factors influencing soil properties.

Although trace elements are minor components of the solid soil phase, they play an important role in soil fertility. A knowledge of the association of trace elements with particular soil phases and their affinity to each soil constituent is the key to a better understanding of the principles governing their behavior in soils.

Two stages are involved in the formation of soil from parent material. The first is the alteration of the primary mineral constituents of the parent rocks by the physical and chemical processes of weathering. The second stage (pedogenesis) results in the formation of a soil profile from the weathered rock material, leading to the development of a mature zonal soil as the end point of the interacting processes. Weathering and pedogenic processes cannot be easily distinguished and separated because they may take place simultaneously at the same sites and most commonly they are closely interrelated. They will be discussed separately, however, in Chapter subsections.

II. WEATHERING PROCESSES

Weathering, the basic soil forming process, has been extensively studied and reviewed as the complex interactions of the lithosphere, the atmosphere, and the hydrosphere that occur in the biosphere and that are powered by solar energy. Weathering can be chemically described as the processes of dissolution, hydration, hydrolysis, oxidation, reduction, and carbonation. All of these processes are based on rules of enthalpy and entropy, and they lead to the formation of mineral and chemical components that are relatively stable and equilibrated in the particular soil environments. Chemical weathering leads to the destruction of parent minerals and to the passing of the elements from the minerals into solutions and suspensions.

Table 7
SOIL UNITS USED IN THE FAO/UNESCO SOIL MAP OF THE WORLD

Name of unit	Symbol	World distribution[a]	Certain old names	Predominating soil forming factor[b]	Short description of soils
Fluvisols	J	2.40	Alluvial soils	a	Recent alluvial deposits, little alteration
Gleysols	G	4.73	Black earths or hydromorphic soils	c	Soils formed from various materials with hydromorphic properties mainly within the top horizon
Regosols	R	10.10	—	a	Little altered, unconsolidated parent rock
Arenosols	Q		—	a	Soils formed from sand, no diagnostic horizons
Lithosols	I	—	—	a	Shallow soils over hard rock
Rendzinas	E	17.17	Rendzinas	a	Shallow soils over limestones
Rankers	U		—	a	Shallow soils formed from recent siliceous deposits, little alteration
Andosols	T	0.76	Volcanic soils	a	Soils formed from volcanic ash
Vertisols	V	2.36	Brown soils	a	Clay soils
Solonchaks	Z	2.03	—	c	Soils, often formed from recent alluvial deposits, with salt accumulation
Solonetz	S		—	c	Soils with hydromorphic properties (high exchangeable Na content)
Yermosols	Y	8.93	Aridisols	c	Desert soils or other formed under aridic regime
Xerosols	X	6.79	Grey soils	c	Similar to above, but better development of a horizon
Kastanozems	K	—	—	b	Soils formed under steppe vegetation
Chernozems	C		—	a,b	Soils developed under prairie vegetation
Phaeozems	H	3.09	Degraded chernozems	b	Similar to K and C soils, but more leached
Greyzems	M	—	—	b	Soils formed under forests in cold temperate climate
Cambisols	B	7.02	Brown soils	c	Highly altered soils, having a cambic B horizon

Luvisols	L	7.00	Lessivage processed	c	Similar to B soils, with clay accumulation and more leached
Podzoluvisols	D	2.00	Grey-podzols	c	Transition soils between L and P soil units
Podzols	P	3.63	—	c	Highly altered profile due to leaching
Planosols	W	0.91	Mineral hydromorphic soils	c	Soils slightly leached due to a slowly permeable horizon
Acrisols	A	7.97	—	c	Highly weathered soils with a clay horizon
Nitosols	N		Red-brown soils	c	Transition soils between A and F soil units
Ferralsols	F	8.11	Oxisols, lateritic soils	c	Soils with sesquioxide rich clay fraction formed under tropical climate
Histosols	O	1.82	Peats, mucks, bog soils	b	Organic soils, having H horizon of more than 40 cm
Anthroposols[c]	—	—	Hortisols, industrosols, regosols	—	Soils formed under a predominant influence of man's activity
Miscellaneous land units	—	3.18	—	—	—

Note: Soil units, definitions, and descriptions are based on data from FAO/UNESCO.[226]

[a] The distribution of major soils, in percent of the world soil area, as given by Dudal.[194]

[b] a, Parent rock; b, vegetation; and c, pedogenic processes stimulated mainly by climate.

[c] Soil unit not given in the FAO/UNESCO system.

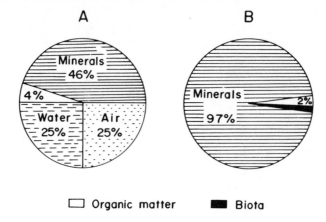

FIGURE 4.   Approximate composition of a representative silty loam surface soil of vertisol soil unit (brown soil). (A) Volume percent of total soils; (B) weight percent of solid phase of soil.

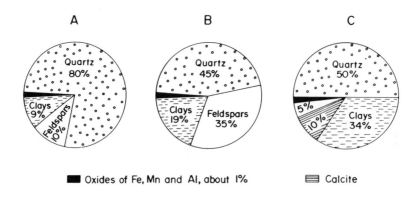

FIGURE 5.   Approximate composition of mineral constituents of surface soils derived from different rocks in a temperate humid climate. (A) Podzoluvisol over sandstone; (B) vertisol over granite; (C) rendzina.

Greatly simplified, basic weathering processes can be characterized as follows:

1. Dissolution — minerals are soluble in the aquatic phase
2. Hydration — minerals increase their water content
3. Hydrolysis — reaction of minerals with water, producing new ions and/or insoluble components
4. Oxidation — incorporation of the oxygen into the chemical components or increase of the element potential
5. Reduction — reactions that are the reverse of oxidation
6. Carbonation — alteration of compounds into carbonates due to the incorporation of CO_2

All of these reactions are controlled by chemical equilibria of the particular earth surface environment. The stability of such a system is often illustrated by Eh -pH diagrams for the given geochemical reaction.[256,477] In spite of the many questions raised and the difficulties in a practical evaluation of these diagrams, they indicate clearly that both factors (Eh and pH values) are significant in geochemical evolution.

The reactions of trace elements in a particular weathering environment are greatly different. Some generalizations of their properties as described by several authors are presented in

Table 8
BEHAVIOR OF TRACE ELEMENTS IN VARIOUS WEATHERING
ENVIRONMENTS[75,381,599]

Degree of mobility	Environmental conditions	Trace elements
High	Oxidizing and acid	B, Br, and I
	Neutral or alkaline	B, Br, I, Mo, Re, Se, U, V, and W
	Reducing	Br and I
Medium	Oxidizing and acid	Cs, Mo, Ra, Rb, Se, Sr, and Zn
	Mainly acid	Ag, Au, Cd, Co, Cu, Hg, and Ni
	Reducing, with variable potential	As, Cd, Co, Cr, F, Fe, Ge, Mn, Nb, Sb, Sn, Tl, U, and V
Low	Oxidizing and acid	Ba, Be, Bi, Cs, Fe, Ga, Ge, La, Li, Th, Ti, and Y
	Neutral or alkaline	Ba, Be, Bi, Ge, Hf, Ta, Te, and Zr
Very low	Oxidizing and acid	Cr, Os, Pt, Rh, Ru, Ta, Te, and Zr
	Neutral or alkaline	Ag, Au, Cu, Co, Ni, Th, Ti, and Zn
	Reducing	Ag, B, Ba, Be, Bi, Co, Cu, Cs, Ge, Hg, Li, Mo, Ni, Re, Se, Zn, and Zr

Table 8. The mobility of these elements during weathering processes is determined firstly by the stability of the host minerals and secondly by the electrochemical properties of the elements.

The pattern of trace element distribution usually is a parameter that is very sensitive to changes of weathering environments. The so-called "chemical nature" of an element reflects mainly its electronegativity parameters and its ionic size. Selected elemental parameters of trace ions given in Table 9 may explain why individual trace elements reveal an affinity for association with major elements in various geochemical environments. The elements with an ionic potential below 3 predominate as free ions, while the elements with an ionic potential between 3 and 12 tend to form hydrolysates or complex ions. Easily mobile elements usually produce smaller hydrate ions in aqueous solutions than do more stable elements; also, the free energy (enthalpy) needed for the formation of their ions seems to be less than the energy uired for ion formation of less mobile elements.

III. PEDOGENIC PROCESSES

Several specific reactions, in addition to those involved in weathering, lead to the formation of a particular soil profile. Although there is great diversity in pedogenic processes, they all include the following similar stages:

1. Addition of organic and mineral materials to the soil
2. Losses of these materials from the soil
3. Translocation of these materials within the soil, both vertically and horizontally
4. Transformation of organic and mineral matter in the soil

These processes can be constructive or destructive in soil formation. Six factors that largely control the kind of soil that finally develops are

1. Climate (temperature, rainfall)
2. Vegetation and other soil biota
3. Parent material (the nature of minerals)
4. Topography (open or closed systems)

Table 9

GEOCHEMICAL ASSOCIATIONS AND SOME PROPERTIES OF TRACE AND MAJOR ELEMENTS

Major elements (boldface) and associated trace element	pH of hydrous oxide precipitation	Ionic radii (Å)	Electronegativity (kcal/g atom)	Ionic potential (charge/radius)	Diameter of hydrated ion in aqueous solution (Å)
K^+	—	1.7—1.6	0.8	0.6	3.0
Na^+	—	1.2—1.1	0.9	0.9	4.5
Cs^+	—	2.0—1.9	0.7	0.5	2.5
Rb^+	—	1.8—1.7	0.8	0.6	2.5
Ca^{2+}	—	1.2—1.1	1.0	1.8	6.0
Mg^{2+}	10.5	0.8	1.2	2.5	8.0
Sr^{2+}	—	1.4—1.3	1.0	1.5	5.0
Ba^{2+}	—	1.7—1.5	0.9	1.3	5.0
Pb^{2+}	7.2—8.7	1.6—1.4	1.8	1.9	4.5
Se^{3+}	—	0.8	1.3	3.7	9.0
Fe^{2+}	5.1—5.5	0.9—0.7[a]	1.8	2.6	6.0
Cu^{2+}	5.4—6.9	0.8	2	2.5	6.0
Ge^{4+}	—	0.5	1.8	8.3	—
Mo^{4+}	—	0.7	—	5.5	—
Mn^{2+}	7.9—9.4	1—0.8	1.5	2.0	6.0
Zn^{2+}	5.2—8.3	0.9—0.7	1.8	2.6	6.0
Fe^{3+}	2.2—3.2	0.7—0.6[a]	1.9	4.4	9.0
Co^{2+}	7.2—8.7	0.8—0.7	1.7	2.6	6.0
Cd^{2+}	8.0—9.5	1.03	—	—	—
Ni^{2+}	6.7—8.2	0.8	1.7	2.6	6.0
Cr^{3+}	4.6—5.6	0.7	1.6	4.3	9.0
Mn^{4+}	—	0.6	—	6.5	—
Li^+	—	0.8	1.0	1.2	6.0
Mo^{6+}	—	0.5	1.8	12.0	—
V^{5+}	—	0.5	—	11.0	—

Al³⁺	3.8—4.8	0.6—0.5ᵃ	1.5	5.6	9.0
Be²⁺	—	0.3	1.5	5.7	8.0
Cr⁶⁺	—	0.4	—	16.0	—
Ga³⁺	3.5	0.7—0.6	1.6	4.9	—
La³⁺	—	1.4—1.3	1.1	2.3	9.0
Sn²⁺	2.3—3.2	1.3	1.8	1.5	—
Y³⁺	—	0.9	1.2	3.1	—
Si⁴⁺	—	0.4	1.8	12.0	—
Ti⁴⁺	1.4—1.6	0.7	1.5	5.8	—
Zr⁴⁺	2.0	—	1.4	4.3	11.0

ᵃ Values given for high and low spin, respectively.

5.   Time
6.   Anthropogenic activity (degradation, contamination, recultivation)

Classification of soil units very commonly is based on the factors predominating in soil forming processes (Table 7). Pedogenic processes stimulated mainly by climate most commonly predominate, but soils influenced most strongly by parent material or vegetation are quite frequent.

Initially, at early stages of weathering and pedogenic processes, the trace element composition of the soil will be inherited from the parent material. With time, however, the trace element status of soil will become different due to the influence of predominating pedogenic processes (Table 10). The fate of trace elements mobilized by dissolution of the host minerals or compounds depends upon the properties of their ionic species formed in the soil solution (Table 8) and may be:

1.   Leached from the soil
2.   Precipitated
3.   Incorporated into minerals
4.   Adsorbed by a soil constituent
5.   Adsorbed onto or into organic matter

Thus, dynamic equilibrium between soil components is governed by various interactions between the soil solid and gaseous phases, biota, and the soil solutions, as is illustrated by the diagram in Figure 6.

A. Dissolution

Chemical reactions leading to solution of each species of ions can be characterized by thermodynamic equations. At each equilibrium state the reaction rates of both directions compensate and keep the composition of the soil phases (solid, liquid, and gaseous) constant.

Chemical equilibria of various soils have been studied and comprehensive mathematical models for the particular soil conditions are presented by Bolt and Bruggenwert.[86] Although many papers have been published on the behavior of trace elements in soils, their chemistry is insufficiently known. The diversity of ionic species of trace elements and their various affinities to complex with inorganic and organic ligands make possible the dissolution of each element over a relatively wide range of pH and Eh. Each element can also be quite readily precipitated and/or adsorbed even under a small change of the equilibrated conditions.

Many textbooks present stability diagrams for ionic species of trace elements as functions of pH and Eh.[166,256,477] In natural soil conditions, pH ranges most often between 5 and 7, and Eh ranges between $+0.5$ and -0.1 values, except where there are high reduction states in waterlogged soils. The properties of ionic species of each element vary, and the pH range for precipitation of their hydrous oxides is flexible (Table 9). One can, however, conclude that usually the most mobile fractions of ions occur at a lower range of pH and at a lower redox potential. It can be anticipated that with increasing pH of the soil substrate the solubility of most trace cations will decrease. Indeed, the concentration of trace elements is lower in soil solutions of alkaline and neutral soils than in those of light acid soils (Table 11).

The solubility of trace elements in soil has great significance in their bioavailability and their migration. Heavy soils, both neutral and alkaline, provide good storage for trace elements and will supply them to plants at a slow rate. This slow release may, however, cause deficiency effects of certain micronutrients to develop in plants. Light soils, on the other hand, can be a source of easily available trace elements during a relatively short period of time. These soils also can lose their pool of available micronutrients at a quite high rate.

Some of the first studies of soil solutions conducted by Hodgson et al.[320] indicated that

Table 10
PRINCIPAL TYPES OF SOIL-FORMING PROCESSES AND TRENDS IN THE BEHAVIOR OF TRACE ELEMENTS

Process	Most favorable climatic zone	Typical soil unit[a]	Behavior of trace elements in surface soil	
			Accumulation	Migration
Lack of chemical alteration	Ice-bound or desert	R, Q, and Y	—	—
Podzolization	Cold nothern	D and P	Co, Cu, Mn, Ni, Ti, V, and Zr (in illuvial horizon)	B, Ba, Br, Cd, Cr, I, Li, Mn, Rb, Se, Sr, V, and Zr
Aluminization	Cool and humid temperate	B, L, M, W, and A	Co, Mn, Mo, and V (in gleyed horizon)	B, Ba, Br, Cu, I, Se, and Sr
Siallitization	Warm temperate and dry tropical	V and K	B, Ba, Cu, Mn, Se, and Sr	—
Lateritization	Humid tropical	A, F, and N	B, Ba, Cu, Co, Cr, Ni, Sr, Ti, and V	—
Alkalization	Warm with dry seasons	Z, S, and (X)	B, Co, Cr, Cu, Mo, Ni, Se, Zn, and V	—
Hydromorphic formations	Intrazonal soils	—	B, Ba, Co, Cu, I, Mn, Mo, Se, Sr, and U (in organic horizon)	B, Br, Co, Cu, Mn, Ni, U, and V

[a] Symbols of the soil units as given in Table 7.

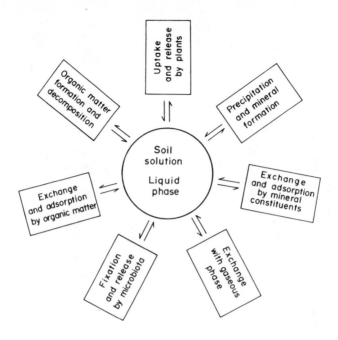

FIGURE 6.   Dynamic equilibria between soil components.

Table 11
TRACE ELEMENTS IN NATURAL SOIL SOLUTIONS OBTAINED BY CENTRIFUGATION FROM SOILS OF VARIOUS pH VALUES[915]

	Soil type and range of pH values				
Element	Acid sandy (2.5—4)	Sandy (4—4.5)	Silty (5.5—6.5)	Loamy (7—7.5)	Calcareous (7.5—7.8)
B	—	—	—	200	800
Cd	107	—	—	—	—
Co	—	—	—	0.5	5
Cu	783	76	20	50	50
Fe	2223	1000	500	200	100
Mn	5965	8000	5000	100	700
Mo	—	—	—	5	3
Pb	5999	—	—	—	—
Zn	7137	1000	5000	100	300

Note: Values are arithmetic means ($\mu g\ I^{-1}$) of 4 to 5 samples.

appreciable quantities of trace elements are present as complexes, mainly with organic ligands. Inorganic complexes may, however, also be of importance for the particular element and condition.

The concentrations of trace elements in soil solutions vary considerably among soils and with time. Great fluctuations have been observed under the influence of the following factors:

1. Time
2. Vegetation
3. Microbial activity
4. Waterlogged states
5. Heterogeneity of the solid soil phase

Moreover, methods used for obtaining solutions from soils differ widely; therefore it is difficult to adequately determine mean concentrations of trace elements. Nevertheless, the range of trace element concentrations, as measured in various soil solutions, is reasonably similar (Table 12).

Rainfall, evaporation, and plant transpiration can change trace element concentrations in soil solutions more than tenfold, whereas the observed variations for major ions (Ca, Mg, K, Na, NO_3, and PO_4) are much less.

Soluble major ions greatly influence the quantities of soluble trace elements. Solutions of most soils contain an excess of Ca, which in many soils constitutes more than 90% of the total cation concentration. Ca is, therefore, the most important cation in governing the soluble stage of trace elements in soils. There are examples, however, of soils in which complexing trace cations prevent precipitation in the presence of Ca^{2+} and in soil solutions having a relatively high pH level. Thus, higher than normal concentrations of dissolved metal ions in solutions and the ready uptake of trace elements by plants usually are related to the formation of complexes. This possibility is suggested also by the observation that half of the calcium in soil solutions is usually present as an organic complex.[356] The anionic composition of soil solutions is also of importance in controlling the trace element status. Little is known, however, about the relation of trace elements to anionic species in soil solutions.

Differing complexing tendencies of cations to interact with ligands can be explained by the rules of coordinating chemistry. It is possible, therefore, to predict that certain cations more readily complex a particular ligand. Thus, some metallic ions such as Be^{2+}, Cr^{3+}, and Co^{3+} will react readily with PO_4^{3-}, CO_3^{2-}, NO_3^{-}, and organic amines, etc.; the group including Ni^{2+}, Co^{2+}, Cu^{2+}, Zn^{2+}, Cd^{2+}, Pb^{2+}, and Sn^{2+} may more easily complex Cl, Br, NO_2, and NH_3, while cations of the chemical nature of Hg^{2+}, Ag^+, and Tl^+ are likely to link with complexes of I, CN, CO, S, P, and As.

In the soil aqueous phase organic compounds and water are the most abundant ligands, therefore, hydrolysis and organic complexing are the most common reactions in soil solutions. These reactions are pH sensitive and can be correlated with the size and charge of the cations. Higher ionic potentials usually indicate a higher degree of hydration in the solution, thus an easier precipitation. The range in pH for the precipitation of hydrous oxides of some cations (Table 9) illustrates that the order of cation mobility in an aqueous phase under an oxidation regime of soil may decrease as follows: $Mg^{2+} = Ca^{2+} > Ag^+ > Hg^{2+} > Mn^{2+} > Cd^{2+} > Ni^{2+} = Co^{2+} = Pb^{2+} > Be^{2+} > Zn^{2+} = Cu^{2+} > Cr^{3+} > Bi^{3+} 3 > Sn^{4+} > Fe^{3+} > Zr^{4+} > Sb^{3+}$. However, the application of data on heavy metal activity in pure systems to soils can be only informative because of various effects of complex ion formation, solid solution, and coating.

The solubility of trace elements in soils evidently depends on complex formation. However, most of the species of trace elements, especially cations, are but slightly soluble, and only small proportions occur in the aqueous phase. The calculation made by Kabata-Pendias[375] showed that the total content of trace cations generally ranges from 10 to 100 $\mu g\ \ell^{-1}$ in normal soil solutions, while in contaminated soils these values can be much higher. When soluble compounds of trace metals are added to soils, their concentrations in equilibrated solutions increase with increasing doses of added metals. In an experiment conducted by Cottenie et al.,[148] the relative solubility of the added metals at the highest dose rate in light sandy soil was as follows: 39% of 1000 mg Zn/kg, 50% of 5000 mg Cu/kg, 30% of 5120 mg Cd/kg, and 26% of 2695 mg Pb/kg.

B. Transport

The transport of dissolved trace elements may take place through the soil solution (diffusion) and also with the moving soil solution (mass flow, leaching). Generally, in soils formed under a cool and humid climate, the leaching of trace elements downward through

Table 12
TRACE ELEMENTS IN NATURAL SOIL SOLUTIONS OF VARIOUS SOILS (μg ℓ^{-1})

Element	Suction from soil paste a	Pump off 0.01 N CaCl$_2$ solution b	Centrifugation			Ceramic plate or porous cup suction			Unspecified technique[a]
			c	d	e	f	g	h	i
Al	400	—	—	460	—	—	4—12	—	—
As	—	—	—	—	—	—	—	—	—
B	3,060	—	67—880	—	—	—	—	—	—
Ba	260	—	—	—	—	—	—	—	—
Cd	—	<0.4—14	0.3—5	6	—	3—5	5—300	0.2	<0.01—0.2
Co	60	—	—	3	—	12—87	—	0.3—1.0	—
Cr	10	—	—	0.4	—	0.6—0.7	Trace	—	—
Cu	40	3—18	28—135	37	78	18—27	14—44	0.5—3.0	<1—3
Fe	50	—	150—549	16	—	36	—	30—40	<50—1,000
Hg	2.4	—	—	—	—	—	—	—	—
Li	111	—	—	—	—	—	—	—	—
Mn	170	—	32—270	243	55	1,000—2,200	—	25—50	2,000—8,000
Mo	730	—	2—8	2	30	—	—	—	—
Ni	20	—	—	150	—	3—15	20—25	3—8	—
Pb	50	—	—	8	—	5—63	Trace	0.6—2.0	<2
Sr	930	—	—	—	—	—	—	—	—
Ti	<100	—	—	—	—	—	—	—	—
V	70	—	—	—	—	—	—	—	—
Zn	70	21—180	73—270	351	22	190—570	40—17,000	4—25	1—15

Note: Sources are as follows: a, 99; b, 320; c, 375; d, 892; e, 905; f, 311; g, 342; h, 292; i, 794.

[a] Data for paddy soils, after flooding for 14 weeks.

the profiles is greater than their accumulation, unless there is a high input of these elements into the soils. In warm, dry climates, and also to some extent in humid hot climates, upward translocation of trace elements in the soil profiles is the most common movement. However, specific soil properties, mainly its cation exchange capacity, control the rates of trace element migration in the profiles.

Although much work has been done on trace element movements within soil profiles, complete knowledge concerning their cycling and balances is still lacking. Theoretical reviews of mechanisms involved in the transport and accumulation of soluble soil components were given recently by Bolt and Bruggenwert[86] and by Lindsay.[447] The equilibria discussed by these authors are useful not only in illuminating fundamental reactions that are important in weathering and soil formation, but also for use in various fields of agricultural and environmental management. However, the models cannot be used for examining the quality of thermodynamic data obtained from a particular soil without making necessary modifications that take into account the variations of soil properties, and even then, some skepticism may remain.

Several detailed studies based on lysimetric experiments, and other research often using isotopic tracers, have yielded much information on element transport. However, each soil profile with developed horizons has its own characteristic trace element movement.

Impoverishment of soils in trace elements is due mainly to their mobility downward with percolating waters through the profiles of freely drained acid soils and also to trace element uptake by plants. On the other side of the balance is the input of trace elements with atmospheric precipitation and their accumulation in particular soil horizons. In acid soils (e.g., with pH below 6.5), several elements, such as Zn, Mn, Cu, Fe, Co, and B, are easily leached. These elements, however, are likely to form quite stable compounds if the pH of the soil rises above 7. Other elements, such as Mo and Se, are mobilized in alkaline soils, while in acid soils they become almost insoluble.

Trace element budgets have been recently calculated for various ecosystems (Table 13). Input/output differences show that for the majority of elements the accumulation rate in the surface soils is positive. Leaching of Mn, Fe, and Be was found to be higher than atmospheric input only in acid forest soils. However, all the data presented in Table 13 are for areas having a possible impact from sources of aerial pollution.

C. Sorption

Soils are considered as sinks for trace elements, therefore, they play an important role in environmental cycling of these elements. They have a great ability to fix many species of trace ions. The term "sorption" used in this chapter refers to all phenomena at the solid-solution boundary, including the following intermolecular interactions:

1. van der Waals' forces
2. Ion-dipole forces
3. Hydrophobic and hydrogen bondings
4. Charge transfer
5. Ion and ligand exchanges
6. Chemisorption
7. Magnetic bonding

Soil components involved in sorption of trace elements are

1. Oxides (hydrous, amorphic) mainly of iron and manganese and, to a much lesser extent, aluminum and silicon
2. Organic matter and biota
3. Carbonates, phosphates, sulfides, and basic salts
4. Clays

Table 13
ANNUAL TRACE ELEMENT INPUT/OUTPUT BALANCE FOR SOILS OF VARIOUS ECOSYSTEMS (g ha^{-1} year^{-1})

Element	Atmosphere input (bulk)		Water flux output or plant uptake	Soil unit	Ecosystem and locality
Van Hook et al.[823]					
Cd	21		7(a)	Acid, light	Deciduous forest,
Pb	286		6	loam derived	Tennessee (U.S.)
Zn	538		149	from dolomite	
Hansen and Tjell[304]					
Cd	3		0.3(b)	(No data)	Agricultural land,
Pb	260		0.3		Denmark
Zn	250		120		
Zöttl et al.[907]					
Be	0.3		5.6(c)	Brown podzol	Pine forest, Schwarzwald,
Cd	4.5		1.4		West Germany
Co	5.6		4.3		
Cu	18		7		
Ni	34		17		
Mn	70		430		
Pb	110		6		
Zn	210		76		
Fe	300		2,000		
Heindrichs and Mayer[310]					
Bi	0.4		0.2(c)	Brown acid	Birch and spruce
Hg	0.4		0.2	silty loam	forest, Solling
Tl	1.2		0.3		Mountains, West
Sb	3		0.3		Germany
Cd	13		9		
Ni	15		14		
Cr	22		2		
Mn	200		6,300		
Cu	224		108		
Zn	3,900		1,900		
Fe	1,600		1,900		
Tyler[815,816]					
Cd	2		5(b)	Podzolic	Spruce forest
Cr	8		10	forest soil	(lysimeter),
Ni	10		9		Hässleholm, Southern
V	12		28		Sweden
Cu	20		29		
Pb	150		81		
Zn	180		270		
Fe	2,000		13,000		
Ruszkowska[671] and Authors' Unpublished Data					
Cd	5	—	3(c)	Podzolic light	Lysimeters and field
Cu	39	18(b)	25	loam	plots, Pulawy,
B	71	29	40		Poland

Table 13 (continued)
ANNUAL TRACE ELEMENT INPUT/OUTPUT BALANCE FOR SOILS
OF VARIOUS ECOSYSTEMS (g ha^{-1} year^{-1})

Element	Atmosphere input (bulk)		Water flux output or plant uptake	Soil unit	Ecosystem and locality
Mn	181	35	90		
Pb	207	—	40		
Zn	547	163	180		

Bubliniec[109]

Cu	40		40(d)	Brown soil	Oak forest, Zvolen,
Mo	70		50		Czechoslovakia
Fe	840		1,040		
B	1,460		80		
Mn	2,480		1,930		

Note: Calculation methods based on (a) stream water, (b) drainage water, (c) seepage water, and (d) uptake by trees.

Of all these components, clay minerals, hydrated metal oxides, and organic matter are considered to be the most important group in contributing to and competing for the sorption of trace elements.

Sorption mechanisms can be based on the valency forces and the process is called "chemisorption". If van der Waals' forces are involved in sorption, the process is called "physisorption". Both sorptions play an important role in the fixation of uncharged complexes. Each trace cation can be sorbed specifically and nonspecifically, as was shown for Cd by Tiller et al.[796]

Reactions of ion exchange (called also "surface exchange") play a basic role in overall sorption processes and are almost exclusively limited to colloidal particles. Various other processes, however, are also involved in sorption phenomena, e.g., precipitation, mineral formation and uptake by mezo- and microbiota, and uptake by plant roots (Figure 6).

D. Adsorption

The term "adsorption" is commonly used for the processes of sorption of chemical elements from solutions by soil particles. Adsorption is thus the kinetic reaction based on thermodynamic equilibrium rules. The forces involved in the adsorption of ionic species at charged surfaces are electrostatic and can be explained by Coulomb's law of attraction between unlike charges and repulsion between like charges. At metal equilibrium concentrations the adsorption by soil particles can be described by either the Langmuir or the Freundlich equations for adsorption isotherms.[86]

Surface charges in soil materials caused primarily by ionic substitutions are exhibited mainly by colloids. At a low pH a positively charged surface prevails, while at a high pH a negatively charged surface develops. The colloids of the majority of soils, therefore, carry negative charges and can be electroneutralized by cations present in the surrounding solutions. In the presence of an excess of cations, the process of exchanging the cations for others maintains the electroneutrality of the system. Thus, the cations adsorbed by the solid phase can be replaced by other cations, most often by H ions. An increase in stability of adsorbed metals may result from dehydration and recrystallization processes that occur on the surface of the colloids, especially in alkaline soils.

The ability of the solid soil phase to exchange cations, the so-called CEC, is one of the most important soil properties governing the cycling of trace elements in the soil. The excess amount of adsorbed cations compared to the amount in solution is interpreted as the buffering capacity of soils, while adsorption capacity defines the amount of ions needed to occupy all adsorption sites per unit of mass.

The CEC of different soils varies widely both in quantity and quality and can range from 1 to 100 meq/100 g of soils.* Surface properties of soil particulates are the most important factor in defining the capacity for adsorption of microcations. Although total adsorption processes cannot be related simply to CEC phenomena, the adsorbed amounts of cations are in accordance with the CEC. Usually the solid soil phase with a large surface area also shows a high CEC value and high adsorption and buffer capacities (Table 14).

The affinity of cations for adsorption, e.g., for anionic exchange sites, is closely related to ionic potential (charge/radius). In some systems the metal ions (Zn, Cd, Mn) occupy nearly the same percentages of the CEC of various minerals.[763] Some cations, however, may have a higher replacing power than others and can be selectively fixed by the sorbing sites. As Abd-Elfattah and Wada[2] stated, the selectivity of adsorption reveals a possible formation of the coordination complexes of heavy metals with deprotonated OH and COOH groups as ligands. This specific sorption is well illustrated by heavy metals having high affinities for organic matter and the surface of oxides, with replacing power over alkali and alkaline earth metals. This phenomenon has great importance in the nutrient supply to plants and in soil contamination.

---

*   CEC of most soils does not exceed a value of 30.

Table 14

SURFACE AND SORPTION PROPERTIES OF SOME SOIL MINERALS[227,287,373a,575,638,699,763,809,833,835,901]

Mineral	Total surface area (m² g⁻¹)	Cation exchange capacity (meq 100 g⁻¹)	Sorption of microcations				
			(µM g⁻¹)	Cd²⁺	Mn²⁺	Zn²⁺	Hg²⁺
Kaolinite	7—30	3—22	30—70	3.1	3.5	3.4	0.46
Halloysite	30	3—57	—	—	—	—	—
Montmorillonite	700—800	69—150	390—460	60—86	72—116	88—108	0.4—2.2
Illite	65—100	10—40	65—95	—	—	—	—
Chlorite	25—40	10—40	—	—	—	—	—
Vermiculite	700—800	100—150	—	98	92	98	9.7
Gibbsite	25—58	—	—	—	—	—	—
Goethite	41—81	—	51—300	—	—	—	—
Manganese oxides	32—300	150	200—1000	—	—	—	—
Imogolite	900—1500	30—135	—	—	—	—	—
Zeolites	720—880	—	—	—	—	—	—
Allophane	145—900	5—50	—	—	—	—	—
Palygorskite	5—30	—	—	—	—	—	—
Sepiolite	20—45	—	—	—	—	—	—
Muscovite	—	15	—	—	—	—	—
Plagioclase	—	7	—	0.47	0.26	1.2	0.14
Quartz	—	7	—	—	—	—	—

Chapter 4

SOIL CONSTITUENTS

I. INTRODUCTION

Quantitatively, trace elements are negligible chemical constituents of soils, but are essential as micronutrients for plants. The first publications on trace elements were devoted to plant nutrition problems. Further, it was recognized that the behavior of trace elements in the soil differs widely for both the element and the soil and that these differences should be understood better for the prediction and effective management of the trace element status of soils. Although trace elements are mainly inherited from the parent rocks, their distribution within the soil profiles and their partitioning between the soil components reflect various pedogenic processes as well as the impact of external factors (e.g., agricultural practices, pollution).

Trace element associations with the particular soil phase and soil component appear to be fundamental in defining their behavior. The trace element composition of soils is relatively well established (Figure 7), although there are still diversities in analytical results, especially in the measurements of very small quantities. Currently, there is also a great deal of work on the distribution of these elements among soil components. One must realize, however, that present-day techniques for soil fractionation are quite drastic and cannot provide very comparative and representative results. Knowledge of the behavior and reactions of separate soil components with trace elements, although fundamental, should not be related directly to overall soil properties, and great caution is needed in using several theoretical models for predicting the behavior of trace elements in soils.

II. MINERALS

The mineral constituents of soils inherited from the parent rocks have been exposed for various periods of time to weathering and pedogenic processes. The soil mineral system, which is not necessarily in equilibrium with the soil solution, is complicated by the processes of degradation and neoformation of minerals, as well as by mineral reactions with organic compounds.

The common primary minerals in soils inherited from the parent material can be arranged in two parallel series, according to their susceptibility to weathering processes: (1) series of felsic minerals; plagioclases (Na $\approx$ Ca) > K-feldspar > muscovite > quartz, and (2) series of mafic minerals; olivine > pyroxenes > amphiboles > biotite. These series are based on broad generalizations, and many exceptions may occur in particular soil environments. The primary minerals occurring in some soils are mostly of a larger dimension and are not involved in sorption processes. They are, however, considered to be the source of certain micronutrient elements.

The approximate composition of mineral constituents of surface soils presented in Figure 5 shows that quartz is the most common mineral in the soils, constituting 50 to more than 90% of the solid soil phase. Even in geochemical conditions favorable for the leaching of silicates, quartz remains as a basic soil mineral. Feldspars are of low relative resistance to weathering in soil environments and their alteration usually provides materials for clay mineral formation. Carbonates (calcite, dolomite) and metal oxides are usually accessory minerals in soils of humid climatic zones, while in soils of arid climatic zones they may be significant soil constituents.

The size and shape of mineral particles determine their ratio of surface to volume and mass, and this ratio determines their physical and chemical properties. Therefore, the grain-size composition (physical composition) of soils is considered to be one of the most important factors in soil characteristics and is included in the systems of soil evaluation and classification.

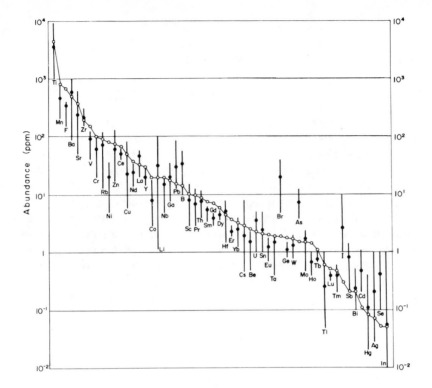

FIGURE 7.   Trace elements in soils compared to their abundance in the lithosphere. Open circles, mean content in the lithosphere; black circles, mean content in topsoils; vertical lines, values commonly found in topsoils.

Sorption properties of the mineral part of soil material are associated principally with the clay and silt-size fractions. These fractions are a mixture of several aluminosilicate clay minerals with lesser amounts of quartz, feldspars, and various oxides and hydroxides. In certain soils carbonate and phosphate minerals are present, and in others, some minerals as sulfides and sulfates may occur.

A. Clay Minerals and Other Aluminosilicates

The boundaries of the mineral group described as "clay minerals" are not well defined, and they tend to enlarge with increasing knowledge. Thus, the division of minerals in this chapter does not relate to any classification system. The common clay minerals in soils can be subdivided into five groups:

1. Kaolinite
2. Montmorillonite, often referred to as smectites
3. Illite
4. Chlorite
5. Vermiculite

Each group includes many compositional and structural varieties; however, they are all 1:1 or 2:1 layer-type aluminosilicates. The structure and chemistry of soil clay minerals have been extensively described in many monographs.[49,281] The surface properties of minerals (area and presence of an electrical charge) seem to be fundamental for the buffer and sink properties of soils.

Table 15
SURFACE AREA OF VARIOUS SOILS

Kind of soil	Surface area $(m^2\ g^{-1})$	
	a	b
Clays and loams	150—250	22—269
Silty loams	120—200	24—117
Sandy loams and loamy sands	10—40	4—67
Rendzinas	—	17—167

Note: Sources are as follows: a,5; b,182.

Clay minerals may contain negligible amounts of trace elements as structural components, but their sorption capacities to trace elements play the most important role. The affinity of trace ions for the clay surface has been examined by many investigators, and while many aspects have been clarified, still much remains to be learned about the adsorption processes.

Although clay mineral samples vary in chemical composition and in their nature, some general surface properties can be given for each mineral group (Table 14) that are responsible for values of the specific surface area of soil materials (Table 15). The capacity values (CEC) vary with the type of clay in the following sequence: montmorillonite, vermiculite > illite, chlorite > kaolinite > halloysite. The ability of the clays to bind the metal ions is correlated with their CEC, and usually the greater the CEC, the greater the amount of cation adsorbed.

The minerals of the montmorillonite group can expand and contract in response to charge and size of the absorbed cation between the clay platelets. Thus, their sorption capacity will differ when saturated with different cations. The microcations sorbed by montmorillonite are also easily released into the liquid phase and, therefore, can be an important pool of micronutrient supply to plants growing in particular soil conditions.

The bonding processes for absorbed microcations, although carefully studied, remain controversial.[2,227,356] It has been well demonstrated, however, that equilibrium and pH values are the most basic qualities in the reactions of sorption and release of microcations by clay minerals.

The chemical nature of transition metals adsorbed on clay minerals has recently been the subject of great interest. Clays containing exchangeable transition metal cations (mainly Cu, Fe, and Co) are known to act as electron or proton acceptors, thus they can be activators in transformations, decomposition, and polymerization of the adsorbed organic species.

Relatively little is known about the adsorption of metal ions on amorphous alumina and silica gels. The mechanisms of coprecipitation and mobilization of certain trace cations with alumina or silica gels may play a significant role in their behavior in the particular soil, especially in those of the tropical climate zone. There are suggestions that metal ions (mainly Cu^{2+}) can substitute for aluminum in the mineral structures, while soluble silicic acid promotes adsorption of Co, Ni, and Zn by clays.[517,793]

The strong adsorption of divalent trace cations (Cu, Pb, Zn, Ni, Co, Cd, and Sr) by freshly precipitated alumina gel is suggested by Kinniburgh et al.[392] to play a role in determining the availability to plants and the movement of some of these cations through the soil.

Some aluminosilicates are known to occur in soils as phyllosilicates (palygorskite, attapulgite, sepiolite) and as zeolites. They all have alternative 2:1-type open structures and are associated with the clay minerals. They can be inherited from parent materials, but also can be of pedogenic origin.[901] Most often these minerals were found in neutral or alkaline soil series, especially in the presence of salts (e.g., solonetz, solochaks, andosols, rendzinas). Some of them are more acid resistant than others in the soil environment. However, the detection of these minerals, especially zeolites, can be questioned due to possible destruction during chemical pretreatments of soil material.

Zeolites exhibit capacities to fix gases, vapors, and liquids, and they are known to be active in the sorption of Ba, Br, F, I, and also Mn and Sr. Barrer[49] has illustrated a high affinity of zeolites to sorb and to complex trace elements, particularly heavy metals and radionuclides.

Amorphous aluminosilicates occurring in soils are often described as allophane and imogolite. Allophanes are present in many soils, and the investigators pointed out their importance in the formation and transformation of noncrystalline clay materials and opaline silica.[833] Both allophane and imogolite make sequences of various types and are formed mainly in the soils developed under a warm humid climate. Allophanes are more stable in acid soils, and imologite is more stable in neutral and alkaline soils. Very often they occur as "gel films" that coat soil particles. The effect of these inorganic coatings may be diverse, both increasing and decreasing trace element sorption, also they may reduce the biological availability of occluded trace elements.[356]

All of these mineral constituents have relatively high CEC values and a great affinity to react with soil organic compounds. Under normal soil conditions, they are important trace element sinks.

B. Oxides and Hydroxides

Several oxide minerals, such as silicon oxides, titanium oxides, aluminum oxides, and hydroxides occur in soils. However, in relation to trace element behavior, the most important are Fe and Mn oxides. Al hydroxides can adsorb a variety of trace elements and in some soils the role of these oxides can be more important than that of Fe oxides in retaining certain trace elements. As Norrish[570] has stated, however, there is little direct evidence to support this view.

Oxides and hydroxides of Fe and Mn are relatively common constituents in soils and, having a high pigment power (mainly Fe oxides), determine the color of many soils. Fe and Mn oxides are present in soils in various mineral forms as well as in crystalline, microcrystalline, and amorphous oxides or hydroxides. Their structure and chemical properties are well described by Hem,[313] Jenne,[356] McKenzie,[524,526] and Schwertmann and Taylor.[699]

Although several minerals of the Fe oxides have been detected in soils, goethite is claimed to be the most frequently occurring form. Norrish[570] reported that simple oxides and hydroxides of manganese do not occur in soils and that the most common mineral forms are lithiophorite and birnessite. Chukhrov et al.[140] on the other hand, have identified vernadite — a simple hydrous oxide of Mn — as the most frequent form in the majority of soils.

These oxides are exposed to reduction and chelation solubility and oxidation-precipitation reactions, in which microbiological processes play an important role. Different nodules of Fe and Mn are known to originate from both chemical and microbial processes; also the formation of some crystalline minerals is known to be effected by microorganisms. The most common Fe-oxidizing bacteria *(Thiobacillium)* and Mn-oxidizing bacteria *(Metallogenium)* are able to tolerate high concentrations of several heavy metals (Zn, Ni, Cu, Co, and Mn). Thus, they are also involved in trace metal cycling in soils.

Fe and Mn oxides occur in soils as coatings on soil particles, as fillings in cracks and veins, and as concretions or nodules. Norrish,[570] using the electron probe analyzer, indicated that many trace elements in soils are concentrated along the deposited oxides in soil material. Fe and Mn oxides have a high sorption capacity, particularly for trace elements, of which large amounts can be accumulated in nodules and at Fe- and Mn-rich points (Table 16). The mechanisms of sorption involve the isomorphic substitution of divalent or trivalent cations for Fe and Mn ions, the cation exchange reactions, and the oxidation effects at the

Table 16
TRACE ELEMENTS IN IRON AND MANGANESE OXIDES (%)

| Element | Iron minerals and concretions | | Manganese minerals and concretions | | | | |
| | Fe-rich points of surface soils (a) | Goethites (a,f) | Manganese nodules | | | Manganese minerals | |
			Mn-rich points of surface soils (a)	(b)	(c)	From soils[a] (d)	From deposits[b] (e)
Fe	7.97—29.65	51.7—61.9	3.92—17.69	—	0.016	0.65—4.60	0.1—4.5
Mn	0.07—1.52	0.26—0.50	5.50—13.59	0.36—7.20	16.0	47.4—59.9	28—61
Ba	0.089—0.179	—	0.573—2.864	0.014—0.23	0.2	3.2—5.4	11.0—12.8
Cd	—	—	—	—	0.0008	—	—
Ce	—	—	—	—	0.072	—	—
Co	0.04—0.07	0.008[c]	0.54—2.44	0.0082—0.038	0.3	0.45—1.20	0.014—1.200
Cr	—	0.1[c]	—	0.003—0.012	0.0014	—	—
Cu	0.004—0.072	0.08	0.039—0.096	—	0.26	—	0.013—1.260
I	—	—	—	—	0.012—0.090	—	—
Li	—	—	—	—	—	0.03—0.07	0.0002—0.534
Mo	—	0.85	—	—	0.041	—	—
Ni	0.026—0.063	0.017[c]	0.086—0.487	0.0039—0.0067	0.49	0.10—0.34	0.012—1.090
Pb	0.046—0.139	—	0.26—2.04	0.0034—0.010	0.087	—	—
Sb	—	—	—	—	0.0004—0.0025	—	—
Sr	—	—	—	—	0.0825	—	—
V	—	1.7	—	0.0088—0.011	0.044	—	—
Zn	0.072—0.257	1.73—2.35	0.032—0.554	0.0030—0.0033	0.071	—	0.005—0.380
Zr	—	—	—	—	0.065	—	—

Note: Sources are as follows: a, 570; b, 524; c, 94; d, 525; e, 837; and f, 881.

[a] Identified minerals: lithiophonite, birnessite, and hollandite.

[b] Identified minerals: psilomelane, cryptomelane, lithiophorite, and pyrolusite.

[c] In magnetite separated from soils (ignited weight).

surface of the oxide precipitates. Variable charges at the surface (mainly of Fe oxides) also promote the adsorption of anions. A high sorption capacity of Fe oxides for phosphates, molybdates, and selenites is most widely observed and is highly pH dependent, being lower at high pH values.[638] The amount of a particular ion that is adsorbed depends mainly on the pH of the equilibrium solution. The maximum adsorption values for various ions on Fe oxides range between pH 4 and 5.[699]

Some investigators give the order of preferential sorption of metals by goethite as Cu > Zn > Co > Pb > Mn,[699] while others have presented metal ion affinities for the oxide surface in the following orders: Cu > Pb > Zn > Co > Cd,[240] and Pb > Zn > Cd > Tl.[252] However, the extrapolation of these results to all soils is difficult. Apparently, hydrous oxides of Fe and Mn are the most important compounds in the sorption of trace metallic pollutants, and they exhibit diverse affinities to cations having approximately the same physical dimensions as Mn^{2+}, Mn^{3+}, Fe^{2+}, and Fe^{3+}; they are Co^{2+}, Co^{3+}, Ni^{2+}, Cu^{2+}, Zn^{2+}, Cd^{2+}, Pb^{4+}, and Ag^+.

C. Carbonates

Carbonates present in soils are often in metastable and polymorphic varieties and thus sensitive to drainage conditions. Carbonates are common constituents in the soils where evapo-transpiration potential exceeds the rainfall. On the other hand, in soils with a high rate of percolating water, carbonates are easily dissolved and leached out. Nevertheless, Ca is usually the predominating cation in solutions of almost all soils.

Calcite is the most widespread and relatively mobile form of Ca carbonates present in soils; it is usually greatly dispersed and has a major influence on the pH of soils and therefore on trace element behavior.

Trace elements may coprecipitate with carbonates, being incorporated in their structure, or may be sorbed by oxides (mainly Fe and Mn) that were precipitated onto the carbonates or other soil particles. Metallic ions may also influence processes of carbonate precipitations.[356] The greatest affinity for reaction with carbonates has been observed for Co, Cd, Cu, Fe, Mn, Ni, Pb, Sr, U, and Zn. However, a wide variety of the elements under various geochemical environments may substitute for Ca in different proportions in nodular calcites. As Vochten and Geyes[828] observed, the secondary calcite crystals show a remarkably high content of Sr and Co — up to 1000 ppm concentrations. Carbonates can be the dominant trace element sink in a particular soil, but the most important mechanisms for regulating the trace element behavior by carbonates are related to variation of the soil pH.

D. Phosphates

Crystalline forms of phosphate minerals rarely occur in soils; however, many varieties of metastable and metamorphous phosphates are of importance in pedogenic processes. There are few data on the occurrence of Ca phosphates (apatite and hydroapatite) or other phosphates in soils. Rather, it has been suggested that an intimate mixture of Ca, Fe, and Al phosphates predominates in soils.[356]

Several of the rock phosphates contain a large amount of trace elements, of which F, and at times Cd, are highly concentrated (Table 17). Some substitutions for Ca by trace elements are known to occur in natural apatites; they are, however, of little importance in soils. Also many trace elements (Ba, Bi, Cu, Li, Mn, Pb, Re, Sr, Th, U, and Zn) can be incorporated, together with Fe^{3+} and Al^{3+}, in hydrated phosphates.[809] Norrish[570] reported extremely high concentrations of lead (1 to 35% PbO) in the phosphate concentrates occurring in ferralsols (lateritic podzolic soils).

E. Sulfides, Sulfates, and Chlorides

Sulfides, sulfates, and chlorides are negligible compounds in soils that developed in a

Table 17
TRACE ELEMENTS IN PHOSPHORITES AND PHOSPHATE
FERTILIZERS (PPM)

| Element | Phosphorites | | Phosphate fertilizers |
	a	b	c
As	30	0.4—188	2—1,200
B	<50	3—33	5—115
Ba	100	1—1,000	<200
Be	<0.5	1—10	—
Cd	0.01—35 (75)(d)	1—10	7—170 (188)(e)
Ce	100	9—85	20
Co	<3—5	0.6—12	1—10
Cr	2—1,000	7—1,600	66—245
Cu	100	0.6—394	1—300
F	31,000	—	8,500—15,500
Hg	0.2	10—1,000	0.01—0.12
I	0.8—280	0.2—280	—
La	—	7—130	—
Li	—	1—10	—
Mn	30	1—10,000	40—2,000
Mo	0.03	1—138	0.1—60
Ni	<2—1,000	2—30	7—32
Pb	2—14	<1—100	7—225
Sb	0.2—7	1—10	—
Se	—	1—10	<0.5
Sn	0.2	10—15	3—4
Sr	1,000	1,800—2,000	25—500
Ti	600	100—3,000	—
U	90	8—1,300	—
V	300	20—500	2—180
Zn	300	4—345	50—1,450
Zr	30	10—800	50

Note: Sources are as follows: a, 94; b, 809; c, 381; d, 399; and e, 554.

humid climate, but in soils of arid climatic zones they can be the dominant controls of the behavior of trace elements. The metallic ions (mainly Fe^{2+}, Mn^{2+}, Hg^{2+}, and Cu^{2+}) may form relatively stable sulfides of acidic or neutral reducing potential in flooded soils. Several other heavy metals (Cd, Co, Ni, Sn, Ti, and Zn) can also be easily coprecipitated with iron sulfides.[356,828]

The precipitation of metallic ions as sulfides is an important mechanism for regulating the solution concentration of both S^{2-} and metallic cations. Sulfides of heavy metals may be transformed into more soluble oxidized sulfates when flooded soil becomes drained and aerated.

Pyrite is the most common mineral of Fe sulfides in soils and other geochemical environments. Some other heavy metals which readily form sulfides may also be, as is Fe, remarkably related to microbial S cycling in soils, as was recently described by Trudinger and Swaine.[809]

Sulfides of heavy metals are not common in soils, especially in soils with good drainage. Sulfates of metals, mainly of Fe (jarosite), but also of Al (alunites) and Ca (gypsum, anhydrite), are likely to occur under oxidizing soil conditions. They are readily soluble and therefore are greatly involved in soil equilibrium processes. Sulfates of heavy metals are also readily available to plants, and their occurrence in soils has practical importance in agriculture.[673] Chlorides as the most soluble salts occur only in soils of arid or semiarid climatic zones.

III. ORGANISMS IN SOILS

Living organisms, often referred to as the soil biota composed of fauna and flora of various dimensions (macro-, mezo-, and microbiota), occur abundantly in soils. At the microbiota level the boundary between plant and animal cells becomes blurred. The importance of living organisms as reflected in biological activity of soils has been discussed in many textbooks.[187,651,856,898]

The abundance of microorganisms in topsoils varies with soil and climatic conditions and may reach as much as 20% of the total biota of a soil system. There is no easy way of knowing with certainty the biomass of microorganisms because this quantity can be determined only indirectly. The maximum weight of soil biota given by Richards[651] for a hypothetical grassland soil corresponds to 7 t ha^{-1} of microbiota (bacteria and fungi) and to 1.3 t ha^{-1} of mezobiota. Kovalskiy et al.[422] calculated that the biomass of bacteria and fungi present in the plow zone of soil (20 cm depth) range from 0.4 to 1.1 t ha^{-1}. The biomass of bacteria varies significantly during the growing season, and may increase about three times from spring to fall.[421]

Microorganisms are very important ecologically because they are the producing, consuming, and transporting members of the soil ecosystem and therefore are involved in the flow of energy and in the cycling of chemical elements. Thus, the microbiota is responsible for many different processes, from mobilization to accumulation of chemical elements, in soils. Although microorganisms are sensitive to both deficiencies and excesses of trace elements, they can adapt to high concentrations of these elements in their environment.

The role of microorganisms in geochemical cycling of the major elements is relatively well understood on the global level. The biogeochemical cycling of trace elements has received much less attention. With the recognition that microbial transformations of compounds of these elements can result in some problems of soil fertility as well as in the formation of some environmental pollution or detoxication processes, the importance of microbiota in cycling of trace elements, especially heavy metals, has been more extensively studied.

The basic microbial phenomena in cycling processes in the soil environment are

1.   Transport of an element into or out of a cell
2.   Charge alteration of an element
3.   Interaction of an element with organic compounds to become a functional part of the system
4.   Complexing an element by organic acids and other compounds produced by microorganisms
5.   Microbial accumulation or mobilization of an element
6.   Microbial detoxication of poisoned soil at a site

The most important microbial function in soil, however, is the degradation of plant and animal residues. It has become apparent that the quantity of trace elements needed or harmful for growth of microorganisms influences also the biological activity of soils.

All available evidence indicates that a low concentration of trace elements stimulates bacterial growth in soil, but a higher content is harmful, being usually most toxic to the bacteria that fix free N and to nitrifying bacteria.[505,812] Of the 19 trace elements studied by Liang and Tabatabai,[471] all inhibited mineral N production in soils. At the concentrations of 5 μM g^{-1} of soil, the most toxic elements were Ag$^+$ and Hg$^+$, while the least toxic were

Co^{2+}, As^{3+}, Se^{4+}, and W^{6+}. Heavy metals especially are known for their toxicity to microbiota, with fungi and actinomycetes having the most resistance. Reduction of microbial growth and enzymatic activity is often reported for soils contaminated by heavy metals.[88,733,814] Mathur et al.[515] showed that the effect of a naturally high Cu content of histosols is most suppressive on levels of accumulated enzyme activities involved in the degradation of the major components of organic debris in soils. A low rate of decomposition of vegetation having a high concentration of Pb and Zn is apparently due to the same processes in nature.[877]

Suppression and/or stimulation of biosynthesis of microorganisms by heavy metals depends upon the nature of the organisms, the kind of metal, and the pH of soils. Even with one species the range of required or inhibitory concentrations of a given metal varies significantly.[421]

Based on data presented in the monograph by Weinberg,[856] it may be generalized that the highest concentration of Fe, Mn, and Zn required by various microorganisms (fungi, bacteria, bacilla, and actinomycetes) was around X00 $\mu M\ \ell^{-1}$. The inhibitory concentrations of these elements on vegetative growth and secondary metabolism of microorganisms have also been established at the above range.

Heavy metals are known to be the most toxic elements, especially to fungi. Somers[746] reported that the fungicidal action of trace cations is due primarily to the formation of an un-ionized complex with surface groups, e.g., phosphate, carboxyl, and sulfhydryl. This author showed that there is a relationship between the toxic concentration of the metal ion and its electronegativity value. The order of toxicity of aqueous solutions of nitrates and sulfates against conidia of *Alternaria tenuis* was given by Somers[746] as follows:

$$Os > Hg > Ag > Ru > Pb \approx Cr > P > Ce > Cu > Ni$$

$$= Be = Y > Mn = Tl > Zn > Li > Sr$$

In soil systems, Hg, Cd, and As seemed to be the most harmful to ammonification processes, while Cu greatly reduced phosphate mineralization rate.[762,813]

Microorganisms can adapt to high concentrations of trace elements. This has been well illustrated by Aristovskaya[35] and Letunova[464] for several elements such as Fe, Mn, Mo, Se, and B (Figure 8). This adaptation is also well shown in various microbiogeochemical processes described in detail by Babich and Stotzky,[41] Gadd and Griffiths,[251] Kowalskiy,[419] and Zajic.[898] The sensitivity of microorganisms (mainly fungi) to different concentrations of trace elements has often been used in the determination of the availability of micronutrients such as Fe, Cu, Zn, and Mo.[531,572]

The physicochemical relationship between bacteria and mineral surfaces leads to diverse effects of dissolution and secondary precipitation of trace metal ions, including changes in their valence and/or conversion into organometallic compounds. A biological oxidation and reduction of Fe and Mn, for example, is one of the most important factor governing the solubility, and thus, the bioavailability, of these metals in soils. Many bacterial species are implicated in the transformation of trace element compounds, including even neoformation of certain Fe and Mn minerals.[140] This effect, however may also at times be an indirect effect. Bacteria also play the most important role in gley formation which affects the mobility of metals in soils.[81]

Microorganisms take up trace elements, several of which play important metabolic functions.[344,614,467] It has been shown by Kokke,[407] however, that cells of microorganisms may show quite variable affinities for radionuclides that are necessarily related to their biological function (Table 18).

A complex balance of trace elements required for microbial activity is of importance in soil productivity. The quantities of specific trace elements available to soil microorganisms can be the critical determinant in the establishment of a disease condition of certain plants. The trace element competition between plants and microorganisms is apparent in various

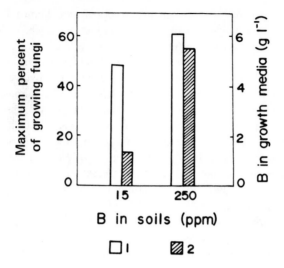

FIGURE 8.   Tolerance of *Actinomycetes* from soils with various B contents to B concentrations in the growth media. (1) Growing fungi, in percent of total number; (2) maximum tolerable concentration of B in solution.[464]

Table 18
RADIONUCLIDE UPTAKE BY THE YEAST *CANDIDA HUMICOLA* AS A FUNCTION OF TIME[407]

Radionuclide	Initial concentration in growth media ($\mu M\ \ell^{-1}$)	Time of growth (days)				
		1	2	4	8	16
		% of initial concentration				
^{89}Ce	40	93	95	98	99	99
^{55}Fe	40	68	83	95	95	95
^{65}Zn	64	60	60	60	75	99
^{89}Sr	40	18	18	18	37	99
^{131}I	20	16	18	25	—	25
^{106}Ru	40	15	30	45	60	75
^{60}Co	40	3	5	5	18	77
^{63}Nd	40	3	3	3	5	12
^{137}Cs	40	1	2	2	3	5

reports, and it may react in different ways. Microbially induced decreases in the availability of trace elements result from a considerably high accumulation of certain elements by microbiota and also from the biological oxidation of compounds of these elements. Microbiological increases of availability, on the other hand, are caused by microorganisms capable of reducing certain compounds (principally, Mn and Fe) and also by their variable bioaccumulation of trace elements (Table 19).

Soil fumigation or steaming and many fungicides kill the fungi and therefore may interfere with the ability of plants to absorb micronutrients. The mechanisms of these phenomena are not yet fully understood; however, they may be related to an imbalance of soil microorganisms and their participation in the transport of ions into or within biological systems. As Martin[510a] reported, variable effects in plants of B, Cu, Li, Mn, and Zn toxicities or deficiencies of Cu, Mn, and Zn were found following soil fumigation.

Gadd and Griffiths[251] concluded that two main types of metal uptake by microorganisms

Table 19
BIOACCUMULATION OF Cu, Mo, and V BY MICROBIOMASS IN TOPSOILS AT VARIOUS SEASONS[421]

	Cu		Mo		V	
Data on soil	Soil (ppm)	Biomass (kg ha^{-1})	Soil (ppm)	Biomass (kg ha^{-1})	Soil (ppm)	Biomass (kg ha^{-1})
Low content of elements	48	0.004 (Sp) 0.006 (Sm) 0.028 (Fl)	6	0.0013 (Sp) 0.002 (Sm) 0.009 (Fl)	66	0.002 (Sp) 0.003 (Sm) 0.013 (Fl)
High content of elements	270	0.60 (Sp) 0.25 (Sm) 0.22 (Fl)	72	0.075 (Sp) 0.031 (Sm) 0.029 (Fl)	840	0.124 (Sp) 0.052 (Sm) 0.049 (Fl)
Control soil (chernozem)	73	0.019 (Sp) 0.069 (Sm) 0.059 (Fl)	10	0.005 (Sp) 0.017 (Sm) 0.013 (Fl)	148	0.005 (Sp) 0.019 (Sm) 0.015 (Fl)

Note: Sp, spring; Sm, summer; Fl, fall.

can occur; the first involves nonspecific binding of the cation to cell surfaces, slime layers, extracellular matrices, etc., while the second involves metabolic-dependent intracellular uptake. The polygalacturonic acid, a common constituent of the outer slime layer of bacterial cells, can complex several trace metals.

The adsorption of trace elements by microorganisms differs widely, as is shown on Tables 18 and 19. Although the mass of microbiota in soils has been calculated to be in the range of 0.X − X t ha^{-1}, the greatest amounts of metals fixed by microorganisms are the following (g ha^{-1}): Ni, 350; Cu, 310; Zn, 250; Co, 150; Mo, 148; and Pb, 8.4, which corresponds to 0.002 to 0.216% of their total abundance in the 20-cm topsoil level of 1 ha.[422,465]

Calculations made by Kovalskiy et al.,[422] and Letunova and Gribovskaya[465] indicated that at the annual rate, the total biocycling of about 11 generations of microbiota may, on the average, involve the following amounts of trace metals (kg ha^{-1}): Ni, 147; Zn, 104; Cu, 78; and Co, 28. Trace elements fixed by microbiomass may be much higher than these values for soils that increase their levels, as has been shown by Krasinskaya and Letunova.[426]

Microbioaccumulation of trace elements may be of great importance both in the cycling of trace elements in soil and in their availability to plants. Fungi and actinomycetes are the most resistant microorganisms to high concentrations of heavy metals, while nitrifying and rhizosphere microorganisms are the most sensitive.

Various waste waters, as well as slurries used for soil irrigation and amendment, can be sources of microbial and other pathogenic organisms of a serious health hazard to humans and animals. This problem has been recently reviewed by Kristensen and Bonde.[429] Bacterial leaching of heavy metals from sewage sludges is, however, a practical application of the biotransformation of the forms of chemical elements.[694]

More than 80% of microorganisms are believed to be adsorbed to soil organic matter and clay minerals.[884] Therefore, comparisons of results obtained for pure cultures of microorganisms may differ from those naturally occurring in soils.

The rhizosphere flora plays a special role in the bioactivity of soils and the availability of nutrients. Although effects of mycorrhizas have been almost always ascribed to an increased phosphate uptake, some observations indicate that they may also influence micronutrient supply. As Lambert et al.[456] and Woldendorp[884] reported, Zn, Cu, and Sr are the chief elements supplied to plants by a given type of mycorrhiza. Some negative effects occurring in rhizospheres may be observed when anaerobiosis around the root surface exists due to a high oxygen demand of microflora leads to the formation of ferrous iron compounds

Table 20
METALS IN SURFACE SOILS AND
EARTHWORMS (PPM)

Metal	Soil	Earthworms	Ratio, worms/soil	Ref.
Cd	2	15	7.5	339
	4	4	1	339
	1.6	11.1	6.9	264
	0.9	14.4	16	264
	1.1	18	16	160
	0.6	12	20	160
	0.1	2.7	27	160
	4.1	10.3[a]	27.6	179
Cu	20	13	0.65	339
	252	11	0.04	339
	335	11	0.03	339
	52	28	0.53	160
	26	18	0.69	160
	9	5	0.55	160
Hg	3.8	1.29[b]	0.33	111
	0.1	0.04[b]	0.40	111
Mn	1330	82	0.06	339
	226	28	0.12	339
	164	27	0.16	339
Ni	26	31	1.19	264
	18	29	1.61	264
	12	32	2.66	264
Pb	1314	3592	2.73	339
	629	9	0.01	339
	700	331	0.47	264
	94	101	1.04	264
	170	62	0.36	160
	20	9	0.45	160
	870	109[a]	0.12	160
Zn	138	739	5.35	339
	992	676	0.68	339
	219	670	3.05	264
	49	400	8.16	264
	275	2000	7.27	160
	40	900	22.50	160
	81	662[a]	8.17	179

Note: Element concentrations expressed on dry weight basis. Organisms
analyzed are *Lumbricus nibellus* or *L. terrestris*, except as indicated.

[a] Other Invertebrata.
[b] FW basis.

which are taken up by plants to concentrations that cause the physiological disorder known as Fe toxicity.[807]

Mezo- and macrobiota occurring in soils are also involved in biocycling of trace elements. Some species of soil fauna, especially earthworms (Annelida), ingest both decaying plant material and mineral soil components. These organisms are known to be able to accumulate certain trace elements in their tissues. Earthworms can selectively take up trace metals and, therefore, are sometimes used as indicators for soil contamination (Table 20). Although heavy metals in earthworms were often found to be significantly related to the corresponding metals in soils, many other factors may also influence this correlation. In soils highly contaminated by trace elements, mezo- and macrobiota are decreased, their metabolism is inhibited, and finally all organisms may vanish.

IV. ORGANIC MATTER

Organic matter of soils consists of a mixture of plant and animal products in various stages of decomposition and of substances that were synthesized chemically and biologically. This complex material, greatly simplified, can be divided into humic and nonhumic substances. Organic matter is widely distributed in soils, miscellaneous deposits, and natural waters. The amount of organic carbon in the earth as humus (50×10^{11} t) has been calculated to exceed that which occurs in living organisms (7×10^{11} t).[201]

The major portion of the organic matter in most soils results from biological decay of the biota residues. The end products of this degradation are humic substances, organic acids of low-molecular and high-molecular weights, carbohydrates, proteins, peptides, amino acids, lipids, waxes, polycyclic aromatic hydrocarbons, and lignin fragments. In addition, the excretion products of roots, composed of a wide variety of simple organic acids, are present in soils. It should be mentioned, however, that the composition and properties of organic matter are dependent upon climatic conditions, soil types, and agricultural practices.

The most stable compounds in soils are humic substances partitioned into the fractions of humic acid, fulvic acid, and humin, which are similar in structure, but differ in their reactions. Humic substances are of a coiled polymer chain structure and contain a relatively large number of functional groups (CO_2, OH, C=C, COOH, SH, CO_2H) having a great affinity for interacting with metal ions. Owing to a particular combination of different groups (mainly, OH and SH), humic substances are able to form complexes with certain cations. Some trace anions, such as B, I, and Se, are also well known to be organically bound in soils. Humic substances are also easily adsorbed by clay and oxide particles in soil and water environments, and these responses are highly dependent on trace cations.[779,799]

Interactions between humic substances and metals have been described as ion exchange, surface sorption, chelation, coagulation, and peptization. It should be emphasized that the existence of a particular site for each cation is not easy to prove because the metal may be bound to two or more ligands from different molecules. All reactions between organic matter and cations lead to the formation of water-soluble and/or water-insoluble complexes.

Sholkovitz and Copland[720] studied the complexing and chelation of trace elements with organic ligands in natural waters. Their studies led to the conclusion that solubilities of humic acid complexes with Fe, Cu, Ni, Cd, Cu, and Mn are the reverse of those predicted from inorganic solubility considerations. The complexing of these ions with humic substances led to the solubilization at high pH (range 3 to 9.5) and precipitation at low pH (range, 3 to 1).

Organic matter is of importance in the transportation (and subsequent leaching) and accumulation of metallic ions known to be present in soils and waters as chelates of various stability and in supplying these ions to plant roots. The ion exchange equilibrium has been extensively studied for determining the stability constant of metallo-organic matter complexes in soils. The values of stability constants determined by several authors described the ability of humic acids to form complexes with metals (Table 21). Metal-fulvic acid complexes with lower stability constants usually are more readily soluble and thus more available to plant roots.

The highest stability-constant values were reported by Takamatsu and Yoshida[771] for Cu^{2+}, Pb^{2+}, and Cd^{2+} complexed with humic acid at pH 5 and by Kitagishi and Yamane[395] for Cu^{2+}, Zn^{2+}, Ni^{2+}, and Cd^{2+} at pH 7. Andrzejewski and Rosikiewicz[28] observed that Mn^{2+}, Co^{2+}, and Ni^{2+} complexes with humic acids were partly soluble, while those of Cu^{2+}, Fe^{2+}, and Cr^{3+} were insoluble. Augustyn and Urbaniak[40] also stated that the higher retention

Table 21

STABILITY CONSTANTS EXPRESSED AS LOG K OF METAL FULVIC AND HUMIC ACID COMPLEXES AT VARIOUS pH LEVELS OF THE MEDIA

| Cation | pH 3 | | pH 3.5 | pH 5 | | | | pH 7 |
	FA (a)	HA (d)	FA (b)	FA (a)	FA (b)	HA (c)	HA (d)	HA (d)
Cu^{2+}	3.3	6.8	5.8	4.0	8.7	8.7	12.6	12.3
Ni^{2+}	3.2	5.4	3.5	4.2	4.1	—	7.6	9.6
Co^{2+}	2.8	—	2.2	4.1	3.7	—	—	—
Pb^{2+}	2.7	—	3.1	4.0	6.2	8.3	—	—
Zn^{2+}	2.3	5.1	1.7	3.6	2.3	—	7.2	10.3
Mn^{2+}	2.1	0	1.5	3.7	3.8	—	0	5.6
Cd^{2+}	—	5.3	—	—	—	6.3	5.5	8.9
Fe^{2+}	—	5.4	5.1	—	5.8	—	6.4	4.8
Ca^{2+}	2.7	0	·2.0	3.4	2.9	—	0	6.5
Mg^{2+}	1.9	0	1.2	2.2	2.1	—	0	5.5
Fe^{3+}	6.1[a]	11.4	—	—	—	—	8.5	6.6
Al^{3+}	3.7[b]	—	—	—	—	—	—	—

Note: Sources are as follows: a, 692; b, 571; c, 771; and d, 395.

[a]   Determined at pH 1.7.
[b]   Determined at pH 2.4.

Table 22

TRACE ELEMENTS IN SOIL ORGANIC MATTER AND IN CLAY FRACTION (PPM DW)[756]

| Surface soil | Element | Content of clay fraction (<1 μm) | | | |
		Total	Organic matter	Humic acid	Fulvic acid
Chernozem	Cu	90	33.0	3.6	29.4
	Zn	116	41.5	3.4	38.1
	Mn	1110	262	Trace	254
	Mo	5	1.7	0.8	0.9
Podzol	Cu	44	17.9	1.2	16.7
	Zn	80	44.7	15.6	29.1
	Mn	1830	307	44	267
	Mo	3	0.7	0.2	0.5

by humic acid was of Fe^{2+}, Cu^{2+}, and Zn^{2+} as compared to other metallic ions. Fe^{3+} and Al^{3+}, however, form the most stable complexes with fulvic acid which greatly interfere with the crystallization of aluminum hydroxide polymorphs.[406]

The stability of metal complexes with fulvic and humic acids increases, in many cases, with increasing pH from 3 to 7 (Table 21). This is best illustrated for Pb, as studied extensively by Hildebrand and Blume.[319] The binding of Fe^{2+} and Fe^{3+} by fulvic acid in solution below pH 5.0 is very strong and, apparently, cannot be exchanged easily by other metals. A relatively high value of the stability constants of Ca^{2+} suggests that this metal can compete with Zn^{2+} and Mn^{2+} in ion exchange processes. Most likely, however, several heavy metals such as Cu^{2+}, Ni^{2+}, Co^{2+}, and Pb^{2+} will readily form stable organic complexes with fulvic acid and, most probably, also with other organic compounds.

Analyses of fractionated organic acids from soils by Stepanova[756] confirm the greater affinity of fulvic acids for heavy metals (Table 22). Mickiewich et al.[537] also reported a much higher concentration mainly of Cu, Pb, and Ti in fulvic acid than was found in humic acid. Heavy metals in soils tend to accumulate in the organic substances, and the lower the metal content, the higher the energy linkage of the metallo-organic groups.[908]

The commonly used value of the stability constant of a complex can be defined as an equilibrium constant of a reaction that forms a soluble complex or chelate. In order to include information about the behavior of insoluble complexes, the value of the stability index has been proposed by Cottenie et al.[148] This index describes the ratio of a given metal fixed with organic substances to its amount in inorganic fractions. The stability index for pure humic and fulvic acids shows that heavy metals (Cu, Zn, Pb, Mn) form complexes several times more readily with humic acid rather than with fulvic acid, and that the highest proportion of Cu is fixed with humic acid over the range of pH 4 to 5, while with fulvic acid the range of pH is limited to 6 to 7. Both acids often show a higher affinity for Cu and Pb rather than for Fe and Mn. These findings agree with those reported by Van Dijk,[820] Stevenson and Ardakani,[757] Förstner and Müller,[242] Pauli,[601] Vlasov and Mikhaylova,[825] and Schnitzer and Khan[692] and indicate that the order of the stability constants of metallo-organic complexes, although quite variable depending on pH and other properties of the medium, can be presented in the following sequence: U > Hg > Sn > Pb > Cu > Ni > Co > Fe > Cd > Zn > Mn > Sr.

Schnitzer and Kerndorff[690] recently established the order of the affinity of metal ions to form water-insoluble complexes with fulvic acid. Although this order depends on the pH of the medium, it may be presented as follows:

$$Fe = Cr = Al > Pb = Cu > Hg > Zn = Ni = Co = Cd = Mn$$

The solubility of fulvic acid-metal complexes is strongly controlled by the ratios FA/metal, therefore, when this ratio is lower than 2, the formation of water-insoluble complexes is favored. There is diversity, however, in the interpretation of metal ion binding by peat, because as Bloom and McBride[80] reported, peat and humic acids are likely to bind, at an acid pH, most divalent cations (Mn, Fe, Co, Ni, and Zn) as hydrated ions. The exception is the Cu^{2+} ion coordinating with functional oxygens of the peat which results in strongly immobilized Cu^{2+} binding.

The index of organic affinities of trace elements in various coal samples was calculated by Gluskoter et al.[269] These authors distinguished three groups of elements:

1. With the highest organic affinity — Ge, Be, B, Br, and Sb
2. With medium organic affinity — Co, Ni, Cu, Cr, Se
3. With the lowest affinity, but occurring in all organic fractions — Cd, Mn, Mo, Fe, Zn, and As

The affinity of humic substances to accumulate trace cations has great importance in their geochemistry. The so-called "geochemical enrichment" factors of humic acid that was extracted from peat can reach a value of 10,000 from very low concentrations of cations in natural waters.[299,691] Trace elements migrating as anions (V and Mo) are reduced by humic acids and fixed in the cationic forms (VO^{2+}, MoO_4^{2+}). Metals complexed by fulvic acid presumably are more available to plant roots and soil biota than are those accumulated by humic acid which can form both water-soluble and water-insoluble complexes with metal ions and hydrous oxides.

Cottenie et al.[148] calculated that the humic acid of a soil containing 4% humus may bind 4500 kg Pb, 17,929 kg Fe, 1517 kg Cu, 1015 kg Zn, and 913 kg Mn per hectare. The ability of humic acid to complex with metals was calculated also by Ovcharenko et al.[587]

and expressed in grams per kilogram of humic acid, as follows: Cu, 3.3; Zn, 3.3; Co, 3.2; Fe, 3.0, and Mn, 2.6. Sapek[681] showed that the ability of humic acids to fix cations differs widely, and that those isolated from the A_o horizon of podzolic soil have about two times lower sorption capacity to metals than those extracted from the B_h horizon of the same soil. In his experiment, the heavy metal content of air-dry humic acid reached more than 29%. All of these values were determined under laboratory conditions; in a natural soil system these proportions would be appreciably smaller. In general, however, it can be expected that up to 50% of total trace-element content is fixed by organic matter in mineral soils. These figures, however, can vary significantly.[756]

Owing to the relatively insoluble complexes of humic acids with heavy metals, especially in an acid medium, these complexes can be considered to be organic storage for heavy metals in soils. The organic matter may act as an important regulator of the mobility of trace elements in soils; however, in the majority of mineral soils, organic matter does not exceed 2% of total soil weight; therefore, it cannot be of the greatest importance in overall controls of trace element behavior in soils.

A high organic matter content of soil has a complex influence on the behavior of trace elements. The deficiency symptoms of plants grown on drained peatland or moorland (histosols) may be the result of a strong retention of Cu, Zn, Mo, and Mn by the insoluble humic acid.[135] A strong fixation of Cu in soils rich in humus is the most common and may result in a high Mo to Cu ratio in forage that is toxic to cattle. Applying organic matter to soil, however, raises the number of microorganisms that can reduce several cations, mainly Fe and Mn, and, in consequence, increase their availability. Increased organic matter content in Pb-amended plots is due to an enhanced preservation of stable humus, perhaps because of newly formed Pb-organic complexes with humic and fulvic acids which are protected from microbial attack.[903] However, there is also evidence that Pb complexes with low molecular weight humic substances were mobilized in the soil solution. On the other hand, some organic compounds present in root exudates and in humus can oxidize and therefore immobilize cupric compounds in soil.[81,189]

Simple organic compounds, such as certain amino acids, hydroxy acids, and also phosphoric acids naturally occurring in soils, are effective as chelating agents for trace elements. Cation chelation is an important factor in soil formation processes, as well as in nutrient supply to plant roots. The solubility of metal complexes depends on both the binding strength and the mobility of the complex thus formed, which is determined mainly by the size of the organic group involved. Strong binding of metal to a low molecular organic substance will appreciably increase its mobility in soil (Figure 9). Organic acids of leaf litter are known to be active in the mobilization of heavy metals in soils. An extract of pine needles dissolved more metals than an extract of oak leaves; in both cases, however, Cu and Zn were more readily complexed than were Co, Ni, and Cd.[87,374] In spite of a high mobilization of heavy metals, forest soil litter is also well known as an important sink of heavy metal and radionuclide pollutants, as reported by Pavlotskaya,[602,603] Van Hook et al.,[823] and Schnitzer and Khan.[692] The ability of simple organic acids to solubilize heavy metals may be of importance in their cycling. Rashid[642] calculated that each gram of amino acids occurring in the sediments may mobilize 4 to 440 mg of various metals, showing the highest affinity to Ni and Co and the lowest to Mn.

Several chelating agents are at present used in diagnostic extraction for available micronutrients in soils (see Norvell,[571] Mengel and Kirkby,[531] and Lindsay).[477] Of those commonly used are ethylenediaminetetraacetic acid (EDTA) and diethylenetriaminepentaacetic acid (DTPA) which have been used for many years for determining plant-available trace elements in soils (Figure 10). Although the results differ when compared with other soil extractants and with uptake by plants, they are applied in many testing methods, and the ranges for critical levels are given for some micronutrients such as Cu, Zn, Mn, and Fe.

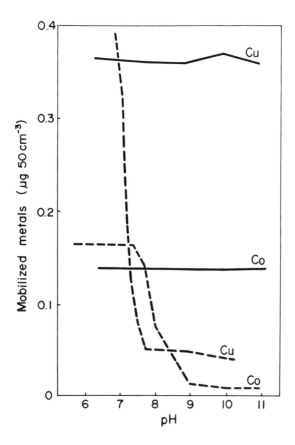

FIGURE 9.   The effect of pH on the solubility of Co and Cu
mobilized by aerobically decomposing alfalfa. Solid lines, complexed metals; broken lines, control solution of CuCl$_2$ or CoCl$_2$
alone.[81]

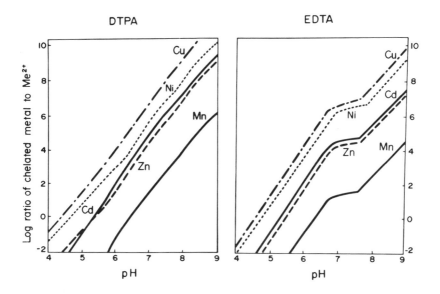

FIGURE 10.   The effect of pH on metal chelating abilities of DTPA and EDTA.[571]

The DTPA soil test has also been developed to assess pollution by heavy metals of soils.[143] Soluble chelates of some trace metals, mainly Mn and Zn (i.e., Mn-EDTA; Na_2Zn-EDTA), are also effective as fertilizers.

Chapter 5

TRACE ELEMENTS IN PLANTS

I. INTRODUCTION

The trace element metabolism of plants has been extensively studied and the basic information on many topics is available in monographs on plant physiology or plant nutrition. The metabolic fate and role of each trace element in plants can be characterized in relation to some basic processes such as:

1. Uptake (absorption), and transport within a plant
2. Enzymatic processes
3. Concentrations and forms of occurrence
4. Deficiency and toxicity
5. Ion competition and interaction

These topics are relatively well understood for certain micronutrients, but further investigations are needed for many other trace elements. The reaction of plants to chemical stresses that are caused by both deficiencies and excesses of trace elements cannot be defined exactly because plants have developed during their evolution and course of life (ontogeny and phylogeny) several biochemical mechanisms that have resulted in adaptation to and tolerance of new or chemically imbalanced environments. Therefore, plant responses to trace elements in the soil and ambient air should always be investigated for the particular soil-plant system.

The chemical composition of plants reflects, in general, the elemental composition of the growth media. The extent to which this relation exists, however, is highly variable and is governed by many different factors. The common concentrations of trace elements in plants growing on various, but nonpolluted, soils show quite a large variation for each element.

II. ABSORPTION

The main sources of trace elements to plants are their growth media, e.g., nutrient solutions or soils. One of the most important factors that determines the biological availability of a trace element is its binding to soil constituents. In general, plants readily take up the species of trace elements that are dissolved in the soil solutions in either ionic or chelated and complexed forms. Much has been written on the absorption of trace elements from solutions by Moore,[548] Loneragan,[489] Mengel and Kirkby,[531] and others, and this absorption can be summarized as follows:

1. It usually operates at very low concentrations in solutions.
2. It depends largely on the concentrations in the solutions, especially at low ranges.
3. The rate depends strongly on the occurrence H^+ and other ions.
4. The intensity varies with plant species and stage of development.
5. The processes are sensitive to some properties of the soil environment such as temperature, aeration, and redox potential.
6. It may be selective for a particular ion.
7. The accumulation of some ions can take place against a concentration gradient.
8. Mycorrhizae play an important role in cycling between external media and roots.

Generalizations of plant processes operative in the absorption of trace elements rest on

the evidence for one or a few of the elements, and most often can represent some approximation to processes acting in a natural plant-soil system. Adsorption by roots is the main
pathway of trace elements to plants; however, the ability of other tissues to readily absorb
some nutrients, including trace elements, also has been observed.

A. Root Uptake

The absorption of trace elements by roots can be both passive (nonmetabolic) and active
(metabolic), but there are some disagreements reported in the literature concerning which
method is involved in certain elements. Despite controversies, in each case the rate of trace
element uptake will positively correlate with its available pool at the root surface.

Passive uptake is the diffusion of ions from the external solution into the root endodermis.
Active uptake requires metabolic energy and takes place against a chemical gradient. Several
data support the suggestion that, at the concentration generally present in soil solutions, the
absorption of trace elements by plant roots is controlled by metabolic processes within
roots.[489,548,788]

Much evidence indicates that roots exhibit great activity in the mobilization of trace
elements that are bound by various soil constituents. The trace elements most readily available
to plants are, in general, those that are adsorbed on clay minerals (especially, montmorillonite
and illite), while those fixed by oxides and bound onto microorganisms are much less readily
available. The depletion of trace elements in solution from the root-soil interface reflects a
higher rate of their uptake by roots than mass-flow and diffusion mechanisms in certain
soils.[531,638] The mechanisms of uptake of trace elements by roots involve several processes:

1.   Cation exchange by roots
2.   Transport inside cells by chelating agents or other carriers
3.   Rhizosphere effects

Ions and other materials released from roots to rooting media control nutrient uptake by
roots. Cation oxidation states around roots are believed to be of great importance in these
processes. Changes in the pH of the root ambient solution may play an especially significant
role in the rate of availability of certain trace elements.[241]

Chaney et al.[128] believed that the reduction step is obligatory in root uptake of Fe. The
reduction of other metals such as Mn, Cu, Sn, or Hg in the uptake step apparently has not
been clearly observed. Rice roots, on the other hand, exhibit a peculiar mechanism to absorb
Si and Se in the form of oxides.[395]

The ability of different plants to absorb trace elements varies greatly; when compared at
a large scale, however, the index of their accumulating ability illustrates some general trends.
Some elements such as Cd, B, Br, Cs, and Rb are extremely easily taken up, while Ba, Ti,
Zr, Sc, Bi, Ga and, to an extent, Fe and Se, are but slightly available to plants (Figure 11).
This trend, however, will differ a great deal for particular soil-plant systems.

Fungi are nongreen plants with quite a diverse mechanism of nutrient uptake, and they
have a specific affinity for some trace elements. They may accumulate Hg and also other
elements such as Cd, Se, Cu, and Zn to high levels (Figure 11).

B. Foliar Uptake

The bioavailability of trace elements from aerial sources through the leaves may have a
significant impact on plant contamination and it is also of practical importance in foliar
applications of fertilizers, especially of elements such as Fe, Mn, Zn, and Cu. Foliar
absorption of radionuclides released into the atmosphere from nuclear weapons testing and
nuclear power installations is of especially great concern.

Foliar uptake is believed to consist of two phases — nonmetabolic cuticular penetration
which is generally considered to be the major route of entry and metabolic mechanisms

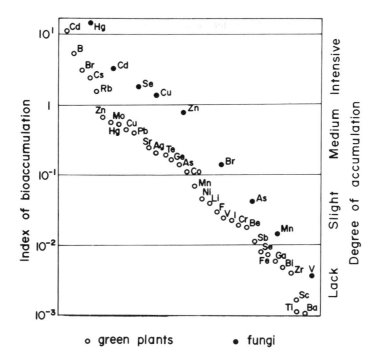

FIGURE 11. Bioaccumulation of trace elements by plants from soils. Index of accumulation was calculated as the ratio of trace elements in plants to their concentration in soils. The calculation is based on data for different plants and soils. Values for fungi are based on data from Byrne and Ravnik.[117]

which account for element accumulation against a concentration gradient. The second process is responsible for transporting ions across the plasma membrane and into the cell protoplast.

Trace elements taken up by leaves can be translocated to other plant tissues, including roots where the excesses of some metals seem to be stored. The rate of trace element movement among tissues varies greatly, depending on the plant organ, its age, and the element involved. Results illustrated in Figure 12 show that Cd, Zn, and Pb absorbed by the tops of brome grass were not likely to move readily to the roots, whereas Cu was very mobile.

A fraction of the trace elements absorbed by leaves may be leached from plant foliage by rainwater. Differences in leaching of trace elements can be related to their function or metabolic association. For example, the easy removal of Pb by washing suggests that the metal was largely a superficial deposit on the leaf surface. In contrast, the small fraction of Cu, Zn, and Cd that can be washed off indicates a greater leaf penetration of these metals than was noted for Pb by Little and Martin[482] and Kabata-Pendias.[374] Moreover, significant absorption of foliar-applied Zn, Fe, Cd, and Hg were reported by Roberts.[657] Foliar leaching by acid rain may involve cation exchange processes, in which the H^+ ion of rainwater replaces microcations held on binding sites in leaf cuticle.[885]

III. TRANSLOCATION

The transport of ions within plant tissues and organs involves many processes:

1. Movement in xylem
2. Movement in phloem
3. Storage, accumulation, and immobilization

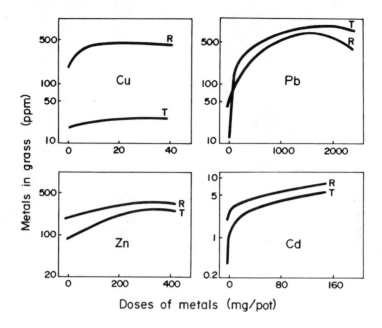

FIGURE 12.   Distribution of heavy metals from aerial sources between the tops (T) and roots (R) of brome grass.[376]

The chelating ligands are most important in the control of cation translocation in plants. However, numerous other factors such as pH, the oxidation-reduction state, competing cations, hydrolysis, polymerization, and the formation of insoluble salts (e.g., phosphate, oxalate, etc.) govern metal mobility within plant tissues.

Tiffin[788] gave a detailed review of the mechanisms involved in the translocation of micronutrients in plants. It can be summarized that long-distance transport of trace elements in higher plants depends upon the vascular tissues (xylem and phloem) and is partly related to the transpiration intensity. Chemical forms of trace metals in phloem exudates differ for each element. Van Goor and Wiersma,[822] for example, reported that Zn was almost all bound to organic compounds, while Mn was only partly complexed.

The distribution and accumulation patterns of trace elements vary considerably for each element, kind of plant, and growth season. As reported by Scheffer et al.,[688a] in the phases of intensive growth of summer barley the amount of Fe and Mn is relatively low, whereas the amount of Cu and Zn is very high. While the first two metals are accumulated mainly in old leaves and leaf sheaths, Cu and Zn seem to be distributed more uniformly through the plant. Differentiation in trace element distribution between various parts of pine trees is also clearly shown by data presented in Table 23. A relatively common phenomenon, however, is the accumulation and immobilization of trace metals in roots, especially when their metal supply is sufficient.

IV. AVAILABILITY

The linear responses of trace element absorption by several plant species in increasing their tissue concentrations from nutrient and soil solutions are illustrated in Figure 13. These responses support the statement that the more reliable methods in diagnosing the available trace element status of soils are those based on element concentrations in the soil solutions rather than methods based on the pool of soluble and/or extractable trace elements.

Different soil-testing methods have been presented by many authors, the most commonly used being those based on specific complexing agents or acid extractants for a given element.

Table 23
VARIATION IN TRACE ELEMENT CONTENTS IN PINE TREES (PPM DW)[186]

Plant part	Al	B	Co	Cr	Cu	Fe	Mn	Ni	Pb	Ti	V
Needles											
1 year old	400	18.0	0.9	4.8	4.2	150	430	6.0	0.2	15	0.6
Older	200	24.0	0.8	4.0	2.5	370	740	2.1	0.5	30	1.2
Branches	400	6.0	0.6	1.6	3.0	650	430	1.1	0.6	25	1.8
Knots	120	4.5	0.2	0.8	1.2	78	185	0.3	0.1	6	0.8
Bark	230	4.5	0.4	1.0	2.0	100	123	0.4	0.3	15	2.8
Wood	7	0.9	0.1	0.3	0.6	5	61	0.3	0.1	1	0.2
Roots											
1-mm diameter	1430	6.5	0.1	0.9	3.5	7171	134	1.1	0.3	46	0.6
5-mm diameter	82	3.2	0.7	0.6	1.2	46	50	0.4	0.1	6	0.5

Note: Samples from a pine forest on old alluvial sands in Ukraina, U.S.S.R.

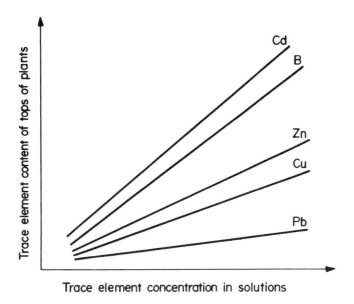

FIGURE 13.   Trace element uptake by plants as a function of their concentrations in nutrient solutions.

Much work has been done on universal extractants for soil tests to assess micronutrient availability. Cox and Kamprath[151] and Walsh and Beaton[847] prepared comprehensive reviews of chemical tests for the determination of water-soluble, acid-extractable, exchangeable, and complexed or chelated trace elements in soils. The interpretation of soil testing results is by no means an easy problem and may give reliable information for only a particular soil-plant system. These methods, nevertheless, are widely used, with various results, in agricultural practices.

The specific properties of plants are very significant in determining the bioavailability of trace elements and are quite variable with changing soil and plant conditions. The availability of different plant species to absorb certain trace elements from the same soil environment is illustrated in Table 24. To provide an effective evaluation of the pool of bioavailable trace elements, therefore, techniques based on both soil tests and plant analyses should be used together.

Table 24

**VARIATION IN TRACE ELEMENT CONTENT AMONG VARIOUS
PLANTS FROM ONE SITE IN THE SAME FOREST ECOSYSTEM[a]
(PPM DW)**

Plants	B	Cd	Cu	Fe	Mn	Pb	Zn	Cr	Ni
Grass	3.4	0.6	4.2	80	740	1.2	59	1.0	1.8
Agrostis alba									
Clover	9.0	0.7	6.0	115	136	2.8	99	1.0	2.2
Trifolium pratense									
Plantain	7.0	1.9	9.8	135	100	2.4	97	1.4	3.0
Plantago major									
Mosses									
Polytrichum juniperinum	3.4	0.8	9.2	800	176	22.4	69	2.0	2.0
Entodon schreberi	3.2	0.7	10.3	425	180	13.0	77	2.8	1.6
Lichens									
Parmelia physodes	2.4	0.4	5.0	1100	62	17.0	78	3.2	4.8
Lobaria pulmonaria	2.4	0.5	7.5	1450	66	28.0	74	3.2	2.4
Edible fungi									
Cantharellus cibarius	4.0	1.0	24.5	49	19	1.2	150	0.4	2.2
Leccinum scabra	0.8	2.7	18.0	44	6	<0.1	125	0.4	1.8
Inedible fungi									
Tylopilus felleus	3.5	1.6	35.0	50	14	0.4	180	0.4	1.2
Russula veternosa	6.4	1.0	32.0	28	18	1.0	175	0.4	1.0

[a]   Pine and birch forest on light sandy soil near Warsaw, Poland.

The sampling procedures for each field, each crop, and a specific plant part in the same stage of growth must be standardized for obtaining compatible results that could be classified as deficient, sufficient, or excessive or toxic for plants. Existing soil and plant tests, however, do not satisfactorily predict trace element deficiencies in crop plants that would respond to application of micronutrients.

Ranges of trace element concentrations and their classification for the mature leaf tissue presented in Table 25 are overall approximations and can differ widely for particular soil-plant systems. It is necessary to emphasize that ranges of concentrations of trace elements required by plants are often very close to the content that exerts a harmful influence on plant metabolism. It is not easy, therefore, to make a clear division between sufficient and excessive quantities of trace elements in plants.

V. ESSENTIALITY, DEFICIENCY, AND EXCESS

Knowledge of the importance of certain trace elements for healthy growth and development of plants dates from the last century. At present, only about ten trace elements are known to be essential for all plants, several are proved necessary for a few species only, and others are known to have stimulating effects on plant growth, but their functions are not yet recognized (Table 26). A feature of the physiology of these elements is that even though many are essential for growth, they can also have toxic effects on cells at higher concentrations. Hypothetical schemes of the reactions of plants to increasing concentrations of the essential and nonessential trace elements are presented in Figures 14 and 15.

The trace elements essential for plants are those which cannot be substituted by others in their specific biochemical roles and that have a direct influence on the organism so that it can neither grow nor complete some metabolic cycle. The elements needing more evidence to establish their essentiality usually are those thought to be required in very low concen-

Table 25
APPROXIMATE CONCENTRATIONS OF
TRACE ELEMENTS IN MATURE LEAF TISSUE
GENERALIZED FOR VARIOUS SPECIES
(PPM DW)[66,171,279,322,369,381,395,531]

Element	Deficient, if less than the stated amounts of essential elements	Sufficient or normal	Excessive or toxic
Ag	—	0.5	5—10
As	—	1—1.7	5—20
B	5—30	10—200	50—200
Ba	—	—	500
Be	—	<1—7	10—50
Cd	—	0.05—0.2	5—30
Co	—	0.02—1	15—50
Cr	—	0.1—0.5	5—30
Cu	2—5	5—30	20—100
F	—	5—30	50—500
Hg	—	—	1—3
Li	—	3	5—50
Mn	15—25	20—300	300—500
Mo	0.1—0.3	0.2—1	10—50
Ni	—	0.1—5	10—100
Pb	—	5—10	30—300
Se	—	0.01—2	5—30
Sn	—	—	60
Sb	—	7—50	150
Ti	—	—	50—200
Tl	—	—	20
V	—	0.2—1.5	5—10
Zn	10—20	27—150	100—400
Zr	—	—	15

Note: Values are not given for very sensitive or highly tolerant plant species.

trations (at $\mu g\ kg^{-1}$ or $ng\ kg^{-1}$ ranges) or that seem to be essential for only some groups or a few species of plants.

Bowen[94] classified the functions and forms of the elements in organisms, based on the current state of knowledge, by dividing the trace elements that occur in plants into the following groups:

1.   Those incorporated into structural materials — Si, Fe, and rarely Ba and Sr
2.   Those bound into miscellaneous small molecules, including antibiotics, and porphyrins — As, B, Br, Cu, Co, F, Fe, Hg, I, Se, Si, and V
3.   Those combined with large molecules, mainly proteins, including enzymes with catalytic properties — Co, Cr (?), Cu, Fe, Mn, Mo, Se, Ni (?), and Zn
4.   Those fixed by large molecules having storage, transport, or unknown functions — Cd, Co, Cu, Fe, Hg, I, Mn, Ni, Se, and Zn
5.   Those related to organelles or their parts (e.g., mitochondria, chloroplasts, some enzyme systems) — Cu, Fe, Mn, Mo, and Zn

In summary, based on extensive literature, trace elements are involved in key metabolic events such as respiration, photosynthesis, and fixation and assimilation of some major

Table 26
FORMS AND PRINCIPAL FUNCTIONS OF TRACE ELEMENTS THAT ARE
ESSENTIAL FOR PLANTS[94,142,349,531,614,859]

Element	Constituent of	Involved in
Al[a]	—	Controlling colloidal properties in the cell, possible activation of some dehydrogenases and oxydases
As[a]	Phospholipid (in algae)	Metabolism of carbohydrates in algae and fungi
B	Phosphogluconates	Metabolism and transport of carbohydrates, flavonoid synthesis, nucleic acid synthesis, phosphate utilization, and polyphenol production
Br[a]	Bromophenols (in algae)	—
Co	Cobamide coenzyme	Symbiotic N_2 fixation, possibly also in non-nodulating plants, and valence changes stimulation synthesis of chlorophyll and proteins (?)
Cu	Various oxidases, plastocyanins, and ceniloplasmin	Oxidation, photosynthesis, protein and carbohydrate metabolism, possibly involved in symbiotic N_2 fixation, and valence changes
F[a]	Fluoracetate (in a few species)	Citrate conversions
Fe	Hemo-proteins and nonheme iron proteins, dehydrogenases, and ferrodoxins	Photosynthesis, N_2 fixation, and valence changes
I[a]	Tyrosine and its derivatives (in angiosperms and algae)	—
Li[a]	—	Metabolism in halophytes
Mn	Many enzyme systems	Photoproduction of oxygen in chloroplasts and, indirectly, in NO_3^- reduction
Mo	Nitrate reductase, nitrogenase, oxidases, and molybdoferredoxin	N_2 fixation, NO_3^- reduction, and valence changes
Ni[a]	Enzyme urease (in *Canavalia* seeds)	Possibly in action of hydrogenase and translocation of N
Rb[a]	—	Function similar to that of K in some plants
Se[a]	Glycene reductase (in *Clostridium* cells)	—
Si	Structural components	—
Sr[a]	—	Function similar to that of Ca in some plants
Ti[a]	—	Possibly photosynthesis and N_2 fixation
V[a]	Porphyrins, hemoproteins	Lipid metabolism, photosynthesis (in green algae), and, possibly, in N_2 fixation
Zn	Anhydrases, dehydrogenases, proteinases, and peptidases	Carbohydrate and protein metabolism

[a] Elements known to be essential for some groups or species and whose general essentiality needs confirmation.

nutrients (e.g., N, S). Trace metals of the transition metal group are known to activate enzymes or to be incorporated into metalloenzymes as electron transfer systems (Cu, Fe, Mn, Zn) and also to catalyze valence changes in the substrate (Cu, Co, Fe, Mo). Some particular roles of several trace elements (Al, Cu, Co, Mo, Mn, and Zn) which seem to be involved in protection mechanisms of frost-hardy and drought-resistant plant varieties are also reported.[511,718]

The requirements of plants and even of individual species for a given micronutrient have been well demonstrated by Hewitt[317] and Chapman.[131] If the supply of an essential trace element is inadequate, the growth of the plant is abnormal or stunted and its further devel-

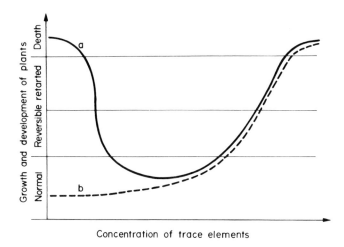

FIGURE 14. Schematic diagram of plant response to stress from deficiency and toxicity of trace elements. (a) Essential trace elements; (b) nonessential trace elements.

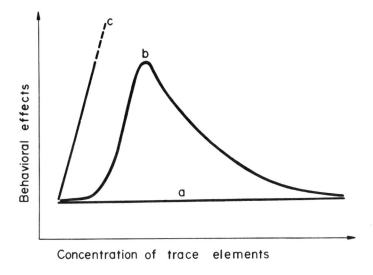

FIGURE 15. Schematic diagram of behavioral plasticity of plants under chemical stress; (a) no behavioral change of entirely tolerant species; (b) development of behavioral tolerance; (c) reaction of nontolerant species leading to damage of organisms followed by no recovery.

opment, especially its metabolic cycles, are disordered. Although deficiency symptoms cannot be generalized, they may be quite characteristic for the particular element. Bergmann and Cumakov[66] presented comprehensive illustrations of deficiency (and some toxicity) symptoms in cultivars. The descriptions of deficiency symptoms summed up in Table 27 indicate that chlorosis is the most frequent symptom. Visible symptoms are important in diagnosis of deficiencies; however, disturbance of metabolic processes and consequent losses in production of biomass may occur before the deficiency symptoms are recognized. In order to develop a better diagnostic method, biochemical indicators based on enzymatic assays were proposed by Ruszkowska et al.,[672] Rajaratinam et al.,[639] and Gartrell et al.[258] as a sensitive test for a hidden deficiency of a given micronutrient. The activity of some enzymes

Table 27
SYMPTOMS OF MICRONUTRIENT DEFICIENCY IN SOME COMMON CULTIVARS[66,114,531,718]

Element	Symptoms	Sensitive crop
B	Chlorosis and browning of young leaves; killed growing points; distorted blossom development; lesions in pith and roots, and multiplication of cell division	Legumes, *Brassica* (cabbage and relatives), beets, celery, grapes, and fruit trees (apples and pears)
Cu	Wilting, melanism, white twisted tips, reduction in panicle formation, and disturbance of lignification	Cereals (oats), sunflower, spinach, and lucerne (alfalfa)
Fe	Interveinal chlorosis of young organs	Fruit trees (citrus), grapes, and several calcifuge species
Mn	Chlorotic spots and necrosis of young leaves and reduced turgor	Cereals (oats), legumes, and fruit trees (apples, cherries, and citrus)
Mo	Chlorosis of leaf margins, "whiptail" of leaves and distorted curding of cauliflower, "fired" margin and deformation of leaves due to NO_3 excess, and destruction of embryonic tissues	*Brassica* (cabbage and relatives) and legumes
Zn	Interveinal chlorosis (mainly of monocots), stunted growth, "little leaf" rosette of trees, and violet-red points on leaves	Cereals (corn), legumes, grasses, hops, flax, grapes, and fruit trees (citrus)

is correlated mainly with Cu, Fe, and Mo levels in plant tissues. The practical use of the enzymatic assays is, however, greatly limited because of a high rate of variation and of technical difficulties in the determination of the enzymatic activity.

The most widely used diagnostic tests are soil and plant analyses. More specific diagnosis of critical levels of some trace metals in plant tissues should also be related to the ratio of antagonistic elements, as described by Nambiar and Motiramani[560] for Fe/Zn ratios in maize. Concentrations of immobile trace elements in old leaves or in whole plants, unlike those of mobile micronutrients, may be misleading in assessing the nutrient status of plants. Nevertheless, plant tissue analysis has been used successfully for assessing deficiencies when based on the normal tissue contents for plant genotypes or species and organs and the stage of plant development. A comprehensive literature has been published in various countries on the diagnosis of trace element deficiencies and their correction through the application of particular micronutrients. Mengel and Kirkby[531] presented the most current information on micronutrients and indicated the need for their application to some cultivars. It should be emphasized, however, that the application of a given micronutrient is effective only if the soil content or the availability of the element is low.

Both deficiencies and toxicities of trace elements for plants most commonly result from complex factors that vary with the specific environment. However, many observations and experiments conducted on various soil types in different countries have clearly demonstrated that soil genesis and soil properties are the main factors controlling micronutrient deficiencies. The data summarized in Table 28 present the general relationship between the occurrence of micronutrient deficiencies in plants and soil properties. The most frequently occurring deficiencies are related to extremely acid soils (light sandy) or to alkaline soils (calcareous) with improper water regimes and with excesses of phosphate, N, and Ca, as well as Fe and Mn oxides.

VI. TOXICITY AND TOLERANCE

Metabolic disorders of plants are effected not only by micronutrient deficiencies, but also

Table 28

SOIL FACTORS CONTRIBUTING TO MICRONUTRIENT DEFICIENCY

Element	Soil units[a]	Contributing soil factors				Critical deficiency limit in soil[b] (ppm)	Plants responding	Some countries of occurrence
		pH range	Organic matter	Water regime	Other factors			
B	Podzols, rendzinas, gleysols, nitosols, and ferralsols	Acid and neutral	Very low or very high	Flooded soils	Light texture, free CaCO₃	0.1—0.3 (HW)	Beets, legumes, crucifers, and grapes	Australia, Egypt, France, Poland, Taiwan, U.S., and U.S.S.R.
Co	Podzols, histosols, rendzinas, and solonetz	Neutral, alkaline, or strongly acid	High	High moisture	Free CaCO₃, high Fe and Mn	0.02—0.3 (AA)[c]	Legumes	Australia, France, Germany, Poland, Sweden, U.S., and U.S.S.R.
Cu	Histosols, podzols, rendzinas, solonchaks, and solonetz	—	Low or high	High moisture	Light leached soil, high N, P, and Zn, free CaCO₃	1—2 (NA, 0.5 N) 0.8—3 (NA, 1 N) 0.8—1 (Ac-ED) 0.2 (AC)	Cereals, legumes, and citrus	Australia, Egypt, U.S., U.S.S.R., and European countries
Fe	Rendzinas, ferralsols, solonetz, arenosols, and chernozems	Alkaline	High with free CaCO₃ or low in acid soil	Poor drainage, moisture extremes	Free CaCO₃, high P, Mn, and HCO₃⁻	0.2—1.5 (DT) 2.5—4.5 (DT) <30—35 (Ac-ED)	Citrus, grapes, pineapples, and tomatoes	Throughout the world, and especially in arid and semiarid regions
Mn	Podzols, rendzinas, and histosols	Strongly acid or alkaline	High, e.g., alkaline peats	Moisture extremes	Free CaCO₃ and high Fe	1—5 (DT) 20—100 (HR) 1—2 (AC) 14—70 (PA)	Cereals, legumes, beets, and citrus	Australia, France, U.S., and U.S.S.R.
Mo	Podzols and ferralsols decalcified	Strongly acid to acid	—	Good drainage	High Fe and Al oxides, and high SO₄-S	0.01—0.6 (AO)[d] <0.1 (HW)	Crucifers, cucurbits, and legumes	Australia, India, U.S., and U.S.S.R.
Se	Podzols, histosols, and ferralsols	Acid	High	Waterlogging	High Fe oxides, and SO₄-S	<0.04 (T)[c]	—	Australia, U.S., and some European countries

Table 28 (continued)
SOIL FACTORS CONTRIBUTING TO MICRONUTRIENT DEFICIENCY

Element	Soil units[a]	Contributing soil factors			Critical deficiency limit in soil[b] (ppm)	Plants responding	Some countries of occurrence
		Organic matter	Water regime	Other factors			
Zn	Podzols, rendzinas, and solonetz	Strongly acid or alkaline	Low	—	Free CaCO₃, high N and P	1—8 (HA) 1.5—3 (ED) 0.4—1.5 (DT) 0.3—2 (AC)	Cereals, legumes, and citrus	Australia, France, India, U.S., and U.S.S.R.

Note: Explanation of symbols used for extraction methods determining contents of soluble elements: AA, 2.5% acetic acid; AC, ammonium acetate; AO, ammonium oxalate; DT, DTPA; ED, EDTA; Ac-ED, ammonium acetate-EDTA; HA, 0.1 *N* hydrochloric acid; HR, hydroquinone reducible method; HW, hot water; NA, 0.5 or 1 *N* nitric acid; PA, 0.1 *N* phosphate acid; and T, total.

a   Generalized soil types assigned to soil units as given in Table 7.
b   Critical deficiency limits vary highly among plant species and soil kinds.
c   Animal response.
d   Highly pH dependent.

by their excesses. In general, plants are much more resistant to an increased concentration than to an insufficient content of a given element. Although many observations have been published on the harmful effects of trace element excesses, the nature of these processes is still poorly understood. Basic reactions, as reviewed by Peterson,[609] Foy et al.,[241] and Bowen[94] related to toxic effects of element excesses are the following:

1.   Changes in permeability of the cell membrane — Ag, Au, Br, Cd, Cu, F, Hg, I, Pb, UO_2
2.   Reactions of thiol groups with cations — Ag, Hg, Pb
3.   Competition for sites with essential metabolites — As, Sb, Se, Te, W, F
4.   Affinity for reacting with phosphate groups and active groups of ADP or ATP — Al, Be, Sc, Y, Zr, lanthanides and, possibly, all heavy metals
5.   Replacement of essential ions (mainly major cations) — Cs, Li, Rb, Se, Sr
6.   Occupation of sites for essential groups such as phosphate and nitrate — arsenate, fluorate, borate, bromate, selenate, tellurate, and tungstate

An assessment of toxic concentrations and effects of trace elements on plants is very complex because it depends on so many factors that it cannot be measured on a linear scale. Some of the most important factors are the proportions of related ions that are present in solution and their compounds. For example, the toxicity of arsenate and selenate is markedly reduced in the presence of excess phosphate or sulfate, and metalloorganic compounds may be either much more toxic than cations or much less so. It should also be noted that certain compounds, e.g., oxygenated anions of metals, may be more toxic than their simple cations.

Several orders of trace element toxicity to plants are presented in the literature and they vary with each experiment and each plant, however, they correlate fairly well with the following factors:

1.   Electronegativity of divalent metals
2.   Solubility products of sulfides
3.   Stability of chelates
4.   Bioavailability

In spite of the reported diversity in toxicity levels, it can be stated that the most toxic metals for both higher plants and certain microorganisms are Hg, Cu, Ni, Pb, Co, Cd, and possibly also Ag, Be, and Sn.

Although plants adapt rather readily to chemical stress, they also may be very sensitive to an excess of a particular trace element. Toxic concentrations of these elements in plant tissues are very difficult to establish. The values presented in Table 25 give very broad approximations of possibly harmful amounts of trace elements in plants. Visible symptoms of toxicity vary for each species and even for individual plants, but most common and nonspecific symptoms of phytotoxicity are chlorotic or brown points of leaves and leaf margins and brown, stunted, coralloid roots (Table 29).

A common feature of plants is their ability to prolong survival under conditions of trace element excesses in their environments, mainly in soils. Lower plants especially, such as microorganisms, mosses, liverworts, and lichens, reveal an extremely high level of adaptation to toxic concentrations of certain trace elements. Zajic,[898] Weinberg,[856] and Iverson and Brinckman[344] presented comprehensive reviews on microorganisms involved in the cycling of trace metals and of their resistance to high metal concentrations.

Although the higher plants are believed to be less tolerant of increased concentrations of trace elements, they are also widely known to accumulate these elements and to survive on soils contaminated by large quantities of various trace elements. Antonovics et al.,[33] Pe-

Table 29
GENERAL EFFECTS OF TRACE ELEMENT TOXICITY ON COMMON CULTIVARS[66,241,381,395,531,731]

Element	Symptoms	Sensitive crop
Al	Overall stunting, dark green leaves, purpling of stems, death of leaf tips, and coralloid and damaged root system	Cereals
As	Red-brown necrotic spots on old leaves, yellowing or browning of roots, depressed tillering	—
B	Margin or leaf tip chlorosis, browning of leaf points, decaying growing points, and wilting and dying-off of older leaves	Cereals, potatoes, tomatoes, cucumbers, sunflowers, and mustard
Cd	Brown margin of leaves, chlorosis, reddish veins and petioles, curled leaves, and brown stunted roots	Legumes (bean, soybean), spinach, radish, carrots, and oats
Co	Interveinal chlorosis in new leaves followed by induced Fe chlorosis and white leaf margins and tips, and damaged root tips	—
Cr	Chlorosis of new leaves, injured root growth	—
Cu	Dark green leaves followed by induced Fe chlorosis, thick, short, or barbed-wire roots, depressed tillering	Cereals and legumes, spinach, citrus seedlings, and gladiolus
F	Margin and leaf tip necrosis, and chlorotic and red-brown points of leaves	Gladiolus, grapes, fruit trees, and pine trees
Fe	Dark green foliage, stunted growth of tops and roots, dark brown to purple leaves of some plants (e.g., "bronzing" disease of rice)	Rice and tobacco
Hg	Severe stunting of seedlings and roots, leaf chlorosis and browning of leaf points	Sugarbeets, maize, and roses
Mn	Chlorosis and necrotic lesions on old leaves, blackish-brown or red necrotic spots, accumulation of MnO_2 particles in epidermal cells, drying tips of leaves, and stunted roots	Cereals, legumes, potatoes, and cabbage
Mo	Yellowing or browning of leaves, depressed root growth, depressed tillering	Cereals
Ni	Interveinal chlorosis in new leaves, gray-green leaves, and brown and stunted roots	Cereals
Pb	Dark green leaves, wilting of older leaves, stunted foliage, and brown short roots	—
Rb	Dark green leaves, stunted foliage, and increasing amount of shoots	—
Se	Interveinal chlorosis or black spots at Se content at about 4 ppm, and complete bleaching or yellowing of younger leaves at higher Se content, pinkish spots on roots	—
Zn	Chlorotic and necrotic leaf tips, interveinal chlorosis in new leaves, retarded growth of entire plant, and injured roots resemble barbed wire	Cereals and spinach

terson,[609] Bradshaw,[101] Woolhouse and Walker,[886] and others attempted to summarize and define what is implied by the term "tolerance" of plants. This term refers to both the species occurring in an area highly contaminated by trace elements, and to individual plants which are able to withstand greater levels of toxicity than are others.

The heavy-metal resistance in plants is of special concern. Practical problems and implications concerning metal-tolerant organisms can be related to:

1. Microbial origin of metal ore deposits
2. Metal cycling in the environment
3. Geobotanical prospecting, i.e., the use of tolerant and sensitive plants to locate natural deposits of metal ores
4. Microbiological extracion of metals from low-grade ores
5. Estabishment of vegetation on toxic waste materials
6. Microbiological treatment of waste waters

Table 30
THE GREATEST ACCUMULATION OF
SOME METALS (PERCENT AW)
REPORTED IN VARIOUS PLANT
SPECIES[609,613]

Element	Plant
	>10%
Ni	*Alyssum bertolonii*
Zn	*Thlaspi calaminare*
	1 to 3%
Cr	*Pimelea suteri* and *Leptospermum scoparium*
Co	*Crotalaria cobaltica*
Ni	*Alyssum bertolonii*
Se	*Astragalus racemosus*
Sr	*Arabis stricta*
U	*Uncinia leptostachya* and *Coprosma arborea*
	0.1 to 1%
Cu	*Becium homblei*
Hg	*Betula papyrifera*
W	*Pinus sibiricus*
Zn	*Equisetum arvense*

7.   Development of resistance in microorganisms to metal-containing fungicides and other pesticides.

The evolution of metal tolerance is believed to be quite rapid in both microorganisms and higher plants and is known to have a genetic basis. Evolutionary changes caused by heavy metals have now been recorded in a large number of species occurring on metalliferous soils that differ from populations of the same species growing on ordinary soils. Species of higher plants that show a tolerance to trace elements belong most commonly to the following families: Caryophyllaceae, Cruciferae, Cyperaceae, Gramineae, Leguminosae, and Chenopodiaceae.

The ranges of highest concentrations of trace elements found in various plant species are presented in Table 30. Various fungi are also well known to be able to accumulate a high proportion of easily soluble and/or easily volatile elements such as Hg, Se, Cd, Cu, and Zn.

Mechanisms of trace element resistance in plants have been the subject of several detailed studies which indicated that both highly specific and multiple metal tolerance may appear as reported by Antonovics et al.,[33] Bradshaw,[101] Simon,[728] Foy et al.,[241] and Cox and Hutchinson.[149] These authors summed up possible mechanisms involved in metal tolerance. They distinguished external factors, such as low solubility and mobility of cations surrounding plant roots, as well as effects of metal ion antagonisms. The real tolerance, however, is related to internal factors. This is not a mechanism of tolerance in a simple sense, but consists of several metabolic processes:

1.   Selective uptake of ions
2.   Decreased permeability of membranes or other differences in the structure and function of membranes
3.   Immobilization of ions in roots, foliage, and seeds

4.   Removal of ions from metabolism by deposition (storage) in fixed and/or insoluble forms in various organs and organelles
5.   Alteration in metabolic patterns — increased enzyme system that is inhibited, or increased antagonistic metabolite, or reduced metabolic pathway by passing an inhibited site
6.   Adaptation to toxic metal replacement of a physiological metal in an enzyme
7.   Release of ions from plants by leaching from foliage, guttation, leaf shedding, and excretion from roots

Considerable evidence was given by Antonovics et al.[33] and Cox and Hutchinson[149] that tolerant plants may also be stimulated in their growth by higher amounts of metal, which reveals a physiological need for an excess of a particular metal by a single plant genotype or species. However, various points still are not clear in the physiology of metal tolerance. Plant resistance to increased concentrations of trace elements and the ability to accumulate extremely high amounts of trace metals may lead to a great health risk by forming a contaminating link in the food chain.

VII. INTERACTION

A chemical balance in living organisms is a basic condition for their proper growth and development. Interactions of chemical elements also are of similar importance to deficiency and toxicity in the physiology of plants. Interactions between chemical elements may be both antagonistic and synergistic, and their imbalanced reactions may cause a real chemical stress in plants.

Antagonism occurs when the combined physiological effect of two or more elements is less than the sum of their independent effects, and synergism occurs when the combined effects of these elements is greater. These interactions may also refer to the ability of one element to inhibit or stimulate the absorption of other elements in plants (Figure 16). All these reactions are quite variable and may occur inside the cells, within the membrane surfaces, and also surrounding plant roots. Interaction processes are controlled by several factors and these mechanisms are still poorly understood, although some data are available.[241,581,840]

Interactions between major and trace elements summarized in Table 31 show clearly that Ca, P, and Mg are the main antagonistic elements against the absorption and metabolism of several trace elements. Some synergistic effects, however, have also been observed for antagonistic pairs of elements, depending on the specific reaction of the plant genotype or species.

Antagonistic effects occur most often in two ways — the macronutrient may inhibit trace element absorption and, in turn, the trace element may inhibit absorption of a macronutrient. These reactions have been observed especially for phosphate, but also have been reported for other macronutrients whose uptake and metabolic activity may be inhibited by several trace elements.[395,463]

Most important for practical application are the antagonistic effects of Ca and P on heavy metals such as Be, Cd, Pb, and Ni that often constitute a health hazard. It is noteworthy that although the antagonistic effects of P and Ca on many trace cations and anions are frequently reviewed in the literature, the antagonistic impact of Mg on trace metals is only occasionally reported.

Interactions observed within plants between trace elements have also indicated that these processes are quite complex, being at times both antagonistic and synergistic in nature, and occasionally are involved in the metabolism of more than two elements (Figure 16). The greatest number of antagonistic reactions have been observed for Fe, Mn, Cu, and Zn which

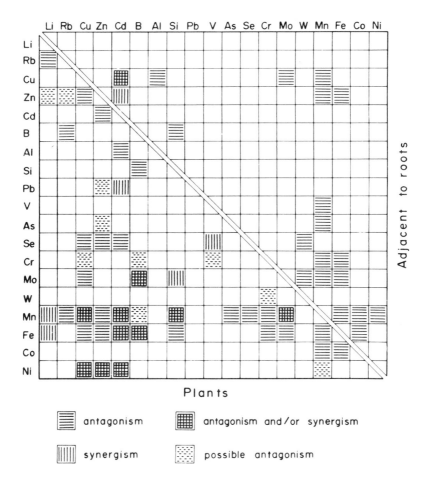

FIGURE 16.   Interactions of trace elements within plant organisms and adjacent to plant roots.

Table 31
INTERACTIONS BETWEEN MAJOR ELEMENTS AND TRACE ELEMENS IN PLANTS[251,381,531,554,663]

Major element	Antagonistic elements	Synergistic elements
Ca	Al, B, Ba, Be, Cd, Co, Cr, Cs, Cu, F, Fe, Li, Mn, Ni, Pb, Sr, and Zn	Cu, Mn, and Zn
Mg	Al, Be, Ba, Cr, Mn, F, Zn, Ni[a], Co[a], Cu[a], and Fe[a]	Al and Zn
P	Al, As, B, Be, Cd, Cr, Cu, F, Fe, Hg, Mo, Mn, Ni, Pb, Rb, Se, Si, Sr, and Zn	Al, B, Cu, F, Fe, Mo, Mn, and Zn
K	Al, B, Hg, Cd, Cr, F, Mo, Mn, and Rb	—
S	As, Ba, Fe, Mo, Pb, and Se	F[b] and Fe
N	B, F, and Cu	B, Cu, Fe, and Mo
Cl	Br and I	—

[a]   Reported for microorganisms.
[b]   Mutual pollution causes significant injury.

are, obviously, the key elements in plant physiology (Table 26). These trace metals are linked to processes of absorption by plants and to the enzymatic pathway. The other trace elements often involved in antagonistic processes with these four trace metals are Cr, Mo, and Se.

Synergistic interactions between trace elements are not commonly observed. Those reported for Cd and other trace metals such as Pb, Fe, and Ni may be artifacts resulting from the destruction of physiological barriers under the stress of excessive concentrations of heavy metals. Moreover, several reactions that occur in the external root media and affect root uptake should not be directly related to metabolic interactions, but the two reactions are not easily separated.

Chapter 6

ELEMENTS OF GROUP I

I. INTRODUCTION

The alkali trace elements of the subgroup Ia are Li, Rb, and Cs. The common characteristic of the alkali elements is the single electron in the outermost energy level, resulting in highly reactive chemical behavior. The relative bonding force holding these monovalent cations is presented most often in the order $Cs^+ > Rb^+ > Li^+$. These cations do not usually form complex ionic species.

The subgroup Ib contains Cu and noble metals Ag and Au. Cu occurs most often as the divalent cation, but can also form single monovalent cations and complex anions. This metal shows many unusual features, including an affinity to combine with S, Se, and Te, and is of great importance in biochemistry. Chemical characteristics of Ag are somewhat similar to those of Cu; however, its role in biochemistry is largely unknown. Au has relatively slight chemical reactivity. Its simple monovalent and trivalent cations exist usually in negligible amounts, thus biogeochemical properties of Au are mostly related to its complex ions.[453] A geochemical relationship is observed between Cu, Ag, and Au.

II. LITHIUM

A. Soils

Li is widely distributed throughout the earth's crust and is likely to be concentrated in acidic igneous rocks and sedimentary aluminosilicates (Table 32). During weathering Li is released from the primary minerals relatively easily in oxidizing and acid media and then is incorporated in clay minerals and is also slightly fixed by organic matter. Thus, the Li content of soils is controlled more by conditions of soil formation than by its initial content in parent rocks. The Li distribution in soil profiles follows the general trends of soil solution circulation; however, it may be highly irregular.[864]

The abundance of Li in surface soils, as presented in Tables 33 and 34 is fairly uniform in various soil units. Grand means for Li vary from 1.2 ppm in light organic soils to 98 ppm in alluvial soils. Lower contents of Li are reported for light sandy soils, especially those derived from glacial drift in a humid climate. In the arid climatic zones Li follows the upward movement of the soil solution and may precipitate at top horizons along with easily soluble salts of chlorites, sulfates, and borates. These reactions explain a relatively higher Li content of soils such as solonchaks, kastanozems, and prairien soils. Also intrazonal young soils derived from alluvium reveal elevated Li concentrations.

In the initial processes of soil formation, Li seems to be highly mobile, while later it may become more stable due to its firm bonding with clay minerals.[864] However, as Shakuri[713] reported, water-soluble forms of Li in the soil profile reach up to about 5% of the total soil content and therefore Li is likely to occur in ground waters of areas having elevated Li contents in rocks and soils. Exchangeable soil Li is reported to be strongly associated with Ca and Mg.[164]

B. Plants

The soluble Li in soils is readily available to plants; therefore the plant content of this element is believed to be a good guide to the Li status of the soil.[279] There are considerable differences in the tolerance of various plant species to Li concentrations, as well as in the plants' ability to take up this element. Borovik-Romanova and Bielova[92] calculated the index of the biological concentrations, based on the ratio Li in plant ash to Li in topsoil. For

Table 32

**LITHIUM, RUBIDIUM, CESIUM, COPPER, SILVER, AND GOLD IN
MAJOR ROCK TYPES**

Rock type	Li (ppm)	Rb (ppm)	Cs (ppm)	Cu (ppm)	Ag (ppb)	Au (ppb)
Magmatic Rocks						
Ultramafic rocks						
Dunites, peridotites, pyroxenites	0.5—X.0	0.1—2	0.X	10—40	50—60	5
Mafic rocks Basalts, gabbros	6—20	20—45	0.5—1.5	60—120	100	0.5—3
Intermediate rocks Diorites, syenites	20—28	100	0.6	15—80	50—70	3.2
Acid rocks Granites, gneisses	25—40	150	2—5	10—30	40	1.2—1.8
Acid rocks (volcanic) Rhyolites, trachytes, dacites	15—45	100—200	3	5—20	50	1.5
Sedimentary Rocks						
Argillaceous sediments	60	120—200	5—10	40—60	70	3—4
Shales	50—75	140—160	6—8	40	70—100	2.5—4
Sandstones	10—40	45	0.5—2	5—30	50—250	3—7
Limestones, dolomites	5—20	5—30	0.5—2	2—10	100—150	2—6

Note: Values commonly found, based on various sources

Table 33

**LITHIUM CONTENT OF SURFACE SOILS
OF THE U.S. (PPM DW)[706]**

Soil	Range	Mean
Sandy soils and lithosols on sandstones	<5—50	16.5
Light loamy soils	9—46	24.5
Loess and soils on silt deposits	9—30	20.5
Clay and clay-loamy soils	10—64	23.5
Alluvial soils	10—120	34.0
Soils over granites and gneisses	10—45	23.5
Soils over volcanic rocks	15—41	25.5
Soils over limestones and calcareous rocks	6—88	26.5
Soils on glacial till and drift	10—30	18.0
Light desert soils	9—69	25.0
Silty prairie soils	10—34	20.5
Chernozems and dark prairie soils	8—40	22.5
Organic light soils	<5—71	13.0
Forest soils	10—57	24.0
Various soils	<5—100	28.0

plants of the Rosaceae family with the highest mean Li content (Table 35), this index is
0.6, while for plants of the Polygonaceae it is 0.04. The highest value of this index, 0.8,
however, was calculated for plants of the Solanaceae, members of which are known to have
the highest tolerance to Li. Some plants of this family when grown in an aridic climatic

Table 34
LITHIUM CONTENT OF SURFACE SOILS OF DIFFERENT COUNTRIES (PPM DW)

Soil	Country	Range	Mean	Ref.
Podzols and	U.S.S.R.	17—60	34	92
sandy soils	New Zealand	5—72	31	864
Loess and silty soils	New Zealand	—	95	864
Loamy and clay soils	New Zealand	1.4—130	45.5	864
Soils on glacial till	Denmark	—	6.2	801
Fluvisols	New Zealand	65—160	98	864
Gleysols	New Zealand	50—100	68	864
Rendzinas	U.S.S.R.	—	42	7
	New Zealand	60—105	80	864
Kastanozems and	U.S.S.R.	31—48	37	7, 92
brown soils	New Zealand	9—175	57.5	864
Ferralsols	U.S.S.R.	10—25	17	92
Solonchaks and	U.S.S.R.	23—53	42	7, 92
solonetz	New Zealand	36—68	55	864
Chernozems	U.S.S.R.	20—65	45	7, 92, 713
Prairien and meadow soils	U.S.S.R.	—	73	713
Histosols and	Denmark	0.5—3.2	1.6	1, 801
other organic soils	New Zealand	0.01—2.8	1.2	864
	New Zealand	0.01—2.8	1.2	864
Forest soils	U.S.S.R.	25—65	50	7, 92, 713
Various soils	Denmark	—	8.6	801
	New Zealand	10—100	64	864
	Great Britain	—	25	818
	38New Guinea	6—28	15	164

Table 35
LITHIUM CONTENT OF PLANT FAMILIES (PPM DW)

Family	U.S.S.R.(a) (mean)	U.S.(b) (range)	New Zealand (c) (range)
Rosaceae	2.9	—	—
Ranunculaceae	2.0	—	—
Solanaceae	1.9	0.01—31 (1120)(d)	—
Violaceae	1.3	—	—
Leguminosae	0.67	0.01—3.1	<0.03—143
Compositae	0.55	—	—
Cruciferae	0.54	—	—
Chenopodiaceae	0.32	—	—
Urticaceae	0.24	—	—
Gramineae	0.24	0.07—1.5	<0.02—13
Polygonaceae	0.10	—	—
Lichenes	—	0.02—0.3(e)	—

Note: Sources are as follows: a, 92; b, 29; c, 864; d, 725; and e, 94 (worldwide data).

Table 36
LITHIUM CONTENT OF PLANT
FOODSTUFFS

Plant	Tissue sample	Mean content (ppm) DW basis	AW basis
Celery(a)	Leaves	6.6(a)	—
Chard(a)	Leaves	6.2(a)	—
Corn(a)	Ears and stover	0.8(a)	—
Corn(b)	Grains	0.05(b)	—
Cabbage	Leaves	0.5	4.9
Carrot	Roots	0.2	2.3
Lettuce	Leaves	0.3	2.0
Onion	Bulbs	0.06	1.6
Potato	Tubers	—	<4
Tomato	Fruits	—	<4
Apple	Fruits	—	<4
Orange	Fruits	0.2	5.3

Note: Sources are as follows: 705; a, 131; b, 197.

zone accumulate more than 1000 ppm Li.[725] The highest uptake of Li was reported for plant species growing on solonetz and solonchak soils or other soils having increased contents of alkali metals.

Li appears to share the K^+ transport carrier and therefore is easily transported in plants, being located mainly in leaf tissues. The Li content of edible plant parts presented in Table 36 shows that some leaves accumulated a higher proportion of Li than did storage roots or bulbs. However, a higher Li content is very often reported for roots. The ratio of root to top for Li in ryegrass was 4.4, while for white clover it was 20.[864] This may suggest that a difference in plant tolerance to Li concentration is related mainly to mechanisms of biological barriers in root tissues.

Wallace et al.[844a] reported that, in most experimental treatments, bush beans accumulated more Li in shoots than in roots while growing in a solution culture with high concentrations of Li. They reported also that higher levels of Li decreased the Zn content in leaves and increased Ca, Fe, and Mn content in all plant tissues.

Although Li is not known to be an essential plant nutrient, there is some evidence that Li can affect plant growth and development.[29] However, stimulating effects of several Li salts reported by various authors have never been confirmed. This observed stimulation may also be related to the influence of other factors, including secondary effects of anions associated with Li.

Increased Li concentrations in soil is toxic to some plant species. Citrus trees are known to be the most susceptible to injury by an excess of Li, which is reported to be toxic at a concentration in leaves of 140 to 220 ppm. Threshold concentrations of Li in plants are highly variable, and moderate to severe toxic effects of 4 to 40 ppm Li concentration in citrus leaves have been observed.[279] In high-Li soils damage to root tips of corn as well as necrotic spots in the interveinal leaf tissues and other nonspecific injury symptoms have been observed.[29]

Ca inhibits Li uptake by plants, therefore the addition of lime to high-Li soil may reduce toxic effects of this element. Li is also toxic to many microorganisms, although the fungi *Penicillium* and *Aspergillus* adapt easily to higher Li concentrations in their growth media.

Table 37
RUBIDIUM CONTENT OF SURFACE SOILS
OF THE U.S. (PPM DW)[706]

Soil	Range	Mean
Sandy soils and lithosols on sandstones	<20—120	50
Light loamy soils	30—100	60
Loess and soils on silt deposits	45—100	75
Clay and clay-loamy soils	45—120	80
Alluvial soils	55—140	100
Soils over granites and gneisses	<20—210	120
Soils over volcanic rocks	20—115	65
Soils over limestones and calcareous rocks	50—100	75
Soils on glacial till and drift	30—80	65
Light desert soils	70—120	95
Silty prairie soils	50—100	65
Chernozems and dark prairie soils	55—115	80
Organic light soils	<20—70	30
Forest soils	<20—120	55

III. RUBIDIUM

A. Soils

Rb abundance in the major rock types reveals its geochemical association with Li, and therefore it has higher concentrations in acidic igneous rocks and sedimentary aluminosilicates (Table 32). In weathering, Rb is closely linked to K; however, its bonding forces to silicates appear to be stronger than those of K; therefore the K/Rb ratio continually decreases in soil-forming processes. The Rb content of soils is largely inherited from the parent rocks, as is indicated by the highest mean Rb contents, 100 to 120 ppm, in soils over granites and gneisses, and in alluvial soils (Table 37). The lowest Rb concentrations (30 to 50 ppm) were reported by Shacklette and Boerngen[706] for sandy soils. The mean Rb content in light loamy soils of Poland was established as 66 ppm by Dobrowolski,[181] while means for soils of various countries range from 33 to 270 ppm as given by Wedepohl.[855] Naidenov and Travesi[558] reported Rb in Bulgarian soils to be within the range of 63 to 420 ppm (mean, 179). Organic matter and micaceous clay minerals increase the sorption capacity of soils for Rb.

B. Plants

Rb apparently is easily taken up by plants, as are other monovalent cations. It may partly substitute for K sites, but cannot substitute for K metabolic roles, therefore, in high concentrations, it is rather toxic to plants. If some plants (e.g., sugar beets) are deficient in K, Rb together with Na can stimulate the growth.[211]

In spite of a chemical similarity of Rb to K, Rb uptake and transportation within plants were reported to be different from those of K.[749,839] The Rb content of green plants differs for each species and for parts of plants, as is shown on Table 38. Zajic[898] reported concentrations of Rb in fungi to vary from 3 to 150 ppm. Some bacteria are known to accumulate Rb and other monovalent cations in subcellular vesicles.[856] Horowitz et al.[325] found the highest Rb content to be more than 100 ppm in some fungi. However, most of the higher plant species analyzed by these authors contained Rb in the range of 20 to 70 ppm (DW).

Table 38
RUBIDIUM CONTENT OF PLANT
FOODSTFFS AND FODDERS
(PPM DW)[381,588,710]

Plant	Tissue samples	Mean content
Cereal	Grains	4
Corn	Grains	3
Onion	Bulbs	1
Lettuce	Leaves	14
Cabbage	Leaves	12
Bean	Pods	51
Soybean	Seeds	220
Apple	Fruits	50
Avocado	Fruits	20
Clover	Tops	44
Lucerne (alfalfa)	Tops	98
Grass	Tops	130

IV. CESIUM

A. Soils

Geochemical characteristics of Cs are similar to those of Rb, but Cs appears to have a greater affinity to be bound to aluminosilicates. Cs is concentrated mainly in acidic igneous rocks and argillaceous sediments (Table 32), as are other monovalent trace cations.

Cs released by weathering in soils is strongly adsorbed, but there is very little information on the Cs status of soils. Using old data cited by Wedepohl,[855] the range of Cs in soils can be calculated as 0.3 to 26 ppm. Koons and Helmke[409] gave the range to be from 0.3 to 5.1 ppm Cs in 4 Canadian reference soils, and Naidenov and Travesi[558] found Cs levels in Bulgarian soils to range from 2.2 to 16.7 ppm, being the highest in surface forest soil and chernozems. These values correspond closely to Cs occurrence in rocks, with the highest values indicating also a possible Cs accumulation in organic horizons of soils.

B. Plants

Cs apparently is not an essential component of plant tissues, and there are few data on its occurrence in plants. As Wedepohl[855] reported from old analytical results, Cs in species of flowering plants was found to range from 3 to 89 ppm (average, 22). Aidiniyan[7] found 0.5 to 1 ppm Cs in tea leaves (AW), and Wallace[840] reported Cs in desert plants to range from 0.03 to 0.4 ppm (DW), with a mean value of 0.1 ppm. Ozoliniya and Kiunke[588] analyzed Cs in different parts of three crops — lettuce, barley, and flax. Each plant accumulated Cs in roots (highest value, 0.32 ppm DW) and in old leaves (highest value, 0.16 ppm DW), while in young leaves Cs reached the highest concentration, 0.07 ppm, in flax. Montford et al.[547] reported Cs to range from 0.2 to 3.3 ppb (FW) in vegetables and from < 0.1 to 2.9 ppb (FW) in fruits.

Souty et al.[749] and Yudintseva et al.[896] reported that Cs is relatively easily taken up by plants, although its absorption by roots appeared not to be parallel to K absorption. An addition of lime and peat to soil greatly inhibited the bioavailability of this metal.

C. [137]Cesium

One of many Cs isotopes, ^{137}Cs is of special environmental concern because it is a by-product of atomic energy production. The geochemical characteristics of this radionuclide are fairly similar to those of nonradioactive Cs, therefore, ^{137}Cs released into the atmosphere becomes strongly adsorbed by clay minerals and also by organic matter in soils. The dis-

Table 39
CONCENTRATIONS OF ^{137}Cs (nCi g^{-1}) IN
VARIOUS PLANTS GROWN ON SOILS WITH
THE ADDITION OF 0.03 nCi kg^{-1} [290,292]

Plant and part	Sandy soil		Loamy soil		Chernozem	
	FW	DW	FW	DW	FW	DW
Cabbage, leaves	4.0	34.9	0.4	6.3	0.6	8.3
Carrot, roots	0.9	6.5	—	—	0.2	1.5
Beet, roots	1.0	5.1	0.4	2.2	0.3	1.6
Potato, tubers	1.3	5.9	1.0	2.3	0.6	2.3
Cucumber, fruits	0.3	8.1	0.1	2.5	0.1	2.4
Tomato, fruits	0.11	2.5	0.02	0.6	0.03	1.2
Oat, straw	—	10—19	—	—	—	—
Oat, grain	—	3—15	—	—	—	—

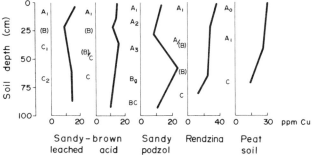

FIGURE 17.   Distribution of Cu in the profiles of different soils developed under humid climate. (Letters indicate genetic soil horizons.)

tribution of ^{137}Cs in soils and plants is a subject of several detailed environmental studies (Table 39).

V. COPPER

A. Soils

Cu in the earth's crust is most abundant in mafic and intermediate rocks and has a tendency to be excluded from carbonate rocks (Table 32). Cu forms several minerals of which the common primary minerals are simple and complex sulfides. These minerals are quite easily soluble in weathering processes and release Cu ions, especially in acid environments. Therefore, Cu is considered among the more mobile of the heavy metals in hypergenic processes. However, Cu is a very versatile trace cation and in soils or depositional material exhibits a great ability to chemically interact with mineral and organic components of soil. The Cu ions can also readily precipitate with various anions such as sulfide, carbonate, and hydroxide. Thus, Cu is a rather immobile element in soils and shows relatively little variation in total content in soil profiles (Figure 17).

Although soluble, therefore mobile and available, forms of Cu in soils are of great importance in agronomic practice, total Cu content of soils gives basic information for geochemical studies. Total Cu distribution in surface soils of various countries is presented in Table 40. In compiling this table an attempt was made to exclude contaminated soils,

Table 40
COPPER CONTENT OF SURFACE SOILS OF DIFFERENT COUNTRIES (PPM DW)

Soil	Country	Range	Mean	Ref.
Sandy soils and podzols	Australia	22—52	—	792
	Israel	—	16[a]	644
	Poland	1—26	8	91, 378, 382
	Nigeria	7—12	9	817
	U.S.S.R.	1.5—29	11	900
Loess and silty soils	Israel	—	25[a]	644
	East Germany	—	21	428
	West Germany	14—31	—	689
	Poland	8—54	19	91, 378, 382
	Nigeria	21—41	31	817
Loamy and clay soils	East Germany	—	14	428
	West Germany	16—70	—	689
	Great Britain	—	37	876
	Poland	4—36	15.5	91, 378, 382
	U.S.S.R.	4—21	12	346, 900
Soils on glacial till	Denmark	—	13.2	801
Fluvisols	Egypt	50—146	80	561
	India	114—160	—	640
	Israel	—	34[a]	644
	Poland	16—28.5	22	378
	U.S.S.R.	11.5—36	25	900
Gleysols	Australia	38—61	—	792
	Great Britain	—	31	876
	Madagascar	35—40	—	557a
	Poland	3—53	12.5	378, 683
Rendzinas	Australia	6.8—43	—	522
	Egypt	11—13	12	561
	East Germany	35—46	—	428
	Israel	—	35[a]	644
	Poland	7—54	16	378, 382, 685
	U.S.S.R.	7—23	15	900
Kastanozems and brown soils	Australia	83—140	—	792
	U.S.S.R.	14—44.5	25.5	343, 900
Ferralsols	Australia	2—96	—	522
	India	44—205	—	640
	Israel	—	60[a]	644
	Ivory Coast	1—100	—	650
	Madagascar	15—40	—	557a
Solonchaks and solonetz	India	55—112	—	640
	Madagascar	15—60	—	557a
	U.S.S.R.	9—37	19	12, 351, 900
Chernozems	Bulgaria	26—38	29	774
	Poland	6.5—53	19	378, 683
	U.S.S.R.	16—70	27.5	346, 351, 900
Prairien and meadow soils	Surinam	1—22.5	—	178
	U.S.S.R.	13—70	35	12, 900
Histosols and other organic soils	Denmark	4—24	15	1, 801
	Israel	27—41[a]	—	644
	Ivory Coast	1—3	—	650
	Poland	1—113	6	682, 683
	U.S.S.R.	5—23	12.5	900
Forest soils	China	—	22	225
	U.S.S.R.	12.5—32	22	346, 900

Table 40 (continued)
COPPER CONTENT OF SURFACE SOILS OF DIFFERENT COUNTRIES (PPM DW)

Soil	Country	Range	Mean	Ref.
Various soils	Canada	5—50	22[a]	521
	Czechoslovakia	23—100[a]	—	64
	Denmark	—	12.5	801
	Great Britain	11—323	23	100, 876
	Japan	4.4—176	34	395
	U.S.S.R.	1—60	28	432, 900

[a] Data for whole soil profiles.

Table 41
COPPER CONTENT OF SURFACE SOILS OF THE U.S. (PPM DW)

Soil	Range	Mean
Sandy soils and lithosols on sandstones	1—70	14
Light loamy soils	3—70	25
Loess and soils on silt deposits	7—100	25
Clay and clay-loamy soils	7—70	29
Alluvial soils	5—50	27
Soils over granites and gneisses	7—70	24
Soils over volcanic rocks	10—150	41
Soils over limestones and calcareous rocks	7—70	21
Soils on glacial till and drift	15—50	21(a)
Light desert soils	5—100	24
Silty prairie soils	10—50	20(a)
Chernozems and dark prairie soils	10—70	27
Organic light soils	1—100	15
Forest soils	7—150	17(a)
Various soils	3—300	26

Note: Sources are as follows: 706, (a) 218, 219.

therefore, the mean values for Cu in the given soil groups can be considered as background Cu contents. The mean levels for Cu vary from 6 to 60 ppm, being highest for the ferralitic soil group and lowest for sandy soils and organic soils. Quite similar Cu background contents were calculated for the surface soils of the U.S. from data given by Shacklette and Boerngen[706] (Table 41). The regularity in large-scale Cu occurrence in soils indicates that two main factors, parent material and soil formation processes, govern the initial Cu status in soils.

The common characteristic of Cu distribution in soil profiles is its accumulation in the top horizons. This phenomenon is an effect of various factors, but above all, Cu concentration in surface soils reflects the bioaccumulation of the metal and also recent anthropogenic sources of the element.

The Cu balance in surface soils of different ecosystems presented in Table 13 shows clearly that the atmospheric input of this metal may partly replace the removal of Cu by biomass production and in some cases may even exceed the total output of the metal from soils. Contemporarily observed soil contamination with Cu can lead to an extremely high Cu accumulation in top soils (Table 42).

Table 42
COPPER CONTAMINATION IN SURFACE SOILS
(PPM DW)

Site and pollution source	Country	Mean or range of content	Ref.
Old mining area	Great Britain	13—129	166
Nonferric metal mining	Great Britain	415—733	166
	Japan	456—2020	395
Metal-processing industry	Australia	847	57
	Bulgaria	24—2015	774
	Canada	1400—3700[a]	245
	Poland	72—125	380
Urban gardens, orchards,	Australia	210	795
and parks	Canada	11—130	332
	Japan	31—300	395
	Poland	12—240	159
	U.S.	3—140	628
	U.S.S.R.	50—83[b]	394
Sludged, irrigated, or fertil-	Holland	265	314
ized farmland	Polland	80—1600	682
	U.S.	90	593
	West Germany	187—280	176
Application of fungicides	West Germany	273—522	652

[a]   For 3 to 6.3 km distance from a smelter.
[b]   Vineyard.

1. Reactions with Soil Components

Although the most common mobile Cu in the surface environment is believed to be the cation with the valence of $+2$, several ionic species may occur in soils (Figure 18). However, Cu ions are held very tightly on both inorganic and organic exchange sites. The processes controlling fixation by soil constituents are related to the following phenomena:

1.   Adsorption
2.   Occlusion and coprecipitation
3.   Organic chelation and complexing
4.   Microbial fixation

Adsorption mechanisms of Cu have been extensively studied by many scientists and comprehensive publications concerning the physical and chemical behavior of the metal in soils are available in papers of McLaren and Crawford,[527] McBride,[518] James and Barrow,[352] and Kitagishi and Yamane.[395]

All soil minerals are capable of absorbing Cu ions from solution and these properties depend on the surface charge carried by the adsorbents. The surface charge is strongly controlled by pH, therefore, the adsorption of Cu ion species can be presented as a function of pH (Figure 19). This type of Cu adsorption is likely to be most important in soils with a large content of variable-charge minerals.[352]

Many authors have reported that Cu can be adsorbed by minerals within the range from 0.001 to 1 μmol dm^{-3} or from 30 to 1000 μmol g^{-1}. The greatest amounts of adsorbed Cu have always been found for Fe and Mn oxides (hematite, goethite, birnessite), amorphous Fe and Al hydroxides, and clays (montmorillonite, vermiculite, imogolite). Harter[307] reported that the most significant correlation was obtained between Cu adsorption and the sum of

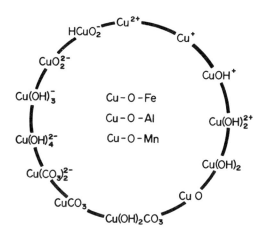

FIGURE 18.   Ionic species, compounds, and bonds of Cu occurring in soils.

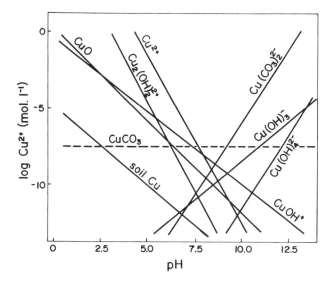

FIGURE 19.   Schematic diagram of solubility of Cu ionic species and Cu compounds in soil.[475, 518]

bases for the surface soils, while in the subsurface horizons the adsorption of Cu was highly related to vermiculite content.

Occlusion, coprecipitation, and substitution are involved in nonspecific adsorption of Cu. Nondiffusible fractions of soil Cu are most probably incorporated in various mineral structures. Some soil minerals, such as Al and Fe hydroxides, carbonates, and phosphates, and to some extent also silicate clays, have a great affinity to bind a part of the soil Cu in a nondiffusible form, which is the most stable portion of the metal in soil. In this chemisorption the formation of oxygen bridge bonds may also be involved (Figure 18).

Chelation and complexing are the key reactions governing Cu behavior in most soils. The ability of the organic soil constituents to bind Cu is well recognized, and there is a tremendous number of comprehensive publications on this subject.

Many kinds of organic substances form both soluble and insoluble complexes with Cu, thus Cu-binding capacities of soils and Cu solubility are highly dependent on the kind and amount of organic matter in soils. Stevenson and Fitch[758] stated that the maximum amount of Cu^{2+} that can be bound to humic and fulvic acids is approximately equal to the content of acidic functional groups. In general, this corresponds to the sorption of from 48 to 160 mg of Cu per gram of humic acid. Sapek[682] reported the maximum sorption capacity of peat-muck soil to be from 130 to 190 meq per 100 g of peat, while Ovcharenko et al.[587] calculated Cu sorption as 3.3 g kg^{-1} of humic acid. These values, however, would differ greatly depending on the physical and chemical properties of organic substances.

Organic binding of Cu in soils differs to some extent from that described for other divalent ions. According to Bloom and McBride[80] and Bloomfield,[81] peat and humic acids strongly immobilize the Cu^{2+} ion in direct coordination with functional oxygens of the organic substances.

To sum up the key role played by organic matter in the behavior of soil Cu, it can be emphasized that humic and fulvic acids are likely to form stable complexes when Cu is present in small amounts and that organic matter can modify several Cu reactions with inorganic soil components.

Microbial fixation plays a prominent role in the binding of Cu in certain surface soils. The amount of Cu fixed by the microbiomass is widely variable and is affected by various factors, such as metal concentration, soil properties, and growing season (Table 19). Microbial fixation of Cu is an important step in the ecological cycling of this metal.

Although Cu is one of the least mobile heavy metals in soil, this metal is abundant in soil solutions of all types of soils. Concentrations of Cu in soil solutions obtained by various techniques from different soils vary from 3 to 135 μg ℓ^{-1}, which corresponds to Cu concentrations of 0.047 to 2.125 μM (Table 12). Overall solubility of both cationic and anionic forms of Cu decreases at about pH 7 to 8. It has been estimated that hydrolysis products of Cu ($CuOH^+$ and $Cu_2(OH)_2^{2+}$) are the most significant species below pH 7, while above pH 8 anionic hydroxy complexes of Cu become important (Figure 19). As Sanders and Bloomfield[677] stated, the solubility of $CuCO_3$ is not pH dependent and this compound seems to be a major inorganic soluble form of Cu in neutral and alkaline soil solutions, while nitrate, chloride, and sulfate do not complex a significant portion of Cu in the soil solution. However, the most common forms of Cu in soil solutions are soluble organic chelates of this metal. Although very little is known of the kinds of soluble organic Cu forms, about 80% of the soluble Cu forms have been estimated to be organic chelates.[320] McBride and Blasiak[519] reported that due to a great affinity of Cu for organic complexing, soluble Cu-organic forms appear to comprise most of the Cu solution over a wide range of pH. Organic complexing of Cu has a prominent practical implication in governing the bioavailability and the migration of Cu in soil. The bioavailability of soluble forms of Cu depend most probably on both the molecular weight of Cu complexes and on the amounts present. Compounds of low molecular weight liberated during decay of plant and animal residues as well as those applied with sewage sludges may greatly increase the availability of Cu to plants. In summary, it should be emphasized that the concentrations of Cu in soil solutions are principally controlled by both the reactions of Cu with active groups at the surface of the solid phase and by reactions of Cu with specific substances.

2. Contamination of Soils

Contamination of soil by Cu compounds results from utilization of Cu-containing material such as fertilizers, sprays, and agricultural or municipal wastes as well as from industrial emissions (Table 42). Some local or incidental Cu input to soils may arise from corrosion of Cu alloy construction materials (e.g., electric wires, pipes).

Major sources of pollution (mainly nonferric metal smelters) present halos in which Cu concentrations in surface soils decrease with distance, which is especially pronounced in a downwind direction (Figure 20). While main point sources of industrial pollution have a

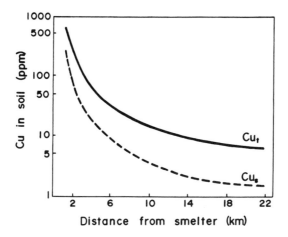

FIGURE 20.   Smoothed patterns for Cu concentrations in soils as a function of downwind distance from the Cu smelter (mean values of 3 years); t, total content; s, soluble content.[380]

local environmental impact, they also contribute to the global long-distance pollution of the atmosphere. Airborne fallout of Cu pollutants differs greatly in a specific area and for parts of the continents, being the greatest in central Europe. Heindrichs and Mayer[310] reported the atmospheric input of Cu to be 224 g ha^{-1} year^{-1} for West Germany, which is, comparatively, the highest value (Table 13).

The addition of Cu to cultivated soils with applied fertilizers, chemicals, and wastes has recently been extensively investigated. Tiller and Merry[795] reviewed all basic problems of Cu behavior in contaminated soils and the assessment of the environmental impacts.

The most important statement on Cu contamination of soils is the great affinity of surface soils to accumulate this metal. As a consequence, Cu content of soils has already been built up to the extremely high concentration of about 3500 ppm Cu from industrial sources of pollution and of about 1500 ppm Cu from agricultural origins of the metal (Table 42). The threshold value of 100 ppm Cu (Table 6) has been exceeded in several contaminated surface soils.

The amelioration techniques based mainly on the addition of lime, peat (organic matter), and phosphate to soils brings variable effects related to the soil and plant factors. It should, however, be kept in mind that Cu stored in surface soils influences their biological activity and may become availabile to plants under various conditions.

B. Plants

1. Absorption and Transport

Numerous studies have greatly increased the present knowledge of Cu absorption mechanisms. The mechanisms of Cu absorption are still far from clear, but it may be stated that, although there is increasing evidence of the active absorption of Cu, passive absorption is likely to occur, especially in the toxic range of this metal in solutions.

In root tissue Cu is almost entirely in complexed forms; however, it is most likely that the metal enters root cells in dissociated forms. Graham[280] compared the results of different studies carried out on the rates of Cu absorption by higher plant roots and stated that these rates are among the lowest of the essential elements, varying from pico- to micromole hr^{-1} g^{-1} (DW) of roots in the physiological concentration range (0.01 to 20 μM of Cu).

In spite of the great complexity of the absorption mechanisms, a relationship between the concentration of the metal measured in either a nutrient solution or a soil solution and in

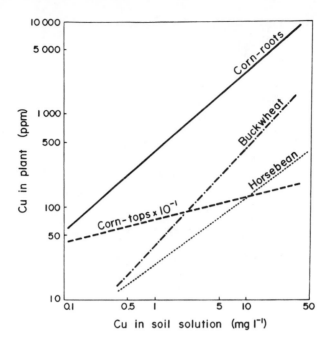

FIGURE 21.   Relationship between Cu content of plants and its concentration in soil solution obtained from Cu-contaminated soils.[379]

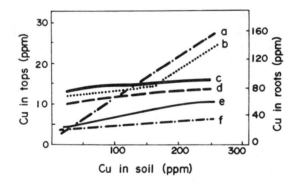

FIGURE 22.   Cu uptake by various plants and organs from Cu-contaminated soils. (a) Valencia orange roots; (b) potato stalks; (c) pasture herbage; (d) lettuce leaves; (e) barley grains; (f) wheat grains.[176,231,783]

the soil that supports plants can be observed, especially in the toxic range (Figures 21 and 22).

The movement of Cu among various parts of plants plays a predominant role in the plant's utilization of Cu. The strong capability of root tissues to hold Cu against the transport to shoots under conditions of both Cu deficiency and Cu excess has been observed. These processes are not yet fully understood; however, Loneragan[490] and Tiffin[788] concluded that excretion of Cu from root cells into the xylem and phloem saps where Cu occurs in mobile forms is a key process in the Cu nutrition of plants.

The concentrations of Cu in xylem and phloem saps range from traces to 140 μM and seem to correlate with the concentrations of amino acids. The Cu mobility within plant tissues strongly depends on the level of Cu supply, being the highest with a "luxury" supply.[490] However, Cu has low mobility relative to other elements in plants and most of this metal appears to remain in root and leaf tissues until they senesce; only small amounts may move to young organs. Therefore, the young organs are usually the first to develop symptoms of Cu deficiency.

The distribution of Cu within plants is highly variable. Within roots, Cu is associated mainly with cell walls and is largely immobile. The highest concentrations of Cu in shoots are always in phases of intensive growth and at the luxury Cu supply level. Scheffer et al.[688a] showed that distribution of Cu in barley leaves is relatively uniform for a given stage of plant growth.

A considerable proportion of the Cu present in green tissues appears to be bound in plastocyanin and in some protein fractions. There is also a tendency to accumulate Cu in reproductive organs of plants; this, however, differs widely among plant species. The highest concentrations of Cu have been found in the embryo of cereal grains and in the seed coat. Loneragan[490] cited Cu concentration in the embryo to be from 2 to 18 ppm and from 8 to 23 ppm in the seed coat, while in whole seeds the highest value was 4 ppm (DW). These proportions do not correspond closely to those reported by Liu et al.,[484] who found a more uniform distribution of Cu throughout the barley grain.

2. Biochemical Functions

Very extensive studies have been made on the forms and behavior of Cu in plants. All findings described in a number of outstanding textbooks can be summarized as follows:

1. Cu is mainly complexed with organic compounds of low molecular weight and with proteins.
2. Cu occurs in the compounds with no known functions as well as in enzymes having vital functions in plant metabolism.
3. Cu plays a significant role in several physiological processes — photosynthesis, respiration, carbohydrate distribution, N reduction and fixation, protein metabolism, and cell wall metabolism.
4. Cu influences water permeability of xylem vessels and thus controls water relationships.
5. Cu controls the production of DNA and RNA, and its deficiency greatly inhibits the reproduction of plants (reduced seed production, pollen sterility).
6. Cu is involved in the mechanisms of disease resistance. This resistance of plants to fungal diseases is likely to be related to an adequate Cu supply. There is also evidence that plants with enriched Cu concentrations are susceptible to some diseases.

These phenomena may indicate that the role of Cu in disease resistance is an indirect one.

The most important practical implications are related to deficiency and toxicity of Cu. Cu deficiency affects physiological processes and therefore plant production. However, as Bussler[115] stated, in most of the processes affected by the Cu deficiency the shortage of Cu operates indirectly. This explains the difficulties in the assessment of Cu needs and Cu availability to plants.

Although Cu deficiency is widespread (Table 28) and has long been known, diagnosis and correction of the deficiency still need more study. Different methods have been applied to diagnosis using plant symptoms, soil testing, or biochemical assays. However, when these approaches are used singly, results are so crop and soil specific that the development of a universal test seems unlikely. Recently Robson and Reuter,[660] Gartrell et al.,[258] and Mengel and Kirkby[531] suggested that the tissue test should be considered together with other relevant information (field observations, soil analysis, etc.) for the most accurate assessment of the need for Cu applications.

The deficiency levels of Cu in plants show large genetic differences; however, some generalizations are presented in Table 25 which indicate that Cu levels less than 2 ppm are likely to be inadequate for most plants. The threshold contents of Cu in soils that were established using different procedures differ widely (Table 28) and their use is not encouraged for the determination of Cu deficiency for various crops.

Removal of Cu by crops is negligible when compared to its content in soil. An average cereal crop removes Cu in amounts of about 20 to 30 g ha^{-1}, whereas forest biomass removes about 40 g ha^{-1} year^{-1}. Gartrell[257] pointed out that comparing the excesses of Cu applied and the amounts removed in farm produce and leached with percolating waters shows that "depletion of soil reserves" is an unlikely explanation for the appearance of Cu deficiency over a short period. Cu application to soils has a long-term effect, and 10 years after Cu fertilization the contents of Cu were still increased in pasture herbage. Repeated Cu applications to soils can lead to toxic concentrations of this metal for some crops.

In spite of the general Cu tolerance of plant species and genotypes, this metal is considered also to be highly toxic. General symptoms of the Cu toxicity listed in Table 29 show that Cu-induced chlorosis and root malformation are the most common characteristic symptoms of this toxicity.

Based on the statements made by Woolhouse and Walker[886] and Sandmann and Boger,[678] the processes induced by an excess of Cu^{2+} and Cu^+ ions may be summarized as follows:

1.   Tissue damage and elongation of root cells
2.   Alteration of membrane permeability, causing root leakage of ions (e.g., K^+, PO_4^{3-}) and solutes
3.   Peroxidation of chloroplast membrane lipids and inhibition of photosynthetic electron transport
4.   Immobilization of Cu in cell walls, in cell vacuoles, and in nondiffusible Cu-protein complexes

Prediction of the Cu content of soil that results in toxic effects on plants is extremely complex. Before toxic symptoms and yield reductions are evident, the nutritive value of the crops having increased Cu levels seems to create the most significant health risk. A number of observations that Tiller and Merry[795] reported on the depressed plant growth, especially retarded seed germination and seedling and root development, resulted from excess Cu concentrations in surface soils.

3. Interactions with Other Elements

For optimal development, the plant must have not only an appropriate amount of active Cu in the cells, but also a balance of chemical elements. Due to the significant functions of Cu in enzymes and its variable valence, ions which have similar affinity as Cu to proteins and other compounds may have antagonistic interrelationships. Many complex interactions of Cu with other elements are observed within plant tissues and also in the external root media, particularly in the uptake-transport processes.

Ce-Zn interactions are commonly observed.[280,654] These metals apparently are absorbed by the same mechanism and therefore each may competitively inhibit root absorption of the other.

Ce-Fe antagonism is indicated as Cu-induced chlorosis. High levels of Cu in the plant decrease the Fe content in chloroplasts.[647] Fe, on the other hand, reduces Cu absorption from soil solutions, especially on peat soils. The optimal Cu/Fe ratio varies for different plant species. The toxic effects of Cu can be decreased by the addition of Fe. However, a synergistic effect of Cu on Fe absorption by rice seedlings was reported by Kitagishi and Yamane.[395]

Cu-Mo interactions are closely related to N metabolism. Cu interferes with the role of

Mo in the enzymtic reduction of NO_3. The mutual antagonism existing between these elements is highly dependent on plant species and kind of N nutrition. Cu aggravates Mo deficiency in plants, especially those using N from NO_3, because Cu interferes with the role of Mo in the enzymatic reduction of NO_3. Some vegetables are quite susceptible to Mo deficiency if growing in soil with a high level of Cu, and in other plants the deficiency of Cu is increased with the application of Mo to the soil.[581]

Cu-Cd interactions are reported by some authors as both antagonistic and synergistic in the element uptake by roots. Synergism may be a secondary effect of the damage to membranes due to the imbalanced proportions of the metals. Cu-Se interactions are observed mainly as inhibited Cu uptake with increased Se level. Cu-Mn interactions are reported to be both synergistic and antagonistic in the uptake processes under defined conditions and at high concentrations. Cu-Ni synergism is observed in similar conditions as Cu-Mn relationships. Cu-Al antagonism leads to the reduction of Cu uptake by roots under Al toxic levels, especially in acid soils.

Cu-Cr interrelationship may occur within plant tissues, as well as in the external root media. Antagonistic reactions apparently are related to the variable valency of Cr. Significant information on synergistic interconnections between Cu-F and Cu-Ag in microbial metabolisms were reported by Gadd and Griffiths.[251]

Interrelationships between Cu and major elements are presented in Table 31. Cu and N interact most strongly in the nutrition of plants. The concentrations of the two elements are highly correlated in shoots of a wide range of species and are related to the formation by protein of strong complexes with Cu. On the other hand, plants with high N levels readily show symptoms of Cu deficiency that results from increased growth.

Cu-P antagonisms occur in root media where phosphates have a strong tendency to adsorb Cu. High phosphate levels in soils also reduce mycorrhizal absorption of Cu. An excess of Cu, in contrast, inhibits activity of phosphatase thereby diminishing the availability of P.[813]

Cu-Ca interactions are highly complex and apparently are cross-linked with the range of pH in the growth media. The affinity of carbonates to precipitate Cu is the most common reaction leading to Cu deficiency in soils within the alkaline range of pH or having free $CaCO_3$. Liming is the most frequent practice in the amelioration of Cu-contaminated soils. The relatively common occurrences of reduced Cu contents in plants that have an increased supply of some nutrients are often related to secondary effects of Cu dilution resulting from enhanced growth rates of the plant.

4. Concentrations in Plants

The appropriate content of Cu in plants is essential both for health of the plant and for the nutrient supply to man and animals. Some plant species have a great tolerance to increased concentrations of Cu and can accumulate extremely high amounts of this metal in their tissues (Table 30).

The concentration of Cu in plant tissues seems to be a function of its level in the nutrient solution or in soils (Figures 21 and 22). The pattern of this relationship, however, differs among plant species and plant parts.

Opinions appear to vary considerably as to which factor, soil or plant, affects concentrations of Cu in plant tissues to a higher degree. Anke et al.[32] reported a significant variation in Cu uptake by red clover from different soils, while Kähäri and Nissinen[388] found fairly uniform Cu levels in timothy from different soils. These authors, however, did not support their results with information on the Cu status of soils (Table 43).

Cu contents of various plants from unpolluted regions of different countries range from 1 to XO ppm (DW) (Tables 44, 45, and 46). Cu in ash of a variety of plant species is reported to range from 5 to 1500 ppm.[710] In several species growing under widely ranging natural conditions, Cu contents of whole plant shoots do not often exceed 20 ppm (DW),

Table 43
VARIATION IN COPPER CONTENT OF RED
CLOVER (BUD STAGE) AND TIMOTHY
(IMMATURE STAGE) AS INFLUENCED BY SOIL
FACTORS (PPM DW)

	Red clover (a)		
Soil parent material	Meadow	Cultivated field	Timothy (b)
Phyllite	11.1	11.3	—
Porphyry	11.5	10.2	—
Basalt	9.6	10.0	—
Loess or silt	9.7	9.8	4.1
Granite	9.2	9.7	—
Alluvial loamy deposit	9.0	9.8	4.8
Diluvial sand	8.4	8.1	3.9
Glacial till	8.5	7.5	4.0
Moor and peat	6.4	6.3	4.0

Note: Sources are as follows: a, 32; and b, 388.

Table 44
MEAN LEVELS AND RANGES OF COPPER IN GRASS AND
CLOVER AT THE IMMATURE GROWTH STAGE FROM
DIFFERENT COUNTRIES (PPM DW)

	Grasses		Clovers		
Country	Range	Mean	Range	Mean	Ref.
East Germany	7.4—15.0	10.1	7.6—15.0	10.5	31, 32, 65
Finland	3.8—4.8	4.3	—	—	388
Hungary	3.6—8.4	5.0	4.2—16.2	10.5	803
Japan	1.3—33.1	6.9[a]	2.0—12.5	6.5[b]	395
New Zealand	7.3—13.4	10.5[c]	8.1—17.5	11.7[d]	536
Poland	2.2—21.0	6.0	4.2—20.9	11.3	381
U.S.	1.5—18.5	9.6	10.2—29.0	16.2	172
U.S.S.R. ecosystems					
Meadow-bog	1.1—3.8	1.8	8.8—20.0	12.5	806
Forest-steppe	1.1—3.9	2.6	5.1—24	14.8	806
Forest	2.2—3.8	3.0	—	6.4	806

[a]   *Dactylis glomerata.*
[b]   *Trifolium repens.*
[c]   *Lolium perenne.*
[d]   *Trifolium pratense.*

and thus this value is most often considered to indicate the threshold of excessive contents (Table 25). However, under both natural and man-induced conditions, the majority of plant species can accumulate much more Cu, especially in root storage tissues (Table 47). The significance of elevated contents of Cu in feed and food plants that reflect man-made pollution needs evaluation from the environmental health point of view.

VI. SILVER

A. Soils

The geochemical characteristics of Ag are similar to those of Cu, but its concentration in

Table 45
MEAN COPPER CONTENT OF PLANT FOODSTUFFS (PPM)

Plant	Tissue sample	FW basis			DW basis		AW basis	
		Data source 547, 574	Data source 395	Data source 705	Data source 354, 381	Data source 354	Data source 705	Data source 354
Sweet corn	Grains	0.60	—	1.4	2.1	—	54	88
Bean	Pods	1.7	10.0[a]	5.1	—	8	73	126
Cabbage	Leaves	0.33	0.3	2.9	3.3	4	31	40
Lettuce	Leaves	0.11	1.7[b]	8.1	—	6	58	42
Beet	Roots	—	—	—	8.1	5	—	87
Carrot	Roots	0.22	—	4.6	8.4	4	65	70
Onion	Bulbs	0.69	—	4.6	6.0	4	110	68
Potato	Tubers	0.38	1.3	3.7	6.6	3	88	105
Tomato	Fruits	0.65	—	8.8	—	6	73	84
Apple	Fruits	0.03	—	1.1	—	—	63	—
Orange	Fruits	0.11	0.4[c]	1.9	—	—	52	—

[a] Pulses.
[b] Spinach.
[c] *Citrus unshiu* (Satsuma orange).

Table 46
COPPER CONTENT OF CEREAL GRAINS
FROM DIFFERENT COUNTRIES (PPM DW)

Country	Cereal	Range	Mean	Ref.
Afghanistan	Barley	—	6.3	446
	Wheat	—	4.1	446
Australia	Wheat	1.3—5.0	3.5	391, 867
Canada	Oats	—	5.5	514
Egypt	Wheat[a]	4.5—10.3	6.7	213
Finland	Wheat[a]	4.7—6.9	5.7	508
	Wheat[b]	3.7—6.8	5.3	508
East Germany	Barley	1.8—6.2	—	65
	Oats	2—4	—	65
	Wheat	6—10	—	65
Great Britain	Barley	2.5—6.0	4.3	783
Japan	Rice	—	2.8[c]	395
	Wheat (flour)	—	1.1[c]	395
Norway	Barley	2.1—9.2	5.5	446
	Wheat[a]	2.1—6.1	4.0	446
Poland	Rye	—	3.7	424
	Triticale	—	3.2	667
	Wheat[b]	2.6—6.5	3.8	267, 355
U.S.	Barley	4—15	—	484
	Rye	4—8	—	490, 492
	Triticale	—	7.8	490, 492
	Wheat	0.6—5.4	4.5	906
U.S.S.R.	Barley	—	5.2	586
	Oats	2.3—4.2	3.2	586
	Wheat[b]	3.8—6.5	5.1	586

[a]   Spring wheat.
[b]   Winter wheat.
[c]   FW basis.

rocks is about 1000 times lower than Cu (Table 32). Ag is easily released by weathering and then precipitated in alkaline reduction-potential media and in media enriched in S compounds.

Ag can form several ionic species, such as simple cations (Ag^+, Ag^{2+}, AgO^+) and complexed anions, AgO^-, $Ag(S_2O_3)_2^{3-}$, $Ag(SO_4)_2^{3-}$). In spite of several mobile complexes, Ag apparently is immobile in soils if the pH is above4. Humic substances are known to absorb and complex Ag leading to an enrichment of the element in surface soils.

The literature published on Ag distribution in the environment has been reviewed by Smith and Carson[739] and shows the common range of Ag in soil to be 0.03 to 0.09 ppm. Soils from mineralized areas are enriched in Ag, but its content does not often exceed 1 ppm.[96] Ag in Canadian soils ranges from 0.2 to 3.2 ppm and for British standard soil samples Ag was reported to average 0.4 ppm.[629,818]

Recent data given by Shacklette and Boerngen[706] show Ag in the plow zone of mineral soil to be 0.7 ppm, and in soils rich in organic matter, to be from 2 to 5 ppm. An average Ag content given for soils by Wedepohl[855] ranges from 0.01 to 5 ppm, and values established by Bowen[94] range from 0.01 to 8 ppm. All higher values for Ag in soils are found in

Table 47
**COPPER CONTENT OF PLANTS GROWN IN CONTAMINATED
SITES (PPM DW)**

Site and pollution source	Plant and part	Mean or range of content	Country	Ref.
Metal processing industry	Lettuce, leaves	64	Australia	57
	Blueberry, leaves	75	Canada	866
	Grass, tops	20—70[a]	Canada	245
	Horsetail, tops	70—250[a]	Canada	245
	Dandelion, tops	73—274[b]	Poland	380
	Dandelion, roots	22—199[b]	Poland	380
Urban garden	Radish, roots	2—14	Great Britain	167
	Leafy vegetables	4—19	U.S.	628
Sludged or irrigated farmland	Grass, tops	14—38	Holland	297
	Rice, grains	4	Japan	395
	Rice, roots	560	Japan	395
	Potato, tubers	5	West Germany	176

[a]   For 1.6- to 5.8-km distance from a smelter.
[b]   For 0.5- to 2.5-km distance from a smelter.

mineralized areas. Davies and Ginnever[168] found up to 44 ppm Ag in soils from the vicinity of old base metal mines. Kiriluk[394] reported the range of Ag from 0.44 to 0.93 ppm in chernozems of vineyards.

B. Plants

Ag concentrations in plants are reported by Smith and Carson[739] to range from 0.03 to 0.5 ppm (DW). Chapman[131] established the intermediate range of Ag in plant foodstuffs as 0.07 to 2.0 ppm (DW). According to Gough et al.[279] and Shacklette[705] the mean Ag content of plant ash is usually less than 5 ppm.

Ag concentrations differ greatly between plant species and between times of sample collection. Horowitz et al.[325] reported that Ag in plants sampled in September was much lower than in plants sampled in May. They found Ag in plants to range from 0.01 to 16 ppm (DW), with the highest values being for fungi and green algae.

The amount of Ag absorbed by several plants (e.g., horsetail, lichens, mosses, fungi, and some deciduous trees) seems to be related to the amount of the metal in soils. Thus, Ag can be concentrated to toxic levels in plants growing in Ag-mineralized areas.[120,417]

Wallace et al.[841] wrote that about 5 ppm Ag in the tops and above 1500 ppm in the roots of bush beans (DW) greatly reduced yields, but the plants grew without symptoms of toxicity. As described by Hendrix and Higinbotham,[315] Ag can substitute for K^+ sites in membranes and thus inhibit the absorption of other cations by roots. Ag compounds are known to precipitate bacterial proteins as well as to form insoluble complexes with ribonucleic acids.[856] Ag ions have a great affinity for binding sulfhydryl groups of some organic compounds.

VII. GOLD

A. Soils

Au is a rare element in the earth's crust and its average concentration in rocks does not exceed the order of magnitude 0.00X ppm (Table 32). Au is relatively stable in hypergenic zones; however, under certain weathering conditions it is known to form several complex ions, $AuCl_2^{2-}$, $AuBr_4^-$, AuI_2^-, $Au(CN)_2^-$, $Au(CNS)_4^-$, $Au(S_2O_3)_2^{3-}$ that are readily mobile. However, Au seems to be transported most often in forms of organometallic compounds or chelates.[97]

Lakin et al.[453] studied Au distribution in soil profiles and showed that, depending on the origin of soil material and its weathering stage, Au can be present in fine soil particulates or in cobbles and pebbles. Au is most often enriched in the humus layer because it can act as a reductant and precipitation medium for mobile forms of Au. However, distribution of Au within soil profiles may follow different trends, depending on its content in parent rocks.

There is a relative paucity of information on the Au content of soils. The Au status in normal soils may be established to be within the range from 1 to 2 ppb. Roslakov[664] gave the range of background contents of Au in soils as ranging from 0.8 to 8.0 ppb, with the greatest amounts in chernozems and kastanozems. Lakin et al.[453] reported that in mineralized areas, Au in forest mull ranged from 0.05 to 5.0 ppm, while mineral-size fractions contained from 0.04 to 0.44 ppm Au.

B. Plants

Plants can absorb Au in soluble forms, and when Au enters the root vascular systems of plants, it can be easily transported to the tops. In reducing media, however, Au precipitates on the cell surface and thus inhibits membrane permeability.

Cyanogenic plants and some deciduous trees are able to accumulate even greater than 10 ppm Au (DW).[453,711] Horsetail is also known as a good indicator for Au; however, Cannon et al.[120] reported Au to be in the range of 0.1 to 0.5 ppm (AW) in horsetail species from Alaska, and only slightly higher Au concentrations were found in plant samples from mineralized aeas. Various plant species collected from mineralized areas of British Columbia contained from 0.7 to 6.5 ppm Au (DW), and the herbaceous plant *Phacelia sericea* (family Hydrophyllaceae) appeared to be the best Au accumulator.[263] Data reviewed by Shacklette et al.[710] indicate that the Au content of plants ranges from 0.0005 to 125 ppm (AW). All anomalous high Au contents of plants are reported for plants of mineralized areas.

Oakes et al.[574] found Au in fruits and vegetables in the range of 0.01 to 0.4 ppb (FW), while Bowen[94] reported 1 to 40 ppb (DW) for other vascular plants. According to Ozoliniya and Kiunke,[588] Au in barley and flax is detectable only in the roots, in the range of 14 to 22 ppb (DW). Several plant species are relatively resistant to higher Au concentrations in tissues. The Au toxicity leads to necrosis and wilting by loss of turgidity in leaves.

Chapter 7

ELEMENTS OF GROUP II

I. INTRODUCTION

Trace elements of Group IIa belong to the alkaline earths (Be, Sr, and Ba) and behave similarly to Ca and Mg. These microcations favor coordination with oxygen donors and do not usually form complex ionic species. The Sr and Ba cations are more similar to Ca than to Mg, and Sr^{2+} in particular is very close in size to Ca^{2+}. These cations can replace each other, while the small size of Be prevents its replacement by other ions. The common characteristic of the alkaline earths is their association with the carbon cycle, therefore, the processes of solution of hydrogen carbonates and the precipitation of carbonates strongly control the behavior of these metals in the terrestrial environment.

Group IIb is composed of three metals of the transition series — Zn, Cd, and Hg. They have comparatively high electronegativity values and easily form bonds with nonmetals of significant covalent character. They occur most often as divalent cations and show a great affinity to combine with S anions and with several organic compounds; thus, they are all of great importance in biochemistry. They are also known to form inorganic complex ions. Unlike the sulfides of the elements of Group IIa, the sulfides of Zn, Cd, and Hg are insoluble in water. However, the compounds of Zn and Cd immediately hydrolyze, but the corresponding compounds of Hg are rather resistant to hydrolysis. The three metals are all relatively mobile in the earth's surface and their cycling may be highly modified by their accumulation by plants and organic debris.

II. BERYLLIUM

A. Soils

Be, the lightest of the alkaline earths, although widely distributed, exists in relatively small quantities, comprising less than 10 ppm of the major rock types. This metal is likely to concentrate in the acid magmatic rocks. Its concentration in argillaceous sediments and shales is also enhanced and ranges from 2 to 6 ppm (Table 48). During rock weathering Be usually remains in the residuum; however, its behavior is much different in various environments because of its small size, relatively high ionization, and electronegativity.

Be occurs most often as the divalent cation, but its complex ions are also known — $(BeO_2)^{2-}$, $(Be_2O_3)^{2-}$, $(BeO_4)^{6-}$, and $(Be_2O)^{2+}$. Thus, Be is present in soils primarily in oxidic-bonded forms. In the alkaline environment Be forms complex anions such as $Be(OH)CO_3^-$ and $Be(CO_3)_2^{2-}$.

The abundance of Be in surface soils of the U.S. is fairly uniform for various soil units (mean, 1.6 ppm), and ranges from <1 to 15 ppm (Table 49). The range for Russian surface soils is reported to be from 1.2 to 13 ppm, while Be in Canadian arable surface soils ranges from 0.10 to 0.89 ppm and averages 0.35 ppm.[45,244] Ure and Bacon[818] gave an average of 2.7 ppm Be for the standard British soils.

Organic substances are known to bind Be easily, therefore, Be is enriched in some coals and accumulates in organic soil horizons. Be may substitute for Al and for some divalent cations, and therefore is strongly bound by montmorillonitic clays. Hädrich et al.[298] reported the range in Be concentrations in soil solutions as 0.4 to 1.0 $\mu g\ \ell^{-1}$. Although Be appears to be rather immobile in soils, its readily soluble salts, such as $BeCl_2$ and $BeSO_4$, can be available and therefore toxic to plants.

Due to some new technologies that employ Be (rocket fuels and light, hard, and high corrosion-resistant alloys), as well as to coal combustion, there is concern that Be may be

Table 48
BERYLLIUM, STRONTIUM, BARIUM, ZINC, CADMIUM, AND MERCURY IN
MAJOR ROCK TYPES (PPM) (VALUES COMMONLY FOUND, BASED ON
VARIOUS SOURCES)

Rock type	Be	Sr	Ba	Zn	Cd	Hg
			Magmatic rocks			
Ultramafic rocks						
Dunites, peridotites, pyroxenites	0.X	2—20	0.5—25.0	40—60	0.03—0.05	0.0X
Mafic rocks						
Basalts, gabbros	0.3—1.0	140—460	250—400	80—120	0.13—0.22	0.0X
Intermediate rocks						
Diorites, syenites	1.0—1.8	300—600	600—1000	40—100	0.13	0.0X
Acid rocks						
Granites, gneisses	2—5	60—300	400—850	40—60	0.09—0.20	0.08
Acid rocks (volcanic)						
Rhyolites, trachytes, dacites	5.0—6.5	90—400	600—1200	40—100	0.05—0.20	0.0X
			Sedimentary rocks			
Argillaceous sediments	2—6	300—450	500—800	80—120	0.30	0.20—0.40
Shales	2—5	300	500—800	80—120	0.22—0.30	0.18—0.40
Sandstones	0.2—1.0	20—140	100—320	15—30	0.05	0.04—0.10
Limestones, dolomites	0.2—2.0	450—600	50—200	10—25	0.035	0.04—0.05

Table 49
BERYLLIUM CONTENT OF SURFACE SOILS
OF THE U.S. (PPM DW)[706]

Soil	Range	Mean
Sandy soils and lithosols on sandstones	<1—3	1.9
Light loamy soils	1—3	1.7
Loess and soils on silt deposits	1—3	1.7
Clay and clay loamy soils	<1—15	1.9
Alluvial soils	1—3	1.6
Soils over granites and gneisses	1—2	1.6
Soils over volcanic rocks	<1—3	1.7
Soils over limestones and calcareous rocks	1—2	1.6
Soils on glacial till and drift	<1—2	1.6
Light desert soils	<1—7	2.1
Silty prairie soils	1—1.5	1.4
Chernozems and dark prairie soils	<1—3	1.5
Organic light soils	<1—1.5	1.2
Forest soils	1—3	1.9
Various soils	<1—5	1.6

increasing in agricultural soils. There is not much information on Be in contaminated soils, which were reported to range from 15 to 50 ppm Be in the vicinity of smelters and coal power stations, while the control soils contained less than 1 ppm Be.[592]

B. Plants

Be apparently is easily taken up by plants when it occurs in soluble forms in soils. Its concentration in plants under natural conditions is reported to range from 0.001 to 0.4 ppm (DW) and to range from <2 to 100 ppm in plant ash.

Be concentrations as high as 250 ppm (AW) are reported for an accumulator plant (*Vaccinium myrtillus*).[855] Some plant species in the Leguminosae and Cruciferae families have a pronounced ability to accumulate Be, particularly in their root tissue.[283]

Although Be is known to be concentrated mainly in roots, Krampitz[425] reported a relatively high Be content of lettuce leaves (0.033 ppm DW) and tomato fruits (0.24 ppm DW), and Bowen[94] stated that Be in lichens and bryophytes ranged from 0.04 to 0.9 ppm (DW). Padzik and Wlodek[592] gave the content of Be in grass from an industrial area as 0.19 ppm (DW). Bohn and Seekamp[84] found that plant Be increased up to 20 ppm (DW) at 100 ppm Be addition to the soil.

Be absorption mechanisms of plants seem to be similar to those involved in the uptake of Mg^{2+} and Ca^{2+}. However, these elements have antagonistic interactions, and Be is able to replace Mg^{2+} in some plants. Gough et al.[279] and Krampitz[425] reported stimulating effects of the dilute solution of $Be(NO_3)_2$ on growth of several plant species, microorganisms in particular (*Aspergillus niger*). Biochemical mechanisms of this phenomenon, however, are not clear.

On the other hand, the toxicity of Be to plants has been frequently reported. Toxic Be concentrations in mature leaves range most often from 10 to 50 ppm (DW). This range is highly variable for each plant species and growth condition.

Relatively low Be concentrations in solution, ranging from 2 to 16 ppm or from 10^{-3} to $10^{-4}M$ Be^{2+}, is highly toxic to plants. Be is known to inhibit the germination of seeds and the uptake of Ca and Mg by roots, to have diverse effects on P absorption, and to degrade some proteins and enzymes. These processes, however, are not fully understood. Although specific symptoms of Be toxicity to plants are not known, the common symptoms are brown, retarded roots and stunted foliage. Although at present there is no evidence that Be in food plants can be a health risk to man, more data are needed to evaluate the risk.[285]

Table 50
STRONTIUM CONTENT OF SURFACE SOIL OF
DIFFERENT COUNTRIES (PPM DW)

Soil	Country	Range	Mean	Ref.
Podzols and sandy soils	Australia	—	118	196
	New Zealand	350—570[a]	—	861
Loess and silty soils	New Zealand	220—380[a]	—	861
Loamy and clay soils	New Zealand	18—86[a]	—	861
	U.S.S.R.	280—310	295	714
Soils on glacial till	Denmark	—	14.7	801
Kastanozems	U.S.S.R.	—	280	714
Chernozems	U.S.S.R.	520—3500	—	714
Prairien and meadow soils	U.S.S.R.	150—500	300	714
Histosols and other organic soils	Denmark	—	92	801
Forest soils	U.S.S.R.	—	675	714
Various soils	Canada	30—500[b]	210	521
	Denmark	—	17.2	801
	Great Britain	—	261	818

[a]   Soil derived from basalts and andesites.
[b]   Data for whole soil profiles.

III. STRONTIUM

A. Soils

Sr is a relatively common trace element in the earth's crust and is likely to concentrate in intermediate magmatic rocks and in carbonate sediments (Table 48). Geochemical and biochemical characteristics of Sr are similar to those of Ca; thus, Sr is very often associated with Ca and, to a lesser extent with Mg, in the terrestrial environment. The Sr to Ca ratio seems to be relatively stable in the biosphere and therefore is commonly used for the identification of built-up concentrations of Sr in a particular environment.

Sr is easily mobilized during weathering, especially in oxidizing acid environments, and then it is incorporated in clay minerals and strongly fixed by organic matter, but most Sr is precipitated as biogenic carbonates, largely in the form of invertebrate shell material. This element is known to occur mainly as Sr^{2+} ions; however, its chelated forms play an important role in cycling which is closely associated with Ca cycling.

Sr content of soils is highly controlled by parent rocks and climate, and therefore its concentrations range in surface horizons from 18 to 3500 ppm (Table 50), being highest in Russian chernozems and in forest soils. Soils derived from glacial till under the humid climate of Denmark are reported to be very poor in Sr, whereas these kinds of soils from the U.S. are relatively rich in this element (Table 51). Mean contents of Sr as calculated for American topsoils range from 110 to 445 ppm, being the highest in desert soils and soils derived from magmatic rocks.

The Sr distribution in soil profiles follows the general trends of soil solution circulation. However, it may also be irregular, depending on soil properties. In acid soils Sr is highly leached down the profile. In calcareous soils Sr may be replaced by various cations and, in particular, hydrogen ions. Displacement of Sr by Ca solutions has practical implications in the reclamation of contaminated soils.[450]

^{90}Sr is produced in many nuclear processes and is considered to be one of the most biologically hazardous radioactive elements for man. Since Ca and Sr are known to be the carrier elements for ^{90}Sr, the cycling of ^{90}Sr is related to that of Ca and Sr.

Numerous studies have been carried out on ^{90}Sr in biological environments. Several authors

Table 51
STRONTIUM CONTENT OF SURFACE SOILS OF
THE U.S. (PPM DW)[706]

Soil	Range	Mean
Sandy soils and lithosols on sandstones	5—1000	125
Light loamy soils	10—500	175
Loess and soils on silt deposits	20—1000	305
Clay and clay-loamy soils	15—300	120
Alluvial soils	50—700	295
Soils over granites and gneisses	50—1000	420
Soils over volcanic rocks	50—1000	445
Soils over limestones and calcareous rocks	15—1000	195
Soils on glacial till and drift	100—300	190
Light desert soils	70—2000	490
Silty prairie soils	70—500	215
Chernozems and dark prairie soils	70—500	170
Organic light soils	5—300	110
Forest soils	20—500	150
Various soils	7—1000	200

stated that ^{90}Sr as a pollutant is easily mobile in light soils and therefore is readily taken up by plants.[603,676,791] Pavlotskaya et al.[604] reported the easy coprecipitation of ^{90}Sr by hydrous Fe oxides, which leads to its accumulation in Fe-rich soil horizons.

B. Plants

The Sr concentration in plants is highly variable and is reported to range from <1 to 10,000 (DW) and to 15,000 ppm (AW). Its common amounts, however, calculated as the mean for different food and feed plants, range from about 10 to 1500 ppm (DW) (Table 52). The lowest mean contents of Sr were found in fruits, grains, and potato tubers, whereas legume herbage contained from 219 to 662 ppm (DW). Bowen gave the broad range in Sr concentrations in lichens as 0.8 to 250 ppm (DW).

Sr uptake by roots is apparently related to both the mechanisms of mass-flow and exchange diffusion.[209] The Ca to Sr ratio of uptake was proposed by some authors to determine the source of these cations and the rate of their uptake. Wallace and Romney[842] reported that Sr is not very readily transported from roots to shoots; however, the highest concentrations of Sr are often reported for tops of plants.[710]

Interactions between Sr and Ca are complex, for they may compete with each other, but Sr usually cannot replace Ca in biochemical functions. Weinberg[856] reported that the enzyme amylase with Sr^{2+} substituted for Ca^{2+} had full activity, but varied in some physical properties. Liming of soil may have both inhibiting and stimulating effects, depending on soil and plant factors.[450]. However, in bush beans grown in solution culture, Ca decreased Sr content, particularly in root and stem (Figure 23). Reported interactions between Sr and P apparently are related to processes in soil; however, opinions vary as to the effects of P on Sr adsorption by plants.[175,439]

There are not many reports of Sr toxicity in plants, and plants vary in their tolerance to this element. Shacklette et al.[710] gave the toxic Sr level for plants as 30 ppm (AW). There is no evidence that stable Sr at levels present in the biosphere may have any deleterious effects on man and animals; however, the accumulation of the radionuclide ^{90}Sr in food and feedstuffs is of the greatest environmental concern.

^{90}Sr is relatively easily taken up by plants, but its availability may be inhibited by the application of Ca, Mg, K, and Na to soils. Beans grown on soil treated with ^{90}Sr accumulated up to 565 nCi g^{-1} (DW) in leaves, while in grains, only 24 nCi g^{-1} (DW).[290] ^{90}Sr in oats

Table 52
MEAN LEVELS AND RANGES OF
STRONTIUM IN FOOD AND FEED
PLANTS (PPM DW)[131,381,514,574,705,853]

Plant	Tissue sample	Range	Mean
Wheat	Grains	0.48—2.3	1.5
Oats	Grains	1.8—3.2	2.5
	Green tops	9—31	20
Corn	Grains	0.06—0.4	—
Lettuce	Leaves	—	74
Spinach	Leaves	45—70	—
Cabbage	Leaves	1.2—150	45
Bean	Pods	1.5—67	18
Soybean	Leaves	58—89	—
Carrot	Roots	1.5—131	25
Onion	Bulbs	10—88	50
Potato	Tubers	—	2.6
Tomato	Fruits	0.4—91	9
Apple	Fruits	0.5—1.7	0.9[a]
Orange	Fruits	—	0.5[a]
Clover	Tops	95—850	219
Lucerne (alfalfa)	Tops	50—1500	662
Grass	Tops	6—37	24

[a]   FW basis.

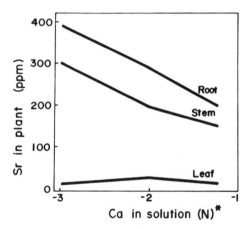

FIGURE 23.   Effects of Ca on Sr distribution in bush beans exposed 48
hr to 10^{-3} N Sr in solution culture.* Ca normality is given in powers of
ten.[842]

in a similar experiment ranged in straw from 17 to 137 nCi g^{-1} (DW) and in grains ranged
from 1 to 11 nCi g^{-1} (DW).[291,292]

IV. BARIUM

A. Soils
In the earth's crust, Ba is likely to concentrate in intermediate and acid magmatic rocks
and commonly ranges in concentration from 400 to 1200 ppm (Table 48). In geochemical

Table 53

BARIUM CONTENT OF SURFACE SOILS OF DIFFERENT COUNTRIES (PPM DW)

Soil	Country	Range	Mean	Ref.
Podzols and sandy soils	Australia	—	207	196
	New Zealand	270—780[a]	—	861
	U.S.S.R.	180—260	220	493
Loess and silty soils	New Zealand	240—590[a]	—	861
	U.S.S.R.	—	960	493
Loamy and clay soils	New Zealand	19—200[a]	—	861
	U.S.S.R.	—	240	493
	Bulgaria	—	402	558
Fluvisols	U.S.S.R.	—	240	493
	Bulgaria	—	691	558
Chernozems	U.S.S.R.	475—620	525	4
	Bulgaria	—	458	558
Histosols and other organic soils	U.S.S.R.	—	84	493
Forest soils	U.S.S.R.	—	560	4
	Bulgaria	397—850	631	558
Various soils	Bulgaria	492—2368	838	558
	Canada	262—867	669	409
	Great Britain	—	672	818

[a] Soils derived from basalts and andesites.

Table 54

BARIUM CONTENT OF SURFACE SOILS OF THE U.S. (PPM DW)[706]

Soil	Range	Mean
Sandy soils and lithosols on sandstones	20—1500	400
Light loamy soils	70—1000	555
Loess and soils on silt deposits	200—1500	675
Clay and clay-loamy soils	150—1500	535
Alluvial soils	200—1500	660
Soils over granites and gneisses	300—1500	785
Soils over volcanic rocks	500—1500	770
Soils over limestones and calcareous rocks	150—1500	520
Soils on glacial till and drift	300—1500	765
Light desert soils	300—2000	835
Silty prairie soils	200—1500	765
Chernozems and dark prairie soils	100—1000	595
Organic light soils	10—700	265
Forest soils	150—2000	505
Various soils	70—3000	560

processes, Ba is usually associated with K^+ due to their very similar ionic radii (Table 9), and therefore its occurrence is linked with alkali feldspar and biotite.

Ba released by weathering is not very mobile because it is easily precipitated as sulfates and carbonates, is strongly adsorbed by clays, and is concentrated in Mn and P concretions and minerals (Tables 16 and 17). Varnishes formed at the surface of aridic soils always have an enrichment in Ba.

Ba in topsoils and rocks has similar ranges in concentration (Tables 53 and 54). The reported range for soil Ba, on a world scale, is from 19 to 2368 ppm (means, 84 to 838

Table 55
MEAN LEVELS AND RANGES OF BARIUM IN FOOD
AND FEED PLANTS (PPM)

| Plant | Tissue sample | DW basis | | FW basis |
		Range	Mean	Mean (a)
Cereal	Grains	4.2—6.6	5.5(b)	—
	Green tops	132—181	160	—
Sweet corn	Grains	—	0.034	—
Snap bean	Pods	—	7	0.4
	Seeds	1—15	8	—
Cabbage	Leaves	—	4.8	1.3
Lettuce	Leaves	—	9.4	0.5
Carrot	Roots	2—50	13	0.14
Onion	Bulbs	3—75	12	—
Potato	Tubers	1.3—35	5	0.7—1.4
Tomato	Fruits	—	2	—
Apple	Fruits	—	1.4	0.03
Orange	Fruits	—	3.1	0.17
Clover	Tops	142—198	170	—
Lucerne (alfalfa)	Tops	—	100	—

Note: Sources are as follows: 381; 705; a, 267; b, 574.

ppm), and the range for Ba in the soils of the U.S. is from 10 to 3000 ppm (means, 265 to 835 ppm). Ba in soils may be easily mobilized under different conditions, therefore, its concentrations in soil solutions show considerable variation.

B. Plants

Although Ba is reported to be commonly present in plants, it apparently is not an essential component of plant tissues. The Ba content ranges from 1 to 198 ppm (DW), being the highest in leaves of cereal and legumes and the lowest in grains and fruits (Table 55). The highest concentrations of Ba, up to >10,000 ppm (DW), are reported to be in different trees and shrubs and in Brazil nuts.[710]

Plants may take up Ba quite easily from acid soils. Weinberg[856] reported a high affinity of Ba^{2+} to be bound to the surface of yeast. There are, however, only a few reports on toxic Ba concentrations in plants. Chaudry et al.[134] gave 1 to 2% (DW) Ba in plants as highly toxic, while Brooks (see Shacklette et al.)[710] stated that 220 ppm (AW) is moderately toxic.

Possible toxicity of Ba to plants may be greatly reduced by the addition of Ca, Mg, and S salts to the growth medium. Antagonistic interactions between these elements and Ba (Table 31) may occur in both plant tissues and soils.

V. RADIUM

Ra occurs in the environment as radioactive nuclides of which ^{224}Ra, ^{226}Ra, and ^{228}Ra are the common products of the uranium and thorium decay chains.^{226} Ra is the most stable and relatively frequent radionuclide in the biosphere.

Bowen[94] reported ^{226}Ra to range in rocks from 0.6 to 1.1 ng kg^{-1} and in soil to be 0.8 ng kg^{-1}. This radionuclide in vegetation ranged from 0.03 to 1.6 ng kg^{-1} (DW).

Megumi and Mamuro[529] studied the concentrations of U-series nuclides in soils derived from granites and found that they increase with decreases in particle size (Figure 24). They

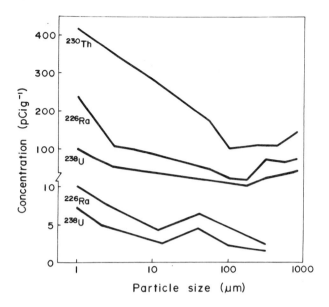

FIGURE 24.   Concentration of some radionuclides in soil particle size from the B horizon of soils derived from two different granites in Japan.[529]

reported that divalent Ra cations seem to be strongly absorbed by hydrous oxides of Mn and Fe.

Taskayev et al.[773] reported stable complexing of ^{226}Ra in soils where this radionuclide ranged from 36 to 351 pg g^{-1}. ^{226}Ra was most mobile under very acid conditions and its solubility ranged from about 1 to 10% of its content in the surface soils.

The ^{226}Ra content of the surface soils is higher than that of deeper soil horizons and this is because the radionuclide has recently been added to soils from anthropogenic sources. As Martell[509] stated, P and K fertilizers increased the ^{226}Ra in the plowed layer of soils, and according to data presented by Moore and Poet,[550] the manufacturing of P fertilizers and cement, as well as coal combustion, are the main sources of the radionuclides of Ra.

VI. ZINC

A. Soils

Zn seems to be distributed rather uniformly in the magmatic rocks, and only its slight increase in mafic rocks (80 to 120 ppm) and its slight decrease in acid rocks (40 to 60 ppm) is observed (Table 48). The Zn concentration in argillaceous sediments and shales is enhanced, ranging from 80 to 120 ppm, while in sandstones and carboniferous rocks concentrations of this metal range from 10 to 30 ppm. The Zn occurs chiefly as single sulfides (ZnS), but is also known to substitute for Mg^{2+} in silicates.

The solubilization of Zn minerals during weathering produces mobile Zn^{2+}, especially in acid, oxidizing environments. Zn is, however, also easily adsorbed by mineral and organic components and thus, in most soil types, its accumulation in the surface horizons is observed (Figure 25).

Mean total Zn contents in surface soils of different countries and of the U.S. range from 17 to 125 ppm (Tables 56 and 57). Thus, these values may be considered as background Zn contents. The highest grand means were reported for some alluvial soils, solonchaks, and kastanozem, while the lowest values were for light mineral and light organic soils.

The Zn balance in surface soils of different ecosystems shows that the atmospheric input

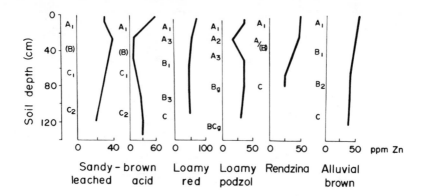

FIGURE 25.   Distribution of Zn in the profiles of different soils developed under humid climate. (Letters indicate genetic soil horizons.)

of this metal exceeds its output due to both leaching and the production of biomass (Table 13). Only in nonpolluted forest regions of Sweden is the discharge of Zn by water flux reported to be higher than its atmospheric input.[816]

1. Reactions with Soil Components

The most common and mobile Zn in soil is believed to be Zn^{2+}, but several other ionic species may occur in soils (Figure 26). The important factors controlling the mobility of Zn in soils are very similar to those listed for Cu, but Zn appears to occur in more readily soluble forms. Many studies of Zn adsorption and retention in soils were reviewed by Lindsay,[476] and it has been shown that clays and soil organic matter are capable of holding Zn quite strongly, thus the solubility of Zn in soils is less than that of $Zn(OH)_2$, $ZnCO_3$, and $Zn(PO_4)$ in pure experimental systems.

Processes involved in Zn adsorption are not yet completely understood; however, some generalizations can be made from studies reported by Lindsay,[475] Farrah and Pickering,[228] Peneva,[605a] Kuo and Mikkelsen,[440] and Wada and Abd-Elfattah.[834] There are two different mechanisms of Zn adsorption: one in acid media related to cation exchange sites and the other in alkaline media that is considered to be chemisorption and is highly influenced by organic ligands.

McBride and Blasiak[519] stated that nucleation of Zn hydroxide on clay surfaces may produce the strongly pH-dependent retention of Zn in soils (Figure 27). The adsorption of Zn^{2+} can be reduced at lower pH (<7) by competing cations and this results in easy mobilization and leaching of Zn from light acid soils (Table 11). At higher pH values, while an increase of organic compounds in soil solution become more evident, Zn-organic complexes may also account for the solubility of this metal (Figure 27).

As Zyrin et al.[911] reported, Zn in soils is associated mainly with hydrous Fe and Al oxides (14 to 38% of total Zn) and with clay minerals (24 to 63%), while its readily mobile fractions and its organic complexes make, respectively, 1 to 20 and 1.5 to 2.3%.

Abd-Elfattah and Wada[2] found the highest selective adsorption of Zn by Fe oxides, halloysite, allophane, and imogolite and the lowest by montmorillonite. Thus, clay minerals, hydrous oxides, and pH are likely to be the most important factors controlling Zn solubility in soils, while organic complexing and precipitation of Zn as hydroxide, carbonate, and sulfide compounds appear to be of much lesser importance. Zn can also enter some layer lattice silicate structures (e.g., montmorillonite) and become very immobile.

Soil organic matter is known to be capable of bonding Zn in stable forms, therefore, the Zn accumulation in organic soil horizons and in some peats is observed. However, stability constants of Zn-organic matter in soils are relatively low. Wada and Abd-Elfattah[834] gave

Table 56
ZINC CONTENT OF SURFACE SOILS OF DIFFERENT COUNTRIES
(PPM DW)

Soil	Country	Range	Mean	Ref.
Podzols and sandy soils	Australia	39—86	—	792
	New Zealand	14—146	42	870
	Poland	5—220	24	378, 382
	U.S.S.R.	3.5—57	31	271, 900
	West Germany	40—76	—	689
Loess and silty soils	New Zealand	—	61	870
	Poland	17—127	47	378, 382
	Rumania	—	73	42
	U.S.S.R.	40—55	48	271
	West Germany	58—100	—	689
Loamy and clay soils	Canada	15—20	17	692
	Great Britain	—	70	876
	New Zealand	31—177	79	870
	Poland	13—362	67.5	378, 382
	U.S.S.R.	9—77	35	271, 900
	West Germany	40—50	—	689
Soils on glacial till	Denmark	—	28	801
Fluvisols	Bulgaria	—	62	752
	Great Britain	67—180	125	166, 786
	New Zealand	53—67	60	870
	Poland	55—124	84.5	378
	U.S.S.R.	34—49	42	900
Gleysols	Australia	31—62	—	792
	Chad	25—300	—	39
	Great Britain	—	54	876
	New Zealand	60—84	73	870
	Poland	13—98	50.5	378
	U.S.S.R.	26.5—79	52.5	900
Rendzinas	Poland	58—150	77	378, 382
	U.S.S.R.	23—71	47	271, 900
	West Germany	—	100	689
Kastanozems and brown soils	Australia	29—79	—	792
	Chad	25—100	—	39
	New Zealand	30—67	53.5	870
	U.S.S.R.	32.5—54	43	343, 900
Ferralsols	Chad	25—145	—	39
	Israel	200—214	—	644
Solonchaks and solonetz	Bulgaria	39—63	—	752
	Chad	25—100	—	39
	New Zealand	54—68	62	870
	U.S.S.R.	44—155	100	12, 900
Chernozems	Bulgaria	63—97	—	752
	Poland	33—82	61.5	378
	U.S.S.R.	39—82	57	712, 900
Prairien and meadow soils	Bulgaria	88—98	—	752
	U.S.S.R.	31—192	105	12, 89, 271, 900
Histosols and other organic soils	Bulgaria	—	80	752
	Denmark	48—130	72.5	1, 801
	New Zealand	21—34	27	870
	Poland	13—250	60	392
	U.S.S.R.	7.5—74	34	271, 900
Forest soils	Bulgaria	35—106	—	752
	China	—	85	225

Table 56 (continued)
ZINC CONTENT OF SURFACE SOILS OF DIFFERENT COUNTRIES
(PPM DW)

Soil	Country	Range	Mean	Ref.
	U.S.S.R.	42.5—118	71.5	900
Various soils	Bulgaria	39—99	65	541
	Canada	20—110	57	629
	Canada	10—200	74[a]	521
	Chad	25—90	—	39
	Denmark	—	31	801
	Great Britain	20—284	80	100, 786, 818, 876
	Japan	10—622	86	395
	New Zealand	45—88	59	870
	Rumania	35—115	61	43
	U.S.S.R.	47—139	78	900

[a]   Data for whole soil profiles.

Table 57
ZINC CONTENT OF SURFACE SOILS OF THE U.S.
(PPM DW)

Soil	Range	Mean
Sandy soils and lithosols on sandstones	<5—164	40.0
Light loamy soils	20—118	55.0
Loess and soils on silt deposits	20—109	58.5
Clay and clay-loamy soils	20—220	67.0
Alluvial soils	20—108	58.5
Soils over granites and gneisses	30—125	73.5
Soils over volcanic rocks	30—116	78.5
Soils over limestones and calcareous rocks	10—106	50.0
Soils on glacial till and drift	47—131	64.0(a)
Light desert soils	25—150	52.5
Silty prairie soils	30—88	54.3(a)
Chernozems and dark prairie soils	20—246	83.5
Organic light soils	<5—108	34.0
Forest soils	25—155	45.7(a)
Various soils	13—300	73.5

Note: Sources are as follows: 706; a, 218, 219.

the range in maximum Zn adsorption by different soils as 16 to 70 μeq g^{-1} for Ca-saturated samples. This finding supports several statements about Ca-exchange sites with high selectivities for Zn. Shukla et al.[721] reported, on the other hand, that the order of Zn adsorption in different cation-saturated soil was the following: H < Ca ≤ Mg < K < Na.

Zn is considered to be readily soluble relative to the other heavy metals in soils. The Zn concentrations in soil solutions range from 4 to 270 μg ℓ^{-1}, depending on the soil and the techniques used for obtaining the solution (Table 12). Itoh et al.[341] reported the maximum of 17,000 μg Zn in 1 ℓ of solution and this value is, apparently, for highly contaminated soils. However, in natural but very acid soils (pH <4), Zn concentration in solutions was reported to average 7137 μg ℓ^{-1}.

Zn is most readily mobile and available in acid light mineral soils. As Norrish[570] stated, the Zn fraction associated with the Fe and Mn oxides is likely to be the most available to plants. Acid leaching is very active in Zn mobilization; thus, losses of this metal are observed

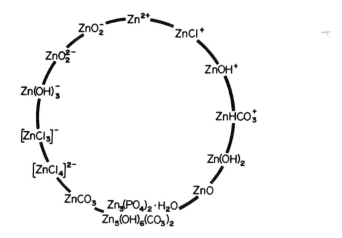

FIGURE 26. Ionic species and compounds of Zn occurring in soils.

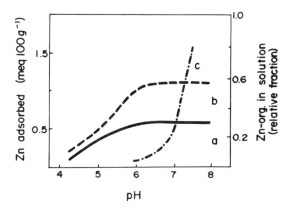

FIGURE 27. Influence of soil pH on adsorption of Zn at (a) 20 ppm and (b) 40 ppm Zn levels in initial solutions and on formation of soluble Zn-organic complexes in the solution of loamy silt soil (c).[519]

in certain horizons, particularly of podsols and brown acid soils derived from sands (Figure 25).

Solubility and availability of Zn is negatively correlated with Ca-saturation and P compounds present in soils. This relationship may reflect both adsorption and precipitation processes, as well as interactions between these elements. However, soluble Zn-organic complexes and complex anionic forms of Zn (Figures 26 and 27) may account for the relative solubility and availability of Zn in soils with a high pH range. This is supported by Bloomfield,[81] who found that Zn was mobilized from the basic carbonates and from the oxides produced by aerobically decomposing plant material.

The immobilization of Zn in soils rich in Ca and P, in well-aerated soils with S compounds, and in soils containing enhanced amounts of certain Ca-saturated minerals such as allophane, imogolite, and montmorillonite, as well as hydrous oxides, has an important practical impact on the Zn deficiency of plants.

2. Contamination of Soils

The anthropogenic sources of Zn are related, first of all, to the nonferric metal industry

Table 58
ZINC CONTAMINATION OF SURFACE SOILS (PPM DW)

Site and pollution source	Country	Mean or range in content	Ref.
Old mining area	Great Britain	220—66,400	808
	Great Britain	455—810	165
Nonferric metal mining	Great Britain	185—4,500	165, 786
	U.S.	500—53,000	615
	U.S.S.R.	400—4,245	467, 567
Metal-processing industry	Canada	185—1,397	363
	Holland	915—3,626	305
	Japan	800—5,400	891, 403
	Poland	1,665—5,567	224
	U.S.	155—12,400	365, 259
	Zambia	180—3,500	573
Urban garden and orchard	Canada	30—117	243, 844
	Poland	15—99	159
	U.S.	20—1,200	628
Sludged farmland	Great Britain	217—525	59
	Great Britain	1,097—7,474	165, 166
	West Germany	190—1,485	397, 176
	Holland	234—757[a]	314
	Sweden	369	24

[a]   6 and 16 tonnes dry matter sludge/ha/year, for 5 years.

and then to agricultural practice. Contemporarily observed soil contamination with Zn has already brought Zn to an extremely high accumulation in topsoils in certain areas (Table 58).

Calculating the first half-life of Zn in contaminated soils in lysimeters showed that Zn decrease was relatively rapid and that soil containing 2210 ppm Zn will reduce Zn content in half during 70 to 81 years.[395] These results, however, were reported for paddy soils with a long drainage period. Based on results of other experiments, the half-life of Zn as a pollutant in soil may be much longer. Amelioration of Zn-contaminated soils is commonly based on controlling its availability by the addition of lime or organic matter or both. Soluble Zn-organic complexes that occur particularly in municipal sewage sludge are very mobile in soils and therefore are easily available to plants.[451] Zn contamination of soils may create an important environmental problem.

B. Plants

1. Absorption and Transport

Soluble forms of Zn are readily available to plants and the uptake of Zn has been reported to be linear with concentration in the nutrient solution and in soils (Figure 28). The rate of Zn absorption differs greatly among both plant species and growth media. The composition of the nutrient solution, particularly the presence of Ca, is of great importance (Figure 29). Disagreement exists in the literature whether Zn uptake is an active or a passive process. Moore,[548] Loneragan,[489] and Hewitt[317] have reviewed this topic, and it may be summed up that several controversial results strongly suggest that Zn uptake is mostly metabolically controlled; however, it can also be a nonmetabolic process.

The form in which Zn is absorbed by roots has not been precisely defined. There is, however, general agreement on the predominating uptake of both hydrated Zn and Zn^{2+}. Several other complex ions and Zn-organic chelates may also be absorbed.[489,788,856] Halvorsen

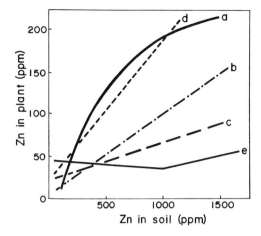

FIGURE 28. Zn uptake by plants from soil contaminated by this metal. (a) Pasture herbage, (b) wheat straw, (c) wheat grains, (d) potato stalks, (e) potato tubers.[176,783]

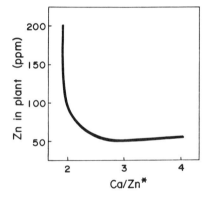

FIGURE 29. Zn content of cereal tops at the boot stage as a function of Ca/Zn ratio in the soil solution. Ca/Zn ratio is given in powers of ten.[915]

and Lindsay[302] concluded that only Zn^{2+} was absorbed by corn roots and that very low concentrations of that ion species is adequate for plant growth.

Several findings support the general statement that Zn generally is bound to soluble low molecular weight proteins; however, the formation of Zn-phytate and other insoluble Zn complexes were also reported by Weinberg[856] and Tinker.[798]

Fractions of Zn bound to light organic compounds in xylem fluids and in other plant-tissue extracts may suggest its high mobility in the plant.[789,822] Tinker[798] reported that the portion of Zn bound in complexes with a negative charge constitutes more than half of the total metal present in the plant.

Some authors regard Zn as highly mobile, while others consider Zn to have intermediate mobility. Indeed, when given luxury supplies of Zn, several plant species have mobilized appreciable quantities of this metal from old leaves to generative organs, but when under Zn-deficiency conditions the same species have mobilized little, if any, Zn from old leaves. Hence, in summary of several findings, it may be stated that Zn is likely to be concentrated in mature leaves. However, Scheffer et al.[688a] reported the highest Zn content of barley leaves, sheaths, and internodes to always be during the phase of intensive growth, which demonstrated a great Zn fluctuation within the plant during the vegetative period. Yläranta et al.[895] on the other hand, reported that the variation in the Zn content of wheat is surprisingly small and that the content increases slightly during the whole growth period. It has been calculated by Baumeister and Ernst[55] that up to 75% of the total Zn that is taken up is in the tops of young plants, whereas 20 to 30% occurs in the tops of old plants.

Roots often contain much more Zn than do tops, particularly if the plants are grown in Zn-rich soils. With luxury levels of soil Zn, this element may be translocated from the roots and accumulated by the tops of the plant. Zn is reported to be concentrated in chloroplasts, especially in those of some plants (e.g., spinach). This metal is also likely to be accumulated in vacuole fluids and in cell membranes.[798]

2. Biochemical Functions

Zn plays essential metabolic roles in the plant, of which the most significant is its activity as a component of a variety of enzymes, such as dehydrogenases, proteinases, peptidases, and phosphohydrolases (Table 26). Lindsay,[476] Price et al.,[630] and Shkolnik[718] have indicated that the basic Zn functions in plants are related to metabolism of carbohydrates, proteins, and phosphate and also to auxins, RNA, and ribosome formations. There is evidence that Zn influences the permeability of membranes and that it stabilizes cellular components and systems of microorganisms.[718,856] Zn is believed to stimulate the resistance of plants to dry and hot weather and also to bacterial and fungal diseases.

Plant species and varieties differ widely in their susceptibility to Zn deficiencies. Although these deficiencies are rlatively common (Table 28), their diagnosis is rather complex, and the best diagnoses are obtained when based on visual symptoms, plant analyses, and soil testing together. It must be emphasized, however, that for certain crops and soil, two chelating extractants, DTPA and EDTA, give linear relationships between Zn in the plant and the soluble Zn pool in the soil.[476,797]

Nambiar and Motiramani[560] reported that Zn levels in tissues used for the prediction of its deficiency often fail as the diagnostic test and that tissue Fe to Zn ratios appear to be more promising for the prediction even of hidden Zn deficiency. The critical Fe to Zn ratio in maize was found to be around 6.0. Lindsay[476] extensively reviewed the patterns of Zn deficiency and its common occurrence throughout the world and listed the most important factors contributing to Zn deficiency as follows:

1.   Low soil Zn content
2.   Calcareous soils and pH above 7
3.   Soils low in organic matter
4.   Microbial inactivation of Zn in soil
5.   Limited Zn uptake by roots due to restricted root zone and to cool spring seasons
6.   Differential responses of plant species and genotypes
7.   Antagonistic effects

Zn toxicity and Zn tolerance in plants have recently been of special concern because the prolonged use of Zn fertilizers, as well as its input from industrial pollution, is reflected in enhanced Zn content of surface soils (Table 58).

Several plant species and genotypes are known to have a great tolerance of Zn and a great selectivity in absorbing Zn from soils. Plants usually reflect changes in the Zn content of growth media and therefore are good indicators in biogeochemical investigations. Some genotypes grown in Zn-rich soils or in areas of heavy atmospheric Zn deposition may accumulate extremely large amounts of this metal without showing symptoms of toxicity. Petrunina[613] and Kovalevskiy[417] listed several species, particularly of the families Caryophyllaceae, Cyperaceae, and Plumbaginaceae, and also some trees, as extremely good Zn indicators that concentrate this metal in the range from about 0.1 to about 1% (DW). Tolerant species may reduce the effect of excessive Zn concentrations either by metabolic adaptation and complexing or by limiting the presence of the metal at cellular locations or by immobilization in storage tissues.

Most plant species and genotypes have great tolerance to excessive amounts of Zn. Chlorosis, mainly in new leaves, and depressed plant growth are the common symptoms of Zn toxicity (Table 29).

Zn phytotoxicity is reported relatively often, especially for acid and heavily sludged soils. Physiology and biochemistry of the toxic effects of Zn in plants are likely to be similar to those reported for other heavy metals; however, Zn is not considered to be highly phytotoxic. The toxicity limit for Zn depends on the plant species and genotypes, as well as on the growth stage. Hence, about 300 ppm Zn is reported to be toxic to young barley, whereas about 400 ppm is toxic to oats at the beginning of tillering.[171,322] However, in root tissues, where Zn is immobilized in cell walls or complexed in nondiffusible Zn proteins, the critical concentration of Zn is much higher.

3. Interactions with Other Elements

Zn is relatively active in biochemical processes and is known to be involved in several biological and chemical interactions with several elements. Zn-Cd interactions appear to be somewhat controversial, since there are reports both of antagonism and synergism between the two elements in the uptake-transport processes. Kitagishi and Yamane[395] explained the observed synergism in rice plants in terms of Zn competition for the Cd sites, resulting in an increase in Cd solubility, and in Cd translocation from the roots to tops. Wallace et al.[846] reported a high Cd accumulation in roots of plants at a high Zn level and at a low pH of the solution. Earlier findings well illustrated by those of Lagerwerff and Biersdorff,[449] however, show antagonism between these cations in the uptake-transport process. It may be stated that the ratio of Cd to Zn in plant media controls the occurrence of synergism and antagonism between these cations.

Zn-Cu antagonistic interactions have been observed in which the uptake of one element was competitively inhibited by the other. This may indicate the same carrier sites in absorption mechanisms of both metals.

Zn-Fe antagonism is widely known and its mechanism is apparently similar to the depressing effects of other heavy metals on Fe uptake. An excess of Zn leads to a marked reduction in Fe concentration in plants. Olsen[581] stated that Zn interfered more with the absorption and translocation of Fe than it did with Cu and Mn. There are two possible mechanisms of this interaction — the competition between Zn^{2+} and Fe^{2+} in the uptake processes and the interference in chelation processes during the uptake and translocation of Fe from the roots to tops. Also, Fe decreases Zn absorption and the toxicity of Zn that was absorbed.

Zn-As interaction was reported by Shkolnik[718] as a possible antagonism observed in the decrease of toxic effects of As excesses after Zn treatments. Zn-P interaction has been widely observed and reported for many crops, especially after phosphate and lime applications.[491] The P-Zn imbalance, resulting from excessive P accumulation, is known to induce Zn deficiency. This antagonism appears to be based to a great extent on chemical reactions in the root media.[581,675] However, Smilde et al.[738] stated that Zn-P antagonism could not be explained by only mutual immobilization and that this interaction is mainly a plant physiological characteristic. Usually the antagonistic effect of P on the concentration of Zn was more pronounced than that of Zn on P. Synergism between P and Zn was also observed in some plants.[654] A balanced P and Zn nutrition was reported by Shukla and Yadav[722] to be essential for the proper activity of *Rhizobium* and for N fixation.

Zn-N interaction is mostly a secondary ''dilution'' effect due to the increase of biomass because of the heavy N treatment. Olsen[581] reported also an enhancement of Zn in tops due to a higher bonding of Zn by proteins and amino acids in root tissues.

Table 59
MEAN ZINC CONTENT OF PLANT FOODSTUFFS (PPM)

Plant	Tissue sample	FW basis Data source 574	FW basis Data source 395	DW basis Data source 705	DW basis Data source 852	DW basis Data source 381	AW basis Data source 705	AW basis Data source 852
Sweet corn	Grains	—	—	25	36	—	980	1060
Bean	Pods	0.3	28.3	38	32	—	550	500
Cabbage	Leaves	0.6	2.2[a]	24	26	31	270	275
Lettuce	Leaves	0.1	11.7	73	44	—	520	240
Carrot	Roots	0.5	—	21	24	27	290	325
Beet	Roots	—	—	—	28	46	—	485
Onion	Bulbs	—	—	22	22	32	530	395
Potato	Tubers	0.3	3.4	14	10	26	340	310
Tomato	Fruits	—	—	26	17	—	220	235
Apple	Fruits	0.03	—	1.2	—	—	67	—
Orange	Fruits	0.09	0.9[b]	5.0	—	—	140	—

[a]   Pulses.
[b]   *Citrus unshiu* (Satsuma orange).

Zn-Ca and Zn-Mg interactions appear to vary for a given plant and media (Table 31). Apparently, several other factors, pH in particular, control the antagonistic and synergistic character of interactions between these elements. Olsen[581] stated that reactions which lowered Zn deficiency by Mg applications occurred within the plant rather than within the soil, but that competition between Mg and Zn in soil exchange sites cannot be precluded.

4. Concentrations in Plants

It is assumed that the Zn content of plants varies considerably, reflecting different factors of the various ecosystems and of the genotypes. However, the Zn contents of certain food-stuffs, cereal grains, and pasture herbage from different countries do not differ widely. Zn content ranged from 1.2 to 73 ppm (DW) in apple and lettuce leaves, respectively (Table 59).

Mean values for Zn in wheat grains ranged from 22 to 33 ppm (DW) and did not show any clear differences in country of origin (Table 60). Rye seems to contain a little less, and barley somewhat more Zn than wheat.

Background content of Zn in grass and clover throughout the world is also relatively stable, and its mean levels in grasses ranged from 12 to 47 ppm (DW), and in clovers ranged from 24 to 45 ppm (DW) (Table 61).

The deficiency content of Zn in plants has been established at 10 to 20 ppm (DW) (Table 25). These values, however, may vary considerably because the Zn deficiency reflects both the requirements of each genotype and effects of the interactions of Zn with other elements within the plant tissues.

Environmental Zn pollution greatly influences the concentrations of this metal in plants (Table 62). In ecosystems where Zn is an airborne pollutant, the tops of plants are likely to concentrate the most Zn. On the other hand, plants grown in Zn-contaminated soils accumulate a great proportion of the metal in the roots. The reported Zn contents of plants from some contaminated sites have already reached the magnitude of 0.X% (DW) and are a real health risk.

Table 60
ZINC CONTENT OF CEREAL GRAINS FROM DIFFERENT COUNTRIES (PPM DW)

Country	Cereal	Range	Mean	Ref.
Afghanistan	Barley	—	20	446
	Wheat	—	25	446
Australia	Wheat	16—35	22	867
Canada	Oats	—	37	514
Egypt	Wheat[a]	19—29	25	213
East Germany	Wheat	6—40	23[b]	65
Finland	Wheat[a]	25—47	37	508
	Wheat[c]	27—35	32	508
Great Britain	Barley	16—49	30	783
Japan	Wheat (flour)	—	5[d]	395
	Brown rice	19—28	23	395
	Unpolished rice	—	21[d]	891
Norway	Barley	15—51	29	446
	Wheat	21—67	33	446
Poland	Wheat[c]	23—38	27	268
	Rye	18—22	21	424
	Triticale	—	22	667
Sweden	Wheat[c]	20—40	34[e]	26
U.S.	Barley	20—23	22	484, 710
	Wheat, soft	—	5	906
	Wheat, hard	20—47	28	906
	Rye	—	34	492
	Triticale	—	26	492
U.S.S.R.	Rye	—	19	501

[a] Spring wheat.
[b] Mean value calculated from the given range.
[c] Winter wheat.
[d] FW basis.
[e] Data calculated from the figures given.

VII. CADMIUM

A. Soils

The abundance of Cd in magmatic and sedimentary rocks does not exceed around 0.3 ppm, and this metal is likely to be concentrated in argillaceous and shale deposits (Table 48). Cd is strongly associated with Zn in its geochemistry, but seems to have a stronger affinity for S than Zn, and exhibits also a higher mobility than Zn in acid environments. Cd compounds are known to be isotypic with corresponding compounds of such cations as Zn^{2+}, Co^{2+}, Ni^{2+}, Fe^{2+}, Mg^{2+}, and, in some cases, of Ca^{2+}.

During weathering Cd goes readily into solution and, though is known to occur as Cd^{2+}, it may also form several complex ions ($CdCl^+$, $CdOH^+$, $CdHCO_3^+$, $CdCl_3^-$, $CdCl_4^{2-}$, $Cd(OH)_3^-$, and $Cd(OH)_4^{2-}$) and organic chelates. However, the most important valence state of Cd in the natural environment is +2, and the most important factors which control the Cd ion mobility are pH and oxidation potential. Under conditions of strong oxidation, Cd is likely to form minerals (CdO, $CdCO_3$) and is also likely to be accumulated in phosphate and in biolith deposits.

The main factor determining the Cd content of soil is the chemical composition of the parent rock. The average contents of Cd in soils lie between 0.07 and 1.1 ppm (Table 63). However, the background Cd levels in soils apparently should not exceed 0.5 ppm, and all higher values reflect the anthropogenic impact on the Cd status in topsoils.

Table 61

MEAN LEVELS AND RANGES OF ZINC IN GRASS AND CLOVER OF IMMATURE GROWTH FROM DIFFERENT COUNTRIES (PPM DW)

Country	Grasses Range	Grasses Mean	Clovers Range	Clovers Mean	Ref.
Bulgaria	24—50	34	—	—	680
Czechoslovakia	15—35	25[a]	—	—	154
East Germany	15—80	47[a]	20—50	24[b]	65
West Germany	27—67	31[c]	—	—	596
Finland	28—39	32	—	—	388
Great Britain	22—54	33	—	—	165
Hungary	21—36	27	30—126	39	803
Japan	18—38	28	23—55	34	770
New Zealand	16—45	28	20—49	27	536
Poland	15—34	25	30—44	37	838
U.S.S.R.	—	—	—	45	501
Yugoslavia	6—11		48—94	62	623, 755

[a]   Mean value calculated from the given range.
[b]   Alfalfa.
[c]   Perennial rye grass and clover mixture.

Table 62

EXCESSIVE LEVELS OF ZINC IN PLANTS GROWN IN CONTAMINATED SITES (PPM DW)

Site and pollution source	Plant and part	Mean, or range of content	Country	Ref.
Old mining area	Grass	65—350	Great Britain	165, 167
	Clover	450	Great Britain	513
Nonferric metal mining	Onion bulbs	39—710	Great Britain	169a
	Lettuce leaves	55—530	Great Britain	169a
Metal industry	Lettuce leaves	316	Australia	57
	Chinese cabbage leaves	1300	Japan	403
	Oat grains	132—194	Poland	224
	Potato tubers	74—80	Poland	224
	Lettuce leaves	213—393	Poland	224
	Carrot roots	201—458	Poland	224
Urban garden	Radish roots	27—708	Great Britain	166, 167
	Leafy vegetables	35—470	U.S.	628
Sludged, irrigated or fertilized farmland	Grass	126—280	Holland	297
	Rice grains	21	Japan	395
	Rice roots	4510	Japan	395
	Oat grains	27—85	U.S.	726
	Soybean leaves	156	U.S.	126
	Soybean seed	114	U.S.	126
	Sagebrush	2,600[a]	U.S.	278
	Potato tubers	36	West Germany	176

[a]   AW basis.

The sorption of Cd species by soil components recently has been widely studied. Farrah and Pickering[228] stated that competitive adsorption by clays is the predominant process in Cd bonding. Also, findings of Tiller et al.[796] and Soon[747] support the opinion that adsorption,

Table 63
CADMIUM CONTENT OF SURFACE SOILS OF DIFFERENT COUNTRIES
(PPM DW)

Soil	Country	Range	Mean	Ref.
Podzols and sandy soils	Poland	0.01—0.24	0.07	378
	Canada	0.10—1.80	0.43	243
Loess and silty soils	Poland	0.18—0.25	0.20	378
Loamy and clay soils	Poland	0.08—0.58	0.26	378
	Canada	0.12—1.61	0.64	243
Soils on glacial till	Denmark	—	0.25	801
	Great Britain	0.49—0.61	—	353
Fluvisols	Austria	0.21—0.52	0.37	6
	Great Britain	0.41—2	1.10	166, 353, 786
	Poland	0.24—0.36	0.30	378
	Bulgaria	0.42	—	612
Gleysols	Poland	0.14—0.96	0.50	378
Rendzinas	Poland	0.38—0.84	0.62	378
Brown soils	Austria	0.22—0.49	0.33	6
Chernozems	Poland	0.18—0.58	0.38	378
	Bulgaria	0.55—0.71	0.61	612
Histosols and other organic soils	Denmark	0.8 —2.2	1.05	1, 801
	Canada	0.19—1.22	0.57	243
	Great Britain	0.56	—	353
Forest soil	U.S.	0.5 —1.5	0.73	87
Various soils	Austria	0.19—0.46	0.29	6
	Denmark	—	0.26	801
	Great Britain	0.27—4	1.00	100, 353, 786, 818
	West Germany	0.3—1.8	0.80	390
	Japan	0.03—2.53	0.44	395
	Poland	0.09—1.80	0.44	158, 378
	Bulgaria	0.24—0.35	0.29	612
	Canada	—	0.56	243
	U.S.	0.41—0.57	—	396
	U.S.S.R.	0.01—0.07	0.06	233

rather than precipitation, controls Cd concentrations in soil solutions until a threshold pH value is exceeded. The pH-solubility diagram (Figure 30) indicates that above pH 7.5 Cd sorbed in soils is not easily mobile, therefore, the solubility of $CdCO_3$, and possibly of $Cd_3(PO_4)_2$, would control the Cd mobility in soils. The solubility of Cd appears to be highly dependent on the pH; however, the nature of sorbent surfaces and of organic ligands are also of importance. As John[361] reported, the coefficient of bonding energy of the Cd adsorption was higher for organic matter than for soil clays. Abd-Elfattah and Wada[2] on the other hand, stated that Fe oxides, allophane, and imogolite reveal the highest affinity for the selective adsorpion of Cd.

Recently, Cd adsorption to organic matter as well as to Fe and Mn oxides has been widely studied by Gadde and Laitien,[253] Forbes et al.,[240] and Street et al.[761] All the findings lead to some generalizations: while in all soil, Cd activity is strongly affected by pH; in acid soils, the organic matter and sesquioxides may largely control Cd solubility, and that in alkaline soil, precipitation of Cd compounds is likely to account for Cd equilibria.

Cd is most mobile in acidic soils within the range of pH 4.5 to 5.5, whereas in alkaline soil Cd is rather immobile. However, as the pH is increased in the alkaline range, monovalent hydroxy ion species are likely to occur (e.g., $CdOH^+$), which could not easily occupy the sites on cationic exchange complexes.

Cd concentration in the soil solution is relatively low and is reported to range from 0.2 to 6 $\mu g \, \ell^{-1}$. The much higher value (300 $\mu g \, \ell^{-1}$) reported by Itoh and Yumura[342] presumably

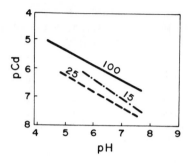

FIGURE 30.   Solubility of Cd^{2+} in soils as a function of soil pH. Equilibration with 15, 25, and 100 μg Cd per 2 g of soil.[747]

Table 64
CADMIUM CONTAMINATION OF SURFACE SOILS (PPM DW)

Site and pollution source	Range in content	Country	Ref.
Ancient mining area	0.6—14.0	Great Britain	166
	2—468	Great Britain	806
Nonferric metal mining	2—144	Belgium	728
	1.5—5.7	Great Britain	166
Metal-processing industry	2—5	Bulgaria	611
	2—36	Canada	363
	9—33	Holland	305
	2.2—88	Japan	336, 403, 891
	6—40	Poland	224, 871
	26—160	U.S.	233
	0.6—46	Zambia	573
Urban garden	0.4—4.5	Poland	159
	0.02—13.6	U.S.	127
	0.1—3.7	U.S.	628
Sludged, irrigated, or fertilized farmland	7.3—8.1	Canada	243
	15—57[a]	Holland	314
	13—35	West Germany	176
Vicinity of highways	1—10	U.S.	705a

[a]   6 and 16 tonnes dry matter sludge/ha/year, for 5 years.

indicates contaminated soil and corresponds to the value (400 μg ℓ^{-1}) given by Kabata-Pendias and Gondek[379] for contaminated soil.

In soils developed under the influence of humid climate, migration of Cd down the profile is more likely to occur than its accumulation in the surface horizon; thus, the enrichment in Cd content observed so commonly in topsoils should be related to contamination effects.

Soil contamination with Cd is believed to be a most serious health risk. Under man-induced conditions Cd is likely to be built up in surface soils (Table 64). Even in the forest or rural regions in various countries, with the exception of one site in Sweden, the atmospheric input of Cd exceeded the output of this metal from the soil profile (Table 13).

The present concentration of Cd in topsoils is reported to be very high in the vicinities of Pb and Zn mines and, in particular, smelting operations (Table 64). Sewage sludges and phosphate fertilizers are also known as important sources of Cd, and there are several comprehensive reviews of this subject, as given by Fleischer et al.,[233] Williams and David,[878,879] Street et al.,[761] Chaney and Hornick,[129] von Jung et al.,[372] and Anderson and Hahlin.[22]

Because of the environmental significance of the Cd accumulation in soil, several techniques for the management of Cd-enriched cropland have been investigated. Similarly, as

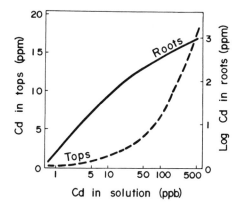

FIGURE 31. The effect of Cd concentration in the culture solution on Cd uptake by grass *Bromus unioloides*.[915]

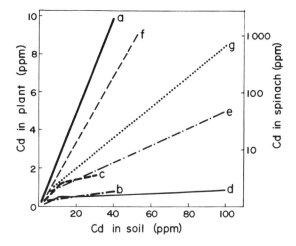

FIGURE 32. Cd uptake by plants from soil contaminated by this metal. (a) Potato stalks, (b) potato tubers, (c) wheat grains, (d) brown rice, submerged during growth period, (e) brown rice, drained after tillering stage, (f) barley grains, value in ng g^{-1}, (g) spinach leaves, data for 0.1 N HCl-soluble Cd in soil.[22,176,336,342]

in the case of Zn-contaminated soils, these techniques are based on increasing the soil pH and CEC. Although liming generally is expected to decrease Cd absorption by raising the soil pH, it is not effective for all soils and plants. Kitagishi and Yamane[395] reported that the best and most reliable results in reducing Cd availability was the layering of unpolluted soil over polluted soil to a depth of 30 cm. The maximum permissible rate of Cd addition to soil should depend strongly on the soil pH.

B. Plants

1. Absorption and Transport

Although Cd is considered to be a nonessential element for plants, it is effectively absorbed by both the root and leaf systems. In almost each case a linear relationship between Cd in plant material vs. Cd in growth medium is reported (Figures 31 and 32). Nevertheless, several soil and plant factors affect the plant uptake of Cd.

In nearly all the publications on the subject the soil pH is listed as the major soil factor controlling both total and relative uptake of Cd (Figure 33). Kitagishi and Yamane[395] reported

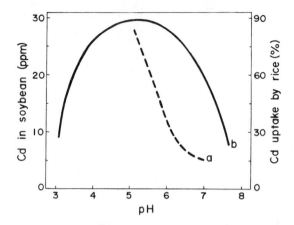

FIGURE 33.   (a) Effect of soil pH on Cd content of soybean
leaves and (b) effect of solution pH on relative Cd absorption
by rice seedlings.[129,395]

results of their experiment indicating that the relative uptake of Cd by rice seedlings was
the greatest within the pH range of 4.5 to 5.5. Bingham et al.[77] also found that the Cd
content of rice grain is highly dependent upon the soil pH and is the highest at pH 5.5.
However, there are contradictory results which show that when Cd becomes more mobile
in alkaline soil due to the formation of complexes or metal chelates, the plant uptake of Cd
may be independent of the pH.[41,129] Kitagishi and Yamane[395] described that when the redox
potential of soils decreases to about -0.14 V, the proportion of soluble Cd decreases,
corresponding to the reduction of sulfate to sulfides. This is well illustrated in the much
lower uptake of Cd by rice grown on submerged soil than on soil drained after the tillering
stage (Figure 32).

Although soil characteristics, other than the pH, can also cause differences in the Cd
absorption by roots, it may be stated that soluble species of Cd in soil are always easily
available to plants. Although an appreciable fraction of Cd is taken up passively by roots,
Cd is also absorbed metabolically.[737]

Chaney and Hornick[129] reviewed plant response to increased levels of Cd in soil and
showed a great difference in the ability of various plant species to absorb this metal. The
difference in Cd content of crops grown on the same soil containing 10 ppm Cd exceeded
100-fold; the lowest concentration was in rice plants, Sudan grass, and clover, and the
highest was in spinach and turnip.

Tiffin[789] found the [109]Cd concentration in tomato xylem exudates to be from 1.5 to 3.5
μM after treatment with this radionuclide. This suggests that Cd, as other heavy metals,
can be easily transported within a plant in the form of metaloorganic complexes.

Mechanisms of the long distance transport of Cd^{2+} have not yet been presented, but Cd
most likely would be transported by the carrier mechanisms similar to those of Zn translocation. Cunningham et al.[155] reported that increased Cd^{2+} treatment progressively sup-
pressed the proportion of Cd translocation to epigeal parts of young leaves and that Cd was
localized mainly in roots, with lesser amounts in stem nodes, petioles, and major leaf veins.

A great proportion of the Cd is known to be accumulated in root tissues, even when Cd
enters the plant via foliar systems.[376] This, however, was not reported for rice plants when
[109]Cd was applied to the leaves.[395] When the amount of Cd is increased in the growth
medium, the concentration of this metal in roots exceeds its content in the tops about 100
times (Figure 31).

It may be concluded that although the roots of several species can take up large amounts of Cd from the growth medium, the translocation of Cd through the plant may be restricted because Cd is easily held mainly in exchange sites of active compounds located in the cell walls.

2. Biochemical Roles

The most important biochemical characteristic of Cd ions is their strong affinity for sulfhydryl groups of several compounds (Cd complexes with metallohionein-like protein are already known). In addition, Cd also shows an affinity for other side chains of protein and for phosphate groups.

Dabin et al.[161] and Braude et al.[102] reported that Cd is likely to be concentrated in the protein fractions of plants. This fact is very important in food production problems.

There are no known enzymes that depend on Cd for their normal activity. Roucoux and Dabin[666] reported that Cd specifically induced cysteine and methionine synthesis in soybeans, depending on the degree of plant resistance to increased Cd levels. Cd is considered to be a toxic element to plants, and the basic cause of its toxicity lies in its disturbing enzyme activities. Cunningham et al.[155] and Baszynski et al.[54] reported inhibition of the formation of anthocyanin and chlorophyll pigments in plants that were treated with Cd.

In general, overt symptoms induced by elevated Cd contents of plants are growth retardation and root damage, chlorosis of leaves, and red-brown coloration of leaf margins or veins. The phytotoxicity of Cd, beyond interfering with normal metabolism of some micronutrients, shows inhibitory effects on photosynthesis, disturbs transpiration and CO_2 fixation, and alters the permeability of cell membranes. Cd is also known to inhibit the DNA-mediated transformation in microorganisms and to interfere with symbiosis between microbes and plants as well as to increase plant predisposition to fungal invasion.

The Cd content of plants is, however, of the greatest concern as a Cd reservoir and as the pathway of Cd to man and animals. Thus, tolerance and adaptation of some plant species to higher Cd levels, although important from the environmental point of view, create a health risk.

3. Interactions with Other Elements

Plants are simultaneously exposed to a variety of pollutants, and thus their integrated effects most often are different from the effect of Cd only. Several elements are known to interact with Cd in both the element uptake by plants and in biochemical roles.

Cd-Zn interactions are commonly observed, but findings appear contradictory, since both depressing and enhancing effects of each have been reported (Chapter 7, Section VI.B.3.). The interaction of Cd and Zn has received much study, and all findings may be summed up by stating that, in most cases, Zn reduces the uptake of Cd by both root and foliar systems. Chaney and Hornick[129] suggested that when the Cd/Zn ratio in plant tissues is limited to 1%, the Cd content is restricted to below 5 ppm, thus below its phytotoxic level (Table 25). Babich and Stotzky[41] described the double antagonistic interactions between Cd-Zn and Cd-Mg in microorganisms. Cd-Cu interactions are also complex (Chapter 6, Section V.B.3.). The inhibitory effect of Cu on Cd absorption is reported most often. Interactions of Cd with other heavy metals such as Mn and Ni are also often reported and appear to be related to their replacement by Cd during the uptake processes.

Cd-Se mutual antagonistic effects were observed in certain crops. Cd-P interaction is exhibited in the P effects on Cd uptake by plants. Both increased and decreased contents of Cd under phosphate treatment are reported. Apparently, these reactions take place in the root media, and thus, the influence of the P supply to soils may differ for various soils and crops. It is most likely that Cd-P interactions may be similar to those in the Zn-P relationship.

Cd-Ca relationship seems to be highly cross-linked with variation in the soil pH. It cannot be precluded, however, that Ca^{2+} cations are able to replace Cd^{2+} in the carrier mechanism and thus Cd absorption by plants may be inhibited by an excess of Ca^{2+} ions. Cd uptake-translocation mechanisms are reported to be influenced by the supply of other nutrients such as K, N, and Al, but the results are not clear at this time and may be related to some secondary effects.

4. Concentrations in Plants

In man and animal nutrition, Cd is a cumulative poison, therefore, its content in food and feed plants has been widely studied. A comparison of the Cd contents of plant foodstuffs produced under uncontaminated conditions of various countries shows the highest Cd concentration in spinach leaves (0.11 ppm, FW) and lettuce leaves (0.66 ppm DW, 3.00 ppm AW) (Table 65). When plants are grown on contaminated soil, however, Cd is very likely to also be concentrated in roots (Table 66). This clearly confirms the statement that leafy vegetables such as spinach, and root vegetables such as turnip, should be considered to be the main routes of Cd supply to man.

The background levels of Cd in cereal grains as well as in common feed plants that are reported for various countries are fairly low and surprisingly similar (Tables 67 and 68). Thus, grand mean values for all cereal grains range from 0.013 to 0.22 ppm (DW), grasses range from 0.07 to 0.27 ppm (DW), and legumes range from 0.08 to 0.28 pm (DW).

Because Cd is readily available to plants from both air and soil sources, its concentration rapidly increases in plants grown in polluted areas. A few data collected for several countries show that both industrial and agronomic practices may create a significant Cd supply to plants (Table 66). The highest concentrations of Cd in polluted plants were always reported for roots and leaves, whereas Cd seems to be excluded from seed crops. The highest Cd values reported for wheat grains (14.2 ppm) and brown rice (5.2 ppm) were less than the amounts of the metal accumulated in the root and leaf tissues of these plants.

The maximum allowable limit of Cd in plant foodstuffs has been widely discussed and should always be calculated on the basis of daily metal intake by a given population group.[395,404] The threshold concentrations in feed plants may be a bit higher than those established for food plants and may differ for each kind of animal.

VIII. MERCURY

A. Soils

In all types of magmatic rocks the Hg content is fairly low and does not exceed the order of XO ppb (Table 48). A much higher concentration of this metal is reported for sedimentary rocks, argillaceous sediments and, in particular, organic-rich shales (common range 40 to 400 ppb).

The most important geochemical features of Hg are

1.   Affinity to form strong bonds with S (e.g., cinnabar, HgS, is the most common Hg mineral)
2.   Formation of organomercury compounds that are relatively stable in aqueous media
3.   Volatility of the elemental Hg

Although Hg may form several ionic species (Figure 34), it is not very mobile during weathering (Table 8). As Jonasson,[367] Shcherbakov et al.,[715a] and Landa[458] have reported, Hg is likely to be strongly bound when added to soils as elemental Hg or as cationic or anionic complexes. Hg is retained by soils mainly as slightly mobile organocomplexes.

The findings of Hem[312] and of Farrah and Pickering[229] clearly indicate that $Hg(OH)_2$ is likely to predominate over other aqueous species at a soil pH near or above neutrality. The sorption of Hg by clays in soil seems to be relatively limited and to vary only a little with pH. However, in acid gley soils, the formation of HgS and even of metallic Hg may take

Table 65
MEAN CADMIUM CONTENT OF PLANT FOODSTUFFS (PPM)

Plant	Tissue sample	FW basis				DW basis		AW basis
		Data source 569	Data source 395, 891	Data source 704, 705	Data source 233	Data source 497, 852	Data source 430	Data source 704, 705
Sweet corn	Grains	—	—	0.007	0.1	—	0.06	1.00
Bean	Pods	—	0.02[a]	0.024	—	0.29[b]	—	0.34
Cabbage	Leaves	0.05	0.02	0.05	0.05	—	—	—
Lettuce	Leaves	—	0.11[c]	0.42	0.4	0.66	0.12[c]	3.00
Carrot	Roots	0.09	0.05	0.15	<0.35	0.24	0.07	2.10
Onion	Bulbs	—	0.01	0.05	—	—	0.08	1.20
Potato	Tuber	0.001—0.09	0.02	0.08	0.05—0.3	0.18	0.03	1.80
Tomato	Fruits	0.02	—	0.11	—	0.23	0.03	1.00
Apple	Fruits	0.008	0.003	0.03	—	—	0.05	0.19
Orange	Fruits	—	0.002	0.005	—	—	—	0.14

[a] Pulses.
[b] Soybean seeds.
[c] Spinach.

Table 66
EXCESSIVE LEVELS OF CADMIUM IN PLANTS GROWN IN
CONTAMINATED SITES (PPM DW)

Site and pollution source	Plant and part	Mean or range in content	Country	Ref.
Ancient mining area	Grass tops	1.0—1.6	Belgium	728
	Lichens	11—22	Belgium	728
	Brussel sprouts	0.10—1.77	Great Britain	168
	Grass tops	1.1—2.0	Great Britain	513
	Clover tops	4.9	Great Britain	513
Metal-processing industry	Lettuce leaves	45	Australia	57
	Silver beet leaves	0.04—0.49[a]	Australia	793a
	Turnip leaves	0.5	West Germany	430
	Brown rice	0.72—4.17	Japan	395, 891
	Lettuce leaves	5.2—14.1	Poland	224
	Carrot roots	1.7—3.7	Poland	224
	Spinach leaves	6.4	Zambia	573
Lignite coal-fired power station	Grass tops	1.1	Czechoslovakia	655
Urban garden	Brussel sprouts	1.2—1.7	Great Britain	786
	Cabbage outer leaves	1.1—3.8	Great Britain	786
	Lettuce leaves	0.9—7.0	U.S.	627
Sludged, irrigated, or fertilized farmland	Cereal grains	0.1—1.1	Finland	748
	Brown rice	5.2 (max.)	Japan	336
	Lettuce[b]	70	U.S.	789
	Corn[b]	35	U.S.	789
	Lettuce leaves	0.5—22.8	U.S.	127
	Carrot roots	0.2—3.3	U.S.	127
	Soybean seeds	2.3	U.S.	102
	Cabbage leaves	130	U.S.S.R.	826
	Wheat grains	5.5—14.2[c]	U.S.S.R.	338
	Wheat leaves	19—47	U.S.S.R.	338
	Wheat roots	397—898	U.S.S.R.	338
Airborne contamination	Lettuce leaves	5.2	Denmark	625
	Spinach leaves	3.9	Denmark	625
	Carrot roots	3.5	Denmark	625

[a] FW basis.
[b] Diagnostic leaf.
[c] Pot experiment.

place. The accumulation of Hg in soil is, therefore, controlled mainly by organic complex formation and by precipitation. Thus, the mobility of Hg requires dissolution processes and biological degradation of organomercury compounds.

The transformation of organomercury compounds, especially the methylation of elemental Hg, plays the most important role in the Hg cycle in the environment. The mechanism of methylation has recently been the subject of many studies because methylated Hg is readily mobile and easily taken up by living organisms, including some higher plants.[58,359,470,856] The mechanism of the methylation of Hg still is not fully understood. It can occur abiotically; also a vast number of organisms (especially microorganisms) may carry out these reactions. The methylation processes evidently have been involved in the environmental catastrophes of Hg poisoning.

Several types of bacteria and yeasts have been shown to effect the reduction of cationic Hg^{2+} to the elemental state (Hg^0); thus the result of these processes is the volatilization of Hg from the medium. The oxidation of elemental Hg to its cationic form can also be mediated by microorganisms.

Table 67
CADMIUM CONTENT OF CEREAL GRAINS FROM DIFFERENT COUNTRIES (PPM DW)

Country	Cereal	Range	Mean	Ref.
Australia	Wheat	0.012—0.036	0.022	878
Canada	Oats	—	0.21	497
Denmark	Oats	—	0.03[a]	625
Egypt	Wheat	0.01—0.09	0.05	213
East Germany	All cereals	0.02—0.06	0.05	430, 488
West Germany	Various cereals	0.01—0.75	0.22	55
	Barley	0.01—0.02	0.02	577
	Wheat	0.03—0.04	0.04	577
	Oats	0.02—0.03	0.02	577
Japan	Rice, unpolished	0.05—0.11	0.08	891
	Rice, unpolished	0.01—0.11	0.05	395
	Wheat (flour)	—	0.03[a]	395
Norway	Barley	0.006—0.044	0.022	446
	Wheat	0.008—0.260	0.071	446
Poland	Oats	—	0.060	915
	Wheat	—	0.056	915
	Rye	—	0.070	915
Sweden	Barley	—	0.013	22
	Wheat	—	0.06	18
U.S.	Corn	—	0.1	704
	Wheat	0.07—0.13	0.10	906
	All cereals	0.1—0.5	—	704
U.S.S.R.	Wheat	0.06—0.07	—	338

[a] FW basis.

Table 68
MEAN LEVELS AND RANGES OF CADMIUM IN GRASSES AND LEGUMES AT THE IMMATURE GROWTH STAGE FROM DIFFERENT COUNTRIES (PPM DW)

Country	Grasses Range	Grasses Mean	Clovers Range	Clovers Mean	Ref.
Canada	—	0.21	—	0.28[a]	497
Czechoslovakia	—	0.6	—	—	655
France	—	0.16	—	0.11[a]	577
East Germany	0.05—1.26	0.27	0.02—0.35	0.16	430, 488
West Germany	0.03—0.14	0.07	0.04—0.18	0.08[a]	577
Iceland	0.07—0.14	0.10	—	—	577
Poland	0.05—0.20	0.08	0.07—0.30	0.10	915
U.S.	0.03—0.3	0.16	0.02—0.2[a]	—	330a, 704

[a] Alfalfa.

The Hg content of virgin soil profiles is inherited mostly from the parent material; however, because Hg is easily volatile, some additional natural sources, such as degassing and thermal activity of the earth, cannot be precluded. The accumulation of Hg generally is related to the organic C and S levels in soils and is distributed in the surface soils at several times the concentration in the subsoils.

Background levels of Hg in soils are not easy to estimate due to widespread Hg pollution.

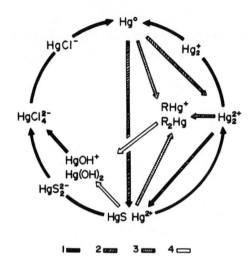

FIGURE 34.   Ionic species and transformations of mercury compounds in soils. (1) Reduction; (2) oxidation; (3) formation of organic compounds; (4) hydration; R: CH_3, CH_3CH_2, C_6H_5.

Table 69
MERCURY CONTENT OF SURFACE SOILS OF DIFFERENT COUNTRIES (PPM DW)

Soil	Country	Range	Mean	Ref.
Podzols and sandy soils	Holland	—	0.04	223
	Canada	0.01—0.70	0.06	243
Desert sands	Israel	0.008—0.03	0.02	724
Cambisols, luvisols, and other loamy soils	Holland	0.45—1.1	—	223
	Canada	0.02—0.78	0.13	243
Soils on glacial till	Canada	0.02—0.10	0.05	368
Gleysols	Canada	0.018—0.22	0.053	520
Histosols and other organic soils	Canada	0.05—1.11	0.41	243
	Switzerland	0.04—0.11	0.08	637
Forest soils	Norway	0.02—0.55	0.19	447
	Japan	0.02—0.20	0.07[a]	395
	Yugoslavia	0.08—0.58	0.23	413, 754a
	Switzerland	0.073—0.10	0.08	637
Paddy soils	Japan	0.15—0.76	0.35[a]	395
	Japan	0.13—0.46	0.27	277
	Viet Nam	0.02—1	0.3	8
Various soils	Canada	0.005—0.10	0.06[b]	521
	Great Britain	0.01—0.09	0.03	164a
	Japan	0.08—0.49	0.28	395
	Norway	0.02—0.35	0.19	367
	Poland	0.02—0.16	0.06	162
	Sweden	0.004—0.99	0.06	19
	West Germany	0.025—0.35	0.09	390
	U.S.S.R.(Europe)	0.04—5.8	—	8

[a]   Calculated from the histogram.
[b]   Data for whole soil profiles.

Nevertheless, data reported for various soils on a world-wide basis show that mean concentrations of Hg in surface soils do not exceed 400 ppb (Tables 69 and 70). The highest mean levels of Hg were reported for histosols of Canada (400 ppb) and for paddy soils of

Table 70
MERCURY CONTENT OF SURFACE SOILS OF THE U.S.
(PPM DW)

Soil	Range	Mean
Sandy soils and lithosols on sandstones	<0.01—0.54	0.08
Light loamy soils	0.01—0.60	0.07
Loess and soils on silt deposits	0.01—0.38	0.08
Clay and clay loamy soils	0.01—0.90	0.13
Alluvial soils	0.02—0.15	0.05
Soils over granites and gneisses	0.01—0.14	0.06
Soils over volcanic rocks	0.01—0.18	0.05
Soils over limestones and calcareous rocks	0.01—0.50	0.08
Soils on glacial till and drift	0.02—0.36	0.07
Light desert soils	0.02—0.32	0.06(a)
Silty prairie soils	0.02—0.06	0.04(a)
Chernozems and dark prairie soils	0.02—0.53	0.10
Organic light soils	0.01—4.60	0.28
Forest soils	0.02—0.14	0.06(a)
Various soils	0.02—1.50	0.17

Note: Sources are as follows: 706; a, 218, 219.

Japan (350 ppb) and Viet Nam (300 ppb). Similarly, in organic and clay soils of the U.S., the highest average concentrations were found to be 280 ppb in histosols and 130 ppb in loamy soils. Apparently, the organic soils and paddy soils are likely to retain more than any other soils, the Hg resulting from vegetable decay and absorption from the atmosphere.

The background content of soil Hg can be approximately estimated as O.X ppm. Thus, Hg contents exceeding this value should be considered as contamination from anthropogenic or other sources.

Sources of the contamination of soil with Hg are related mainly to base metal processing industries and some chemical works (chlor-alkali, in particular), as well as to the use of fungicides containing Hg (which has recently been discontinued in many countries). Sewage sludges and other wastes may also be sources of Hg contamination (Table 71).

In soils, the migration processes involving Hg are rather limited, therefore, the Hg content of surface soils is slowly built up, even under a low input of this metal.[205] However, Landa[459] reported Hg losses by volatilization from soils, which increased with higher soil temperature and with higher soil alkalinity, and Kulikova and Nurgaleyeva[438] described a short life of Hg residues in chernozems from Hg seed dressing.

The behavior of Hg in contaminated soils is of great interest since the ready bioavailability of this metal creates an important health hazard. Kitagishi and Yamane[395] have widely reviewed this topic, giving special concern to the paddy soil. It can be generalized that inorganic compounds of Hg added to soils are likely to be absorbed well by humus and partly by clays. Various organic compounds of Hg (methyl, ethyl, and phenyl) added to soil are partly decomposed or adsorbed by soil constituents. However, all these compounds, having a relatively small degree of dissociation and adsorbability, are readily taken up by plants. The authors reported that methyl-Hg was the most available, while phenyl- and sulfide-Hg were the least available to plants and that residue of organomercuries varied, being the highest for phenyl-Hg and methyl-Hg iodide.

The potenial for microbial methylation of Hg by bacteria and fungi exists under both aerobic and anaerobic soil conditions. Soil contamination with Hg itself is usually considered not to be a serious problem, although there is the possibility that a large amount of methyl-Hg will result. Nevertheless, even simple Hg salts or metallic Hg create a hazard to plants and soil biota from the toxic nature of Hg vapor.

Table 71
MERCURY CONTAMINATION OF SURFACE SOILS (PPM DW)

Site and pollution source	Maximum or range of content	Country	Ref.
Ancient mining area	0.21—3.4	Great Britain	164a, 165
Hg mining or ore deposit	0.2 —1.9	Canada	368
	0.6 —4.2	Yugoslavia	413, 414
	8.2 —40	U.S.	79, 703
	0.1 —2.4	U.S.	198
Chlor-alkali or chemical works	3.8	Great Britain	111
	0.32—5.7	Canada	776
	0.10—0.43	Switzerland	637
Urban garden, orchard, and park	0.03—1.14	Canada	243
	0.25—15.0	Great Britain	488a, 850
	0.6	Great Britain	634
	0.6	U.S.	208
	0.04—0.08	Israel	724
	0.06—0.24	Japan	277
Sludged or irrigated farmland	10.0[a]	Holland	223
	1.5[b]	West Germany	205
	0.29—0.71	Japan	395
	0.8	Sweden	25
	0.12—0.35	Poland	162
Application of fungicides	9.4 —11.5	Canada	498

[a]   Soil flooded by Rhine River water.
[b]   Soil irrigated with sewage water for 80 years.

Lagerwerff[447a] stated that Hg uptake by plant roots may be minimized by neutralizing the soil pH with lime. Also sulfur-containing compounds and rock phosphates have been proposed to inactivate mercurial fungicides or elemental Hg in soils. Sorterberg,[748] in contrast, reported that heavy liming was ineffective in reducing toxic effects of Hg excesses in the soil.

B. Plants
1. Biochemical Roles
The information on the biochemistry of Hg is concerned mainly with biological transformation of Hg compounds, but it is not yet clear which processes are most important in the cycling of Hg in the environment. Weinberg[856] reviewed all topics related to Hg transformation and to Hg resistance in organisms.

Plants seem to take up Hg easily from solution culture (Figure 35). There is also much evidence that increasing soil Hg generally causes an increase in the Hg contents of plants. The rate of increase of the Hg content in plants when the soil was the only source of this metal was reported to be highest for roots, but leaves and grains also accumulated much Hg.[321,481,748] These findings show that Hg is easily absorbed by the root system and is also translocated within the plants. Blanton et al.,[79] on the other hand, reported that Hg levels in plants bear little relationship to the Hg content of soils from an Hg mining district and that this reflects strong Hg bonds to soil components. Also in the official report on Environmental Mercury and Man[214] it was stated: "For most plants, even when grown on soils having much higher concentrations of mercury, there is very little additional uptake".

Plants are known to directly absorb Hg vapor. Browne and Fang[107] reported that the rate of Hg vapor uptake is particularly influenced by illumination, but is unaffected by ambient temperature.

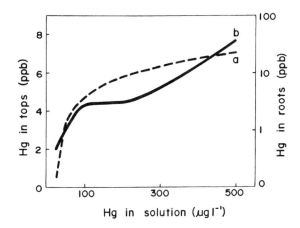

FIGURE 35.   Uptake of Hg by 7-day-old oat seedlings from the culture solution of HgNO$_3$ concentration. (a) Tops; (b) roots.[735]

Translocation of Hg occurs in various plant tissues, including apple leaves to fruits, from potato leaves to tubers, from rice leaves to grains, and also from the wheat and pea seed treated with mercurial fungicides into the first generation seed.[928] The content of Hg in rice grains harvested from plants treated with phenylmercuric acetate was reported to be the highest in the bran, but the fraction of Hg accumulated in whole grains apparently was associated mainly with gluten.[395]

Although Hg is known to bind strongly to the amino acid S atoms contained in many proteins and enzymes, this metal seems to be relatively easily transported in plants. The Hg affinity to sulfhydryl groups apparently is the key reaction in disrupting metabolic processes of plants.

Toxic effects in young barley were observed at an Hg level of 3 ppm (DW), and Hg at a concentration of 0.01 ppm (AW) was reported to be severely toxic.[171,710] The toxicity of volatilized elemental Hg and of some methylated compounds are known to be the most serious for plants.

The symptoms of Hg toxicity are, most commonly, stunting of seedling growth and root development and inhibition of photosynthesis and, as a consequence, a reduction in yield. Hg accumulated in root tissue inhibits K$^+$ uptake by plants, although the stimulating effect of low Hg concentration on K$^+$ uptake was also reported by Hendrix and Higinbotham.[315]

The resistance to Hg compounds in fungi and bacteria has received most studies, as it is of practical importance. Weinberg[856] reviewed recent results and demonstrated that in most cases the resistant strain was capable of volatilizing Hg. Plasmids are believed to carry determinants for the Hg resistance.

The tolerance to Hg in higher plants has also been reported, and although the mechanism of the physiological barrier is not known, it is most probably related to the inactivation of Hg at the membrane sites. The affinity of Hg to form insoluble compounds with S-rich proteins was reported for several plant species.

2. Concentrations in Plants

The distribution of Hg in plants has recently received the most study because of the Hg pathway into the food chain. Therefore, most information is at present related to the Hg content of plant foodstuffs. The background levels of Hg in vegetables and fruits vary from 2.6 to 86 ppb (DW) and from 0.6 to 70 ppb (FW) (Table 72).

Gracey and Stewart[279a] reported Hg to average 39 ppb (DW) in alfalfa, and other observations (author's unpublished data) show that natural Hg levels in grass and feed legumes

Table 72
MEAN MERCURY CONTENT OF PLANT
FOODSTUFFS (PPB)

Plant	Tissue sample	FW basis	DW basis
Sweet corn	Grains	—	4.6(d), 3(f)
Bean	Pods	70(e), 17(g)	3(d), 11(i)
Beet	Roots	3(a)	—
Carrot	Roots	—	86(c), 5.7(d)
Lettuce	Leaves	<0.6(a)	8.3(d)
Cabbage	Leaves	10(e)	6.5(d)
Potato	Tubers	3(b), 12(g)	47(c), <10(d)
Onion	Bulbs	7(g)	<10(d)
Cucumber	Unpeeled fruits	1(a), 11(g)	—
Tomato	Fruits	1(a)	34(c), 3.1(d)
Apple	Fruits	10(g)	<10(d)
Orange	Fruits	—	2.6(d)
Lemon	Fruits	43(g)	—
Mushroom	Caps and stalks	—	3.5(h)

Note: Sources are as follows: a, 776; b, 710; c, 496; d, 705; e, 395; f,
126; g, 163; h, 754a; and i, 373.

Table 73
MERCURY CONTENT OF CEREAL GRAINS FROM
DIFFERENT COUNTRIES (PPB DW)

Country	Cereal	Range	Mean	Ref.
Canada	Barley	5—17	12	279a
	Oats	4—19	9	279a
	Wheat	7—15	11	279a
Egypt	Wheat	11—28	21	213
West Germany	Wheat	—	<10	400
Japan	Wheat (flour)	—	20[a]	395
	Buckwheat (flour)	—	10[a]	395
Norway	Barley	0.2—17.2	3.4	446
	Wheat	0.2—2.7	0.9	446
Poland	Barley	7—82	19	373
	Oats	7—42	20	373
	Rye	3—18	9	373
	Wheat	4—33	13	373
Sweden	Oats	<4—45	14	748
Switzerland	Wheat	6—10	7	637
U.S.	Barley	—	19	110
	Oats	—	12	110
	Wheat	10—16	14	110, 447a
U.S.S.R.	Wheat	7—12	10	163

[a]   FW basis.

do not exceed 100 ppb (DW). The Hg contents of cereal grains seem to be fairly similar
for various countries and for certain kinds of cereals, with mean values ranging from 0.9
to 21 ppb (DW) (Table 73). However, grain crops from land where mercuric compound
dressings of seeds were used show some elevation (up to 170 ppb DW) in the Hg content.[162]
A similar statement was made by Kulikova and Nurgaleyeva,[438] although without analytical
data. The latter reports do not support the earlier findings indicating that the use of Hg in
the U.K. for the treatment of cereal seeds had very little effect on Hg levels in the grains.[214]

Table 74
MERCURY CONTENT OF PLANTS GROWN IN CONTAMINATED SITES
(PPM DW)

Site and pollution source	Plant and part	Maximum or range in content	Country	Ref.
Metal-processing industry	Edible mushrooms	37.6	Yugoslavia	414
	Carrot, roots	0.5—0.8[a]	Yugoslavia	116
	Apple, flesh	0.04—0.13[a]	Yugoslavia	116
	Apple, pips	0.33—1.32[a]	Yugoslavia	116
Soil overlying Hg deposit	Labrador tea, stems	1—3.5	U.S.	703
	Carrot, roots	0.05—0.1[a]	Yugoslavia	116
Chlor-alkali or chemical works	Lettuce, leaves	0.15—0.36	Switzerland	637
	Spinach	0.11—0.59	Switzerland	637
	Corn, grains	0.074—0.136	Switzerland	637
	Wheat, grains	0.007—0.025	Switzerland	637
	Festuca rubra	4.0	Great Britain	111
	Lichens	36.0	Finland	485
	Lettuce leaves	0.1[a]	Canada	776
Urban vicinity and parks	Bryophytes	1.4	U.S.	893
	Edible mushrooms	33.6	Switzerland	635
Sludged or irrigated farmland	Brome grass, tops	0.09—2.01[b]	Canada	321
	Brown rice	4.9	Japan	336
Application of fungicides or Hg salts	Potato, foliage	1.1—6.8	Canada	585
	Lettuce, leaves	0.1—0.3	Canada	496
	Oat, grains	631[c]	Sweden	748
	Oat, straw	99[c]	Sweden	748
	Wheat, grains	0.05—0.17[d]	Poland	162

[a] FW basis.
[b] Different Hg compounds added to soil columns.
[c] Pot experiment.
[d] After Hg treatment of seeds.

Plants differ in their ability to take up Hg and can also develop a tolerance to high Hg concentrations in their tissues when grown in soils overlying Hg deposits. Shacklette et al.[710] reported Hg to range from 0.5 to 3.5 ppm (DW) in trees and shrubs from areas of Hg mineralization.

Plants grown in contaminated sites may accumulate much higher than normal amounts of Hg (Table 74). Certain plant species, lichens, carrots, lettuce, and mushrooms in particular, are likely to take up more Hg than other plants grown at the same sites. Also, some parts of plants have a greater ability to adsorb Hg, as is the case of apple flesh and apple pips.

Several authors have made an attempt to estimate a permissible limit for Hg in food plants and have proposed 50 ppb (FW), although the background Hg levels for plants have been estimated by Kosta et al.[415] to range from 1 to 100 ppb DW. The allowable limit of Hg in plant foodstuffs should always be calculated on the basis of daily Hg intake by a given population group.

Chapter 8

ELEMENTS OF GROUP III

I. INTRODUCTION

Geochemical and biochemical properties, as well as the abundance in the biosphere, of all the trace and rare elements of Group III are highly divergent. The geochemistry of Group IIIa elements is especially complicated, reflecting a wide range in occurrence and behavior from B, the lightest nonmetal, to amphoteric Al, which is one of the basic constituents of the lithosphere. Ga, In, and Tl (the latter two being widely distributed elements in the lithosphere and biosphere) also belong to this group. A strong affinity for oxygen and the predominant +3 state are common characteristics of these elements. In Group IIIb, Sc and Y are very rare in the environment. Other rare elements are subdivided into the lanthanides and actinides, of which many are natural or artificial radionuclides.

II. BORON

A. Soils

B, the only nonmetal among the elements of Group III, is not uniformly distributed in the earth's crust. The B content of magmatic rocks increases with the acidity of the rocks, while in sedimentary rocks, the element is associated with the clay fraction (Table 75). The largest quantities of B are concentrated in marine evaporites and in marine argillaceous sediments, therefore, their B content can serve as a paleosalinity indicator. It should be emphasized, however, that the geochemistry of B is characterized by an abnormally large range of variation in its concentration in rocks. In the terrestrial environment B is likely to occur in chemical combination with oxygen and is known to form several minerals, mainly hydroxides and silicates, of which the tourmaline group is the most common in soils.

During chemical weathering of rocks B goes easily into solution, forming several anions such as BO_2^-, $B_4O_7^{2-}$, BO_3^{3-}, $H_2BO_3^-$, and $B(OH)_4^-$. Although B is likely to be retained by clays (illitic minerals in particular and also by sesquioxides and organic substances), its concentration in soil solutions is relatively high, ranging from 67 to 3000 $\mu g \ell^{-1}$ (Table 12). The most common forms of B in soil solutions are, apparently, its undissociated acid H_3BO_3 and, in part, $B(OH)_4^-$. Only at pH above 7 are other anions such as $H_2BO_3^-$ and $B_4O_7^{2-}$ likely to occur in soil solutions.

The behavior of B in soils has been widely studied, and the basic results have been summed up by Ellis and Knezek.[207a] It has been shown that B is sorbed more strongly by soils than are other anions (e.g., Cl^- and NO_3^-), and the manner of B sorption by the clay surface is somewhat similar to that of cations rather than anions.

In general, retention of B is greater on sesquioxides than on clay minerals, and the hydrous oxide of Al is more effective than that of Fe. Lindsay[475] stated that the B adsorption on oxides of Fe and Al is believed to be an important mechanism governing B solubility in soils. Organic matter also exercises a powerful influence on B mobility and availability, particularly in acid soils.[166]

There are several descriptions in the literature of mechanisms of B reactions with soil components; however, the nature of B adsorption in soils is still not well understood. These reactions are highly pH dependent, with the maximum always occurring at pH above 7. The B adsorption by oxy and hydroxy bonds by surface coatings and by incorporation into interlayer or structural positions of aluminosilicates are the mechanisms likely to predominate in acid and neutral soils. On the other hand, in arid-zone soils B is likely to be coprecipitated with Mg and Ca hydroxides as coatings of soil particles, and B may also occur as Na-metaborate. In sodic soils tourmaline is reported to be the major source of B.[72]

Table 75
BORON, ALUMINUM, GALLIUM, INDIUM, THALLIUM, SCANDIUM, AND YTTRIUM IN MAJOR ROCK TYPES
(VALUES COMMONLY FOUND, BASED ON VARIOUS SOURCES)

Rock type	B (ppm)	Al (%)	Ga (ppm)	In (ppb)	Tl (ppm)	Sc (ppm)	Y (ppm)
Magmatic Rocks							
Ultramafic rocks: dunites, peridotites, pyroxenites	1—5	0.45—2.0	1—3	10—60	0.05—0.2	5—15	0.5
Mafic rocks: basalts, gabbros	5—20	7.8—8.8	15—20	20—220	0.1—0.4	20—35	5—32
Intermediate rocks: diorites, syenites	9—25	8.8	15—24	40—130	0.5—1.4	3—10	20—35
Acid rocks: granites, gneisses	10—30	7.2—8.2	16—20	40—200	0.6—2.3	3—14	30—40
Acid rocks (volcanic): rhyolites, trachytes, dacites	15—25	6.9—8.1	20	30—150	0.5—1.8	3—8	28—44
Sedimentary rocks							
Argillaceous sediments	120	7.2—10.0	19—25	70	0.5—1.5	12—15	25—35
Shales	130	7.8—8.8	15—25	50	0.5—2.0	10—15	30—40
Sandstones	30	2.5—4.3	5—12	XO	0.4—1.0	1	15—50
Limestones, dolomites	20—30	0.43—1.30	1—3	XO	0.01—0.14	0.5—1.5	4—30

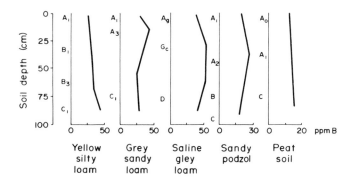

FIGURE 36. Distribution of B in the profiles of different soils developed under humid climate. (Letters indicate genetic soil horizons.)

In soils, B is considered to be the most mobile element among the micronutrients. Thus, the B movement in soils follows the water flux, and in cool humid-zone soils B is leached downward in soil profiles, whereas in soils of warm humid, or arid and semiarid regions, B is likely to concentrate in surface horizons (Table 10). B may also be concentrated in other soil horizons that are enriched in illitic clays or sesquioxides (Figure 36).

The B status of arable soils has been extensively investigated throughout the world. The total B content in surface soil ranges from 1 to 467 ppm, and its average content ranges from 9 to 85 ppm. The lowest amounts of B were found in sandy and loamy soils of Poland and New Zealand, while the highest concentrations were reported for lateritic soils of India, for solonchaks of the U.S.S.R., and for calcareous soils of Israel (Table 76). Levels of total B reported for soils of the U.S. seem to be fairly stable, with calculated means from 20 to 55 ppm (Table 77). It should be pointed out, however, that due to the low detectability of the analytical method that was used, some values of B in soils could be overestimated.

Although B is a rather deficient micronutrient in most soils, some soils of arid or semiarid regions and soils overfertilized with B may contain hazardous amounts of this element. Some sewage sludges and fly ash may also be significant sources of B contamination of soils.

Light acid soils with an excessively high level of B may be easily improved by irrigation. The B hazard in sodic soils is proposed by Bhumbla and Chhabra[72] to be ameliorated by the addition of gypsum, which converts readily soluble Na-metaborate to sparingly soluble Ca-metaborate. Heavy applications of $Ca(H_2PO_4)_2$ also resulted in a lower availability of B, especially in acid soils. As Prather[626] reported, sulfuric acid can effectively aid in reclaiming soil high in B, but the enhanced desorption of B by silicate ions is believed to be the main soil reaction.

B. Plants

B is important in plants metabolically and is believed to play the most significant role in the translocation of sugars, because the borate-polyhydric complex is more mobile than polar sugar molecules. Most studies have been done on effects of B on the metabolism of sugar beet, and it has been shown that adequate B supply is necessary for sugar synthesis (Figure 37).

1. Absorption and Transport

Soluble forms of B are easily available to plants which can take up the undissociated boric acid as well as other B species present in the ambient solution. The property of boric acid to complex with polysaccharides is believed to play an important role in passive sorption.[531]

Table 76
BORON CONTENT OF SURFACE SOILS OF DIFFERENT COUNTRIES (PPM DW)

Soil	Country	Range	Mean	Ref.
Podzols and sandy soils	Israel	29—43	—	644
	New Zealand	1—56	15.5	865
	Poland	5—134	9	91, 382
	U.S.S.R.	—	15.5	912
Loess and silty soils	New Zealand	—	37	865
	Poland	14—48	35	91, 382
Loamy and clay soils	New Zealand	<1—32	10.5	865
	Poland	3—75	15	91, 382
Fluvisols	Israel	50—85	—	644
	New Zealand	14—37	29	865
	India	4—9	—	772
Gleysols	New Zealand	13—60	31	865
Rendzinas	Israel	100—145	—	644
	New Zealand	25—64	45	865
	Poland	1—194	25	91, 382
	U.S.S.R.	—	10.5	912
Kastanozems and brown soils	New Zealand	11—70	34.5	865
	U.S.S.R.	—	40.5	912
Ferralsols	Israel	30—60	—	644
	India	14—467	—	772
Solonchaks and solonetz	New Zealand	28—67	44	865
	U.S.S.R.	49—105	85	12, 351, 912
	India	12—81	34[a]	772
Chernozems	Yugoslavia	—	32	412
	U.S.S.R.	47—68	54	4, 351, 912
Prairien and meadow soils	Yugoslavia	—	38	412
	U.S.S.R.	27—50	38.5	12
Histosols and other organic soils	New Zealand	4—15	8.8	865
	Poland	17—48	—	91, 382
	U.S.S.R.	8—47	26.5	912
Forest soils	China	—	46	225
	U.S.S.R.	—	32	4, 912
Various soils	New Zealand	2.5—47	15.5	865
	Romania	21.5—68.5	43	43
	Great Britain	4.7—21	13	818

[a]   Data for alkali, saline, and calcareous alluvial soils.

There is still controversy as to the extent to which the uptake process is either passive or active. From several recent reviews presented by Moore,[548] Price et al.,[630] Shkolnik,[719] and Loneragan,[489] it may be concluded that B uptake by roots consists of different phases. Moore[548] described three processes, whereas Bowen[95] observed six phases of B absorption by barley roots. The metabolically controlled process seems to be relatively minor; the B absorption mainly follows water flow through the roots. The B uptake is, therefore, proportional to its concentration and to the water flow.

Soil pH is one of the most important factors affecting the availability of B to plants. The lowest ratio of B uptake occurs when the soil pH is approximately 7. In alkaline soils the availability of B increases with an increase in soil pH. This affects B hazard problems, particularly in irrigated saline-alkaline soils.[72] The absorption of B is temperature dependent and increases during warm periods.

Table 77
BORON CONTENT OF SURFACE SOILS OF
THE U.S. (PPM DW)

Soil	Range	Mean
Sandy soils and lithosols on sandstones	<20—100	35
Light loamy soils	7—150	40(a)
Loess and soils on silt deposits	<20—70	40
Clay and clay loamy soils	<20—150	55
Alluvial soils	<20—70	40
Soils over granites and gneisses	30—50	40
Soils over volcanic rocks	<20—50	20
Soils over limestones and calcareous rocks	<20—70	35
Soils on glacial till and drift	<20—50	31(b)
Light desert soils	<20—100	35
Silty prairie soils	20—70	35(b)
Chernozems and dark prairie soils	<20—70	35
Organic light soils	<20—100	30
Forest soils	30—70	35(b)
Various soils	<20—150	45

Note: Sources are as follows: 706; a, 250; b, 218, 219.

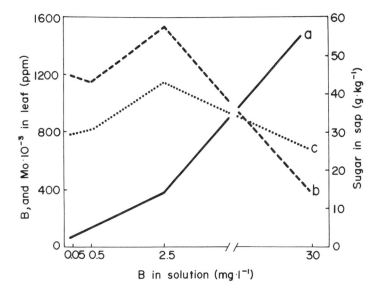

FIGURE 37.   The effect of B levels in the nutrient solution on sugar beet. (a) B concentration in leaves; (b) Mo concentration in leaves; (c) sugar content of sap.[90]

B is relatively immobile in plants, but because it is translocated mainly through the xylem, it is largely accumulated in old leaves in which the highest B content is in the tips and margins. Epigeal parts of plants usually contain more B than do roots.

Van Goor[820a] compared the B concentrations in phloem exudate (10 μg g^{-1}) and in leaf tissue (34 μg g^{-1}) and showed that B content relative to other micronutrients is higher. This may indicate that the absorption of B during its transport within veins is less than that of other elements. Although B may become quite immobile within plant tissues, there are indications that B can be transported from leaves to developing fruits and seeds.

2. Biochemical Functions

The physiological role of B differs from other micronutrients in that this anion has not been identified as a component of any specific enzyme. In spite of the essentiality of B for higher plants, the biochemical role of this element is still not well understood. There have been many physiological experiments, usually with B-deficient plants, and the results are extensively reviewed by Price et al.,[630] Shkolik,[719] Jackson and Chapman,[348] and Mengel and Kirkby.[531]

The functions of B are related to some basic processes such as:

1.   Carbohydrate metabolism and transport of sugars through membranes
2.   Nucleic acids (DNA and RNA) and phytohormone synthesis
3.   Formation of cell walls
4.   Tissue development (involvement in a messenger role in plants is suggested)

B deficiency in several plant species is common on a world scale (Table 28), and its deficiency in some commercial crops such as sugar beet and mangels, celery, sunflower, legumes, and apples is of great agronomic concern. Specific symptoms of B deficiency (Table 27) first appear as retarded and abnormal development of growing points, blue-green color of young leaves, and impairment of fruit formation. Jackson and Chapman[348] reported that there is a general similarity of B deficiency symptoms and the response of plant tissue to treatment with growth hormones (e.g., auxin, gibberellic acid).

Strangely, B seems not to be essential for some fungi and algae, although this element was reported to stimulate N fixation by bacteria.[504,630] Mycorrhizal plants have a greater need for a B supply than do nonmycorrhizal plants.[457]

The assessment of B availability and its requirement by plants have been extensively studied and it has been found that analysis of water-soluble B in soils and plant tissue tests are adequate for the diagnosis and prediction of B deficiency (Table 28). Some plant species have a low B requirement and may also be sensitive to elevated B levels even only slightly above those needed for normal growth. Therefore, toxic effects of B are likely to arise by excessive use of B fertilizers. The toxicity is usually more common in arid and semiarid regions on soils with naturally high levels of B. Sensitive crop plants (e.g., cereals, cotton) may be affected with a B concentration in soil solutions as low as 1 mg ℓ^{-1}, but 5 mg ℓ^{-1} may be tolerated by various plant species, whereas 10 to 15 mg ℓ^{-1} was toxic to tolerant plants.[11,90,648] The increased B content of irrigation water may be especially toxic to crops grown in an arid region.[133,674]

If there is an excess of B, a very high concentration (1000 to 1500 ppm) of this element often occurs in leaf tips or margins. These parts of leaves become necrotic, while at the early stage of B toxicity the leaves are dark green and wilted. The growing points of such plants becomes dark and decay.

3. Interactions with Other Elements

Interactions of B in the uptake of other nutrients by plants apparently are related to changes in membrane permeability and in the status of cell colloids. Physiological mechanisms of these reactions are, however, still not well understood.

Several interactions reported for B and other trace elemens, as shown in Figure 16, have not yet been confirmed. The possible antagonisms with Cu, Cr, Mo, and Mn may be effects of indirect influence through the increase of growth, hence increased demands for a given micronutrient. However, Lambert et al.[457] reported that B-deficient alfalfa also contained a lower amount of Cu. Leal et al.[463] explained Fe-B antagonism as a result of increased B accumulation in roots with a higher soil supply of Fe. B-Si antagonism is an effect of possible competition by silicate ions for adsorption sites of B, and this reaction has been observed in both soil environments and root tissues.

Table 78
BORON CONTENT OF GRASSES AND LEGUMES
(PPM DW)

Country	Grasses		Clovers		Ref.
	Range	Mean	Range	Mean	
Great Britain	—	26[a]	—	—	112
Czechoslovakia	14—30	22[a]	—	—	154
Finland	3.9—6.3	4.9	—	—	590
East Germany	—	—	20—50[b]	—	65
	—	—	20—60	—	65
West Germany	—	—	30—100[b]	—	65
Hungary	1.0—7.9	5.8	20—35	33	803, 804
Japan	1.6—12	4.9	12—35	21	770
Poland	1.0—15.6	5.6	11.3—16.5	14	915
New Zealand	1.7—10.0	5.2	6—120	26[b]	865
U.S.	<5—20	7.4	10—70	22[c]	710
U.S.S.R.	2—10	5	32—50	40[b]	337
	—	—	10—40	26	337
Yugoslavia	—	—	70—97	78	623

[a] Pasture herbs.
[b] Alfalfa.
[c] Calculated from AW basis.

B-Ca interrelationship is reported most often. Plants grow normally if a certain balance exists in the intake and in the tissue concentrations of Ca and B. Lime-induced B deficiency has frequently been observed in acid soils. However, it has also been shown that at equivalent amounts of Ca, tissue B concentrations are much higher if $CaSO_4$, rather than $CaCO_3$, is applied to the soil.[626,772] Liming is believed to result in decreased B adsorption; thus, toxic effects of B may be reduced or prevented by adding Ca to soils. These phenomena have been ascribed both to reactions within soil media and to metabolic processes.

B and P have similar reactions with OH^- groups; thus, the uptake of these elements by plants is likely to follow similar patterns.[52] The uptake and distribution of P are reported to be dependent on B concentrations because B increases P immobilization in roots.[463] B-P interactions in soils are related to the interference of phosphate ions with B mobility. Other effects of nutrients, such as K and N, on the B content of plants are, presumably, secondary results of increased plant growth or of some physiological disorders.

4. Concentrations in Plants

The B content of plants grown under natural conditions is widely varied for plant species and kinds of soil. In general, however, dicot plants have a higher B requirement and thus a higher B content than do monocot plants. The average B content of forage plants from various countries supports this statement (Table 78). This table gives the average B concentration in grasses as 5.7 ppm, whereas the content of alfalfa and clover is 37 ppm. Using these averages, the ratio of B in dicots to monocots is 6.5, which is higher than that calculated by other authors.[166]

The range of the B content in vegetables and fruits is 1.3 to 16 ppm (DW) or 58 to 455 ppm (AW) (Table 79). Shacklette et al.[710] reported that trees and shrubs (B content, 50 to 500 ppm AW) generally contain 2 to 10 times as much B as do vegetables. The lowest B amounts, however, have always been found in seeds and grains, cereal grains in particular (Table 80). The highest B level was found in leaves of sugar beets (Figure 37).

Table 79
MEAN BORON CONTENT OF
PLANT FOODSTUFFS (PPM)[381,710]

Plant	Sample	DW basis	AW basis
Sweet corn	Grains	1.5	58
Bean	Pods	13	180
Cabbage	Leaves	14	140
Lettuce	Leaves	1.3	93
Carrot	Roots	9.9	140
Onion	Bulbs	10	250
Potato	Tubers	6.1	58
Tomato	Fruits	6	84
Apple	Fruits	8.3	455
Orange	Fruits	9.4	260

Table 80
BORON CONTENT OF CEREAL GRAINS FROM
DIFFERENT COUNTRIES (PPM DW)

Country	Cereal[a]	Range	Mean	Ref.
Canada	Oats	—	0.7	293
Great Britain	Barley	—	3.4	112
	Oats	—	3.3	112
Poland	Oats	1.9—2.3	2.0	915
	Rye	1.1—1.6	1.3	915
	Wheat	0.3—1.5	0.8	267
Finland	Barley	0.7—1.1	1.0	829
	Oats	0.9—1.3	1.2	829
	Wheat	1.1—1.2	1.2	829
U.S.	Barley	0.9—2.3	1.6	200
	Oats	1.6—3.8	2.2	200
	Wheat (s)	0.8—4.3	1.9	200
	Wheat (w)	0.8—3.5	1.8	200
U.S.S.R.	Barley	5—9	6.6	337
	Oats	5—8	6.8	337
	Rye	5—12	7.3	337
	Wheat	1—15	6.8	337

[a]   Spring, s; winter, w.

It has been relatively easy to establish the critical B levels in plant tissues as 5 to 30 ppm (DW) (Table 25). The toxic B contents, on the other hand, have been reported as follows: John et al.[362] found that spinach, though with reduced yield, could grow when containing 348 to 990 ppm B, whereas corn could tolerate a content of 1007 to 4800 ppm; Chapman[131] reported toxic concentrations for alfalfa to be from 283 to 333 ppm; Alikhanova[11] observed B toxicity in cotton at 283 to 333 ppm; and Davis et al.[171] stated that 80 ppm B is toxic to barley seedlings.

III. ALUMINUM

A. Soils

As one of the main constituents of the earth's crust, Al in rocks commonly ranged from 0.45 to 10% (Table 75). The only stable and frequently occurring ion, Al^{3+}, is known to coordinate with oxygen-bearing ligands.

Table 81
ALUMINUM CONTENT OF FOOD AND FORAGE
PLANTS (PPM)

Plant	Tissue sample	FW(a) basis	DW basis	AW basis
Barley	Grains	—	135(b), 10(c)	—
Oats	Grains	—	82(b)	—
Sweet corn	Grains	—	2.6	100
Cabbage	Leaves	1.5	8.8	95
Spinach	Leaves	—	104(d)	—
Lettuce	Leaves	0.1	73	520
Carrot	Roots	0.4	7.8	110
Onion	Bulbs	—	63	1,500
Potato	Tubers	3.8	76(d), 13	310
Tomato	Fruits	—	20	170
Apple	Fruits	0.9	7.2	400
Orange	Fruits	1.4	15	430
Legumes	Tops	—	85—3,470(e)	—
Grass, timothy	Tops	—	6.5—23.5(f)	—
Grasses[a]	Tops	—	60—14,500(g)	—
Grasses	Tops	—	50—3,410(e)	—

Note: Sources are as follows: 705; a, 547; b, 113; c, 484; d, 55; e, 536; f, 590; g, 15.

[a] Sample from grass tetany pasture.

During weathering of primary rock minerals the series of Al hydroxides of variable charge and composition, from $Al(OH)^{2+}$ to $Al(OH)_6^{3-}$, are formed and they become the structural components of clay minerals. In general, the solubility of Al hydroxides is low, especially at the pH range of 5 to 8, and solubility decreases with aging. Freshly precipitated solid Al hydroxide species and colloidal species have a potential for anion adsorption, as well as the ability to flocculate negatively charged particles. Thus, Al hydroxides contribute greatly to various soil properties.

The total Al content of soils is inherited from parent rocks; however, only that fraction of Al which is easily mobile and exchangeable plays an important role in soil fertility.[528] In acid soils with pH below 5.5, the mobility of Al increases sharply and very actively competes with other cations for exchange sites. The solutions of neutral soils contain Al in the order of about 400 $\mu g\ \ell^{-1}$ (Table 12), while in the soil solution at pH 4.4 Al concentration was reported to be 5700 $\mu g\ \ell^{-1}$.[279] The mobile Al in acid soils can be taken up rapidly by plants and it creates a problem of chemical stress in the plants.

B. Plants

Al is a common constituent of all plants and is reported to occur in higher plants in the order of about 200 ppm (DW). However, the content of this element in plants varies greatly, depending on soil and plant factors (Table 81). Some species of Al-accumulating plants may contain more than 0.1% (DW) of Al.

The physiological function of Al in plants is not clear, although there is some evidence that low levels of Al can have a beneficial effect on plant growth, especially in Al-tolerant plant species.[141,241] Al injury or toxicity is often reported for plants grown in acid soils. Several recent reviews have been published on this subject which emphasized that a high availability of Al in acid soils is one of the limiting factors in the production of most field crops.[44,241,646] In fact, the reduced yield of crops on acid soils is often due to increased availability of Al rather than high H^+ concentrations.

Plant species and even cultivars of the same species differ considerably in their ability to take up and translocate Al, which affects the tolerance of plants to excesses of Al. In most plants the symptoms of Al injury first appear in the roots, and Al is likely to be concentrated in the roots of several plant species. The amount of Al passively taken up by roots and then translocated to tops reflects the Al tolerance of plants, but the ability to accumulate Al in roots is not necessarily associated with Al tolerance.

As Foy et al.[241] stated, the physiological mechanisms of Al toxicity are still debated; however, they are related mainly to impaired nutrient uptake and transport and to an imbalanced ratio of cations to anions. The Al excess in plants is also likely to interfere with cell division and with properties of the protoplasm and cell walls. Al is known to form organic complexes and therefore to precipitate nucleic acids.

The complex physiology of Al toxicity in plants is reflected in several interactions with the uptake of nutrients such as P, Ca, Mg, K, and N. In general, cation uptake by plants is reduced with an excess of Al. Al toxicity is also frequently associated with increased levels of Fe and Mn, and possibly other heavy metals, which are readily available in acid soils. However, Al-induced chlorosis due to impaired Fe metabolism in some plants was also reported by Foy et al.[241] It is to be expected that the toxicity would be accompanied by lower levels of Ca and Mg in both soils and plants.

The interaction of Al and P is related to the formation of sparingly soluble Al-phosphates in soils and to other coreactions of internal adsorption or precipitation of Al and P, as well as to Al interference with normal P metabolism, mainly in root tissues. Hence, Al toxicity is often manifested as a P deficiency, and P is an effective agent for detoxifying excess Al.

Al excess in plants is known to induce Ca deficiency or reduce Ca transport. Also the Mg content of plants is greatly decreased by Al, and this decrease in Mg may be an important response of plants sensitive to Al. The addition of both Ca and Mg to soil greatly reduces Al toxicity. Al tolerance in plants seems to be associated with NH_4 tolerance because nitrification is strongly inhibited in acid soils. The mechanisms of Al tolerance in plants are known to be genetically controlled, therefore, the selection of plants having genetic adaptability may be a solution to the problem of Al stress for crops grown in acid soils.

IV. GALLIUM

A. Soils

Ga is distributed rather uniformly in the major types of rocks and its common values in both magmatic and sedimentary rocks range from 5 to 25 ppm, but in ultramafic and calcareous rocks the concentration of this metal is about 3 ppm (Table 75). Only two minerals of Ga are known — sulfide and hydroxide. Somewhat elevated concentrations of this metal are reported to be in feldspars and amphiboles.

In weathering, Ga behaves like Al and is usually strongly associated with Al minerals (e.g., bauxites). This general tendency of Ga is reflected in the fact that in soil profiles Ga is positively correlated with the clay fraction. The distribution of Ga in soils also shows a relation to Fe and Mn oxides.

The ionic form of this element in natural environments is Ga^{3+}, but the low solubility of $Ga(OH)_3$ seems to be most responsible for its limited migration. Ga is likely to be accumulated in the organic matter of soil. Moreover, the higher concentrations of Ga are reported for bioliths, but whether Ga forms organometallic complexes has not been confirmed.[855]

The Ga status in soils has not been intensively studied. The data summarized by Wedepohl[855] for soils from various countries show that the Ga content ranges from 1 to 70 ppm and that the grand mean content is 28 ppm. Wells[863] reported Ga for soils derived from basalts and

Table 82
GALLIUM CONTENT OF SURFACE SOILS OF
THE U.S. (PPM DW)[145,706]

Soil	Range	Mean
Sandy soils and lithosols on sandstones	<5—30	11.0
Light loamy soils	5—50	20.5
Loess and soils on silt deposits	5—30	16.5
Clay and clay loamy soils	5—70	18.5
Alluvial soils	5—30	18.0
Soils over granites and gneisses	15—50	29.5
Soils over volcanic rocks	15—30	22.5
Soils over limestones and calcareous rocks	<5—30	12.0
Soils on glacial till and drift	7—30	15.0
Light desert soils	7—30	17.0
Silty prairie soils	10—20	14.5
Chernozems and dark prairie soils	7—30	15.0
Organic light soils	<5—50	13.5
Forest soils	<5—50	17.0
Various soils	<5—50	12.0

andesites of New Zealand to range from 16 to 48 ppm, and Gribovskaya et al.[283] gave the range from 6 to 17 ppm in various soils of the U.S.S.R. The British standard soil sample contained 21 ppm Ga, as reported by Ure and Bacon.[818]

Average Ga content has been calculated for different soils of the U.S. to range from 11 to 30 ppm, being the lowest in sandy and calcareous soils and the highest in soils derived from granitic and volcanic rocks (Table 82).

Ga as a pollutant is emitted from aluminum works and during coal combustion. However, elevated concentrations of Ga in surface soils has not yet been reported.

B. Plants

There is insufficient evidence to demonstrate either the necessity for or the toxicity of Ga in plants, although some earlier studies suggested a beneficial role of this element in the growth of microorganisms. Nevertheless, Ga is commonly found in plant tissues, and its concentration is reported to range from 3 to 30 ppm (AW) in a variety of native species from the U.S., as reported by Shacklette et al.[710] and from 0.02 to 5.5 ppm (DW) in the native herbage from the Soviet Union as described by Gribovskaya et al.[283] and Dvornikov et al.[199] The highest Ga contents were given by Wedepohl[855] and Bowen[94] for lichens (2.2 to 60 ppm, DW) and bryophytes (2.7 to 30 ppm, DW). A higher ratio of Ga to Al in land plants than in the soils in which the plants grow can reflect the selective uptake of Ga by plants.

V. INDIUM

A. Soils

The geochemistry of In is not well known. Concentrations of In in magmatic rocks exceed its occurrence in sedimentary deposits about ten times (Table 75). The grand mean In content of rocks is 0.1 ppm. It exhibits a chalcophilic behavior in the earth's crust and therefore forms mostly sulfide minerals, but selenide and telluride minerals are also common. Most recent geochemical studies of In have been made because it seems to be associated with base metal deposits. The current information on In in the environment has been extensively discussed by Smith et al.[742] During weathering, In oxidized to In^{3+} follows Fe^{3+}, Mn^{4+}, and, partly, Al^{3+}, and usually precipitates under conditions which form hydrous Fe oxide.

In in acid solutions may form several ionic species, e.g., $InCl^{2+}$ and $In(OH)^{2+}$, which are precipitated in the pH range of 5 to 9. Above pH 9.5 the anion $In(OH)_4^-$ is likely to occur.

In is commonly found in coals and crude oil and is reported to be combined with the organic substances. In soils In also seems to be associated with organic matter and therefore its concentration is increased in surface-soil horizons. This concentration may also reflect pollution.

Natural In content of various soils in the U.S. ranged from <0.2 to 0.5 ppm (average, 0.2 ppm), whereas in soils of other countries In is reported to average 0.01 ppm. Chattopadhyay and Jervis[132] reported that the In content of cultivated surficial organic soil increased up to 2.6 ppm. Somewhat elevated concentrations of In (up to 4.2 ppm) in the top soil near Pb and Zn works are described by Smith et al.[742] Some sewage sludges may also be a source of In.[249]

B. Plants

In is known to be readily available to plants, although it is not significantly concentrated by most plants. Physiological effects on plants are reported mainly in relation to In-induced toxicity in roots, which were described by Smith et al.[742] to occur in various plants at 1 to 2 ppm In concentrations in culture solutions. More results were obtained from studies of the effects of In on microorganisms, which reveal a greater resistance to In concentrations in solution than do higher plants. However, concentrations of 5 to 9 ppm In were reported to inhibit activity of nitrate-forming bacteria in soil.

A few data collected by Smith et al.[742] show that the In content of vegetation from unpolluted sites ranged from 30 to 710 ppb (FW) (mean, 210 ppb), whereas in unwashed plants (chiefly grass) from the industrial region these values were 0.008 to 2.1 ppm (FW). Furr et al.[249] gave the range of In in beets grown in soil amended with sewage sludge to be 80 to 300 ppb (DW). Much lower values reported for In content (0.64 to 1.8 ppb DW) of the standard samples of orchard and tomato leaves show that either the In content of plants is highly variable or that determinations are not very precise.

VI. THALLIUM

A. Soils

The distribution of Tl in the earth's crust shows that its concentration seems to increase with increasing acidity of magmatic rocks and with increasing clay content of sedimentary rocks (Table 75). Common Tl contents of mafic rocks range from 0.05 to 0.4 ppm and in acid rocks from 0.5 to 2.3 ppm. Calcareous sedimentary rocks contain as little as 0.01 to 0.14 ppm Tl.

In geochemical environments Tl is known to occur in three oxidation states, $+1$, $+2$, and $+3$. The cation Tl^+ is highly associated with K and Rb, and also with several other cations, and is incorporated into various minerals, mainly sulfides.

During weathering Tl is readily mobilized and transported together with alkaline metals. However, Tl is most often fixed *in situ* by clays and gels of Mn and Fe oxides. The sorption of Tl by organic matter, especially under reducing conditions, is also known.

Smith and Carson[740] widely reviewed environmental occurrences of Tl and cited its concentration to range from 0.02 to 2.8 ppm in surface soils of the U.S. and enriched contents of Tl (up to about 5 ppm) in soils over sphalerite veins. Dvornikov et al.[199] reported the Tl content of soil within Hg mineralization to be 0.03 to 1.1 ppm. In garden soil analyzed by Chattopadhyay and Jervis[132] Tl occurred in concentrations from 0.17 to 0.22 ppm, with the highest value in the surface samples. Ure and Bacon[818] found 0.27 ppm Tl in the standard soil sample. The largest anthropogenic source of Tl is related to coal combustion, but also

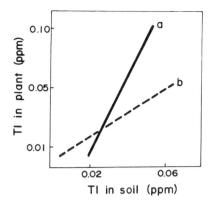

FIGURE 38. Tl content of two herbs, (a) Wormwood and (b) euphorbia, as a function of its concentrations in soil.[199]

heavy metal smelting and refining processes may release some amounts of Tl into the environment.

B. Plants

The Tl content of plants seems to be a function of Tl concentrations in soils, as is illustrated in Figure 38. Herbage and woody plants apparently contain higher amounts of Tl than do other plant species. Dvornikov et al.[199] found Tl in herbage to range from 0.02 to 1.0 ppm (DW), and Shacklette et al.[710] cited Tl for pine trees to range from 2 to 100 ppm (AW), being higher in needles than in stems. Smith and Carson[740] gave Tl levels in edible plants to range from 0.02 to 0.125 ppm (DW), in clover from 0.008 to 0.01 ppm (DW), and in meadow hay from 0.02 to 0.025 ppm (DW).

Zýka[909] analyzed herbaceous plants grown in soil over Tl mineralization and showed accumulations as high as 17,000 ppm (AW) in flowers of *Galium* sp. (Rubiaceae family), while other plants accumulated Tl in leaves and stalks at about 100 ppm (AW). Increased Tl levels in plant tissues are highly toxic to both plants and animals. Some plant species, e.g., wormwood (*Artemisia* sp., Compositae family), are likely to accumulate Tl, and the concentration factor may be high. As Smith and Carson[740] described, at potash fertilizer works and smelter and bituminous coal plant sites, plants contained elevated Tl amounts up to 2.8 ppm (DW).

Microorganisms are reported to be relatively sensitive to Tl, and therefore the inhibition of nitrate formation in Tl-polluted soils may have an agronomic impact. There is also an opinion that Tl is likely to be involved in microbial cycling by possible methylation.[856]

VII. SCANDIUM

A. Soils

The lithospheric abundance of Sc presented in Table 75 shows that the element is likely to be enriched in mafic rocks and also in argillaceous sediments, whereas the Sc content of sandstones and limestones is low. Sc is known to occur in natural environments as Sc^{3+} which can substitute for Al^{3+}, Fe^{3+}, Y^{3+}, and also Ti^{4+}; thus, the element is mainly associated with ferromagnesian minerals and biotite. However, the simple Sc^{3+} ion probably does not exist in solutions. The complexes such as $Sc(H_2O)_6^{3+}$ and $Sc(H_2O)_5OH^{2+}$ are likely to occur in aqueous environments. Sc shows also an affinity for complexing with PO_4, SO_4, CO_3, F, and amines. PO_4 especially is most effective in the precipitation of Sc compounds; thus, its enrichment in phosphorites can be expected (Table 5).

The Sc content of surface soils ranges from 0.5 to 45 ppm (Tables 83 and 84). Erdman

Table 83
SCANDIUM CONTENT OF SURFACE SOILS OF
DIFFERENT COUNTRIES (PPM DW)

Soil	Country	Range	Mean	Ref.
Podzols	Poland	0.8—3.5	1.5	915
Loamy and clay soils	Bulgaria	11.7—13.9	12.9	558
	Poland	2.4—3.5	3.0	915
Fluvisols	Bulgaria	—	7.6	558
	Poland	—	3.1	915
Black earth	Poland	—	2.3	915
Rendzinas	Poland	—	5.8	915
Chernozems	Bulgaria	10.9—13.7	12.0	558
Forest soils	Bulgaria	4.2—24.8	12.5	558
Various soils	Bulgaria	3.4—46.4	16.6	558
	Canada	4.9—17.8	10.5	409
	Great Britain	—	12.7	818
	West Germany	0.5—9.0	4.3	325

Table 84
SCANDIUM AND YTTRIUM CONTENTS OF SURFACE SOILS OF
THE U.S. (PPM DW)[145,706]

Soil	Sc Range	Sc Mean	Y Range	Y Mean
Sandy soils and lithosols on sandstones	<5—30	5	<10—100	22
Light loamy soils	<5—15	7	10—70	29
Loess and soils on silt deposits	<5—20	8	10—50	27
Clay and clay loamy soils	<5—20	10	10—100	28
Alluvial soils	<5—15	8	10—50	23
Soils over granites and gneisses	5—30	11	10—150	30
Soils over volcanic rocks	7—30	16	10—70	33
Soils over limestones and calcareous rocks	<5—15	7	<10—70	27
Soils on glacial till and drift	5—15	8	10—20	16
Light desert soils	<5—30	6	10—100	31
Silty prairie soils	<5—10	8	10—30	20
Chernozems and dark prairie soils	<5—20	9	10—70	24
Organic light soils	<5—15	5	<10—50	21
Forest soils	5—20	7	<10—150	25
Various soils	5—30	11	10—150	26

et al.[218] calculated the mean Sc content of uncultivated soils in the U.S. to be 7.1 ppm and in cultivated soils to be 5.1 ppm. Laul et al.[462] gave the range of Sc concentration in soils as 2.9 to 17 ppm.

The soil Sc content is governed mainly by the parent material, and its lowest concentrations are reported for sandy and light organic soils, whereas somewhat higher amounts have been found in soils derived from granitic and volcanic rocks.

Wedepohl[855] reported that the ash residues of some peats, coal, and crude oil carry significant amounts of Sc (5 to 1000 ppm, AW), therefore, environmental enrichment of Sc due to coal and oil combustion processes should be expected.

B. Plants

There is a paucity of data on Sc distribution in plants. Connor and Shacklette[145] gave the mean content of Sc in some shrubs and trees as <5 ppm (AW) and reported that in about 3% of the analyzed samples Sc was at detectable concentrations. Duke[197] gave the range in Sc for several plant foods of tropical forest Indians to be 0.002 to 0.1 ppm (DW). Laul et al.[462] found 0.005 ppm Sc in vegetables and 0.07 ppm in grass.

Ozoliniya and Kiunke[588] reported high concentrations of Sc in barley roots (up to 0.63 ppm, DW) and observed that the greatest amounts of Sc were taken up by plants from sandy soils. The Sc content seemed to be higher in old leaves than in young leaves, and its highest concentrations (0.014 to 0.026 ppm DW) were reported for flax plants, while in lettuce leaves Sc ranged from 0.007 to 0.012 ppm. Bowen[94] reported the range of Sc in lichens and bryophytes to be from 0.3 to 0.7 ppm (DW), whereas in fungi the amounts of Sc were lower (<0.002 to 0.3 ppm).

VIII. YTTRIUM

A. Soils

The occurrence of Y in the earth's crust is relatively common, and its abundance does not show any great differences between various rock types (Table 75). Ultramafic rocks contain somewhat smaller amounts of Y (0.5 to 5 ppm) than do acid rocks and sandstones (28 to 50 ppm).

Geochemical properties of Y are similar to those of the lanthanides. Y is known to be incorporated mainly as Y^{3+} into several minerals, of which silicate, phosphate, and oxide forms are most frequent. The Y content of coal (range, 7 to 14 ppm) does not indicate its sorption by organic substances.

Soils have not often been analyzed for Y. The most comprehensive data are reported for Y in the soils of the U.S. (Table 84). The Y content of these soils ranged from <10 to 150 ppm and averaged 25 ppm. Erdman et al.[218] gave the mean Y content of uncultivated soils as 23 ppm and of cultivated soils as 15 ppm. Similar values for Y in soils are given by Ure and Bacon[818] for Great Britain (22 ppm), by Duddy[196] for Australia (17 ppm), and by Dobrowolski[181] for Poland (10 ppm in sandy soil).

B. Plants

The most data for the Y content of plants are given by Connor and Shacklette[145] and Shacklette et al.[170] These authors found Y at the detectable levels in about 10% of the plants studied. In edible plants, the Y content ranged from 20 to 100 ppm (AW), the highest amount being reported for cabbage. Duke[197] gave the range from 0.01 to 3.5 ppm (DW) for the Y content of food plants from a tropical forest region.

Woody seed plants can accumulate Y to as much as 700 ppm (AW). Data collected by Bowen[94] show that lichens accumulated Y in the range of 0.2 to 2 ppm (DW), whereas bryophytes concentrated 1.3 to 7.5 ppm (DW). Erämetsä and Yliroukanen[217] gave the range in Y contents in mosses and lichens as 2 to 200 ppm (DW).

IX. LANTHANIDES

A. Soils

Lanthanides, also called rare earth metals, comprise a group of 15 elements, of which one, promethium (Pm), does not occur naturally in the earth's crust (it has not yet been detected), while the others occur in all types of rocks (Tables 85 and 86). The terrestrial abundance of the rare earth metals shows a general peculiarity; their contents decrease with the increase in their atomic weights, and the element with the even atomic number is more frequent than the next element with the odd atomic number (Figure 39 and Table 87).

The geochemical properties of the lanthanides are fairly similar — they occur mainly as $^+3$

Table 85

LANTHANUM, CERIUM, PRASEODYMIUM, NEODYMIUM, SAMARIUM, EUROPIUM, AND GADOLINIUM IN MAJOR ROCK TYPES (PPM) (VALUES COMMONLY FOUND, BASED ON VARIOUS SOURCES)

Rock type	La	Ce	Pr	Nd	Sm	Eu	Gd
Magmatic Rocks							
Ultramafic rocks Dunites, peridotites, pyroxenites	0.X—1.8	0.X—3.3	0.6	0.X—2.4	0.X	0.01— 0.X	0.X
Mafic rocks Basalts, gabbros	2—27	4—50	1—15	5—30	0.9—7.0	0.8—3.5	2—8
Intermediate rocks Diorites, syenites	30—70	60—160	7—15	30—65	6—18	1.3—2.8	7—18
Acid rocks Granites, gneisses	45—60	80—100	7—12	33—47	8—9	1.1—2.0	7.4—10.0
Acid rocks (volcanic) Rhyolites, trachytes, dacites	30—150	45—250	6—30	18—80	6—11	1.0—1.9	4.3—8.7
Sedimentary Rocks							
Argillaceous sediments	30—90	55—80	5.5—9.5	24—35	6.0—6.5	1.0—1.8	6.4—7.4
Shales	34—50	30—90	5—10	18—41	5—7	1.0—1.4	5.0—6.5
Sandstones	17—40	25—80	4—9	16—38	4—10	0.7—2.0	3—10
Limestones, dolomites	4—10	7—20	1.0—2.5	4.7—9.0	1.3—2.1	0.2—0.4	1.3—2.7

Table 86
TERBIUM, DYSPROSIUM, HOLMIUM, ERBIUM, THULIUM, YTTERBIUM, AND LUTETIUM IN MAJOR ROCK TYPES (PPM) (VALUES COMMONLY FOUND, BASED ON VARIOUS SOURCES)

Rock type	Tb	Dy	Ho	Er	Tm	Yb	Lu
Magmatic Rocks							
Ultramafic rocks Dunites, peridotites, pyroxenites	0.X	0.05—0.95	0.X	0.X	0.X	0.X	0.X
Mafic rocks Basalts, gabbros	0.5—1.2	0.9—6.9	1.0—1.5	0.9—3.9	0.2—0.6	0.8—3.4	0.2—0.6
Intermediate rocks Diorites, syenites	1.1—2.8	6—13	1.5—3.5	3.9—7.0	0.6	3.8—7.0	0.6—2.0
Acid rocks Granites, gneisses	1.0—2.5	5—7	1.3—2.0	3.5—4.2	0.3—0.7	3.5—4.3	0.5—1.2
Acid rocks (volcanic) Rhyolites, trachytes, dacites	1.0—1.2	5—8	1.3—1.7	3.1—4.6	0.5—0.7	2.9—4.6	0.7
Sedimentary rocks							
Argillaceous sediments	0.9—1.1	4.6—5.4	1.0—1.6	2.5—3.8	0.2—0.6	2.6—3.6	0.7
Shales	1	4.0—5.8	1.0—1.8	2.5—4.0	0.6	2.2—3.9	0.2—0.8
Sandstones	1.6—2.0	2.6—7.2	2	1.6—4.9	0.3	1.2—4.4	0.8—1.2
Limestones, dolomites	0.2—0.4	0.8—2.1	0.3	0.4—1.7	0.04—0.16	0.3—1.6	0.2

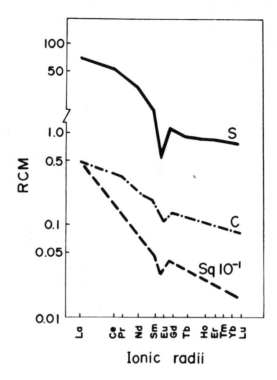

FIGURE 39.   The relative occurrence of rare earth elements in surface soil and plants as a function of their ionic radii and expressed as a ratio (RCM) of the abundance in samples to the content of chondrite metorites. S, soil; C, cheatgrass; Sq, squash.[462]

cations, show an affinity for oxygen, and are likely to be concentrated in phosphorites and in argillaceous sediments. Most often their lowest concentrations are reported for ultramafic and calcareous rocks. Two subgroups of lanthanides are distinguished: the first, composed of the more basic and more soluble metals from La to Gd, and the second, composed of less basic and less soluble metals from Tb to Lu.

The concentrations of lanthanide metals of the basic subgroup are reported to range from X to XO ppm, and concentrations of the less basic subgroup are reported to range from 0.X to X ppm (Table 87). The concentrations of lanthanides do not show any obvious differences among soil types (Table 88).

More detailed studies are presented for the occurrence of La and Yb in soils of the U.S. (Table 89). The grand mean for La in surface soils has been calculated to be 47 ppm, and for Yb, the grand mean is 3 ppm. Elevated concentrations of some lanthanides (La, Ce, Sm, Eu, and Tb) found in the air of industrial and urban areas (Table 3) indicate that these elements are likely to be released into the environment mainly from coal burning and nuclear energy materials processing.

B. Plants

Neither the distribution of lanthanides in plant tissue nor their physiological functions have received much attention. Laul et al.[462] calculated the relative abundance of lanthanides in both soil and plants and showed that the concentrations of these elements in plants followed their occurrence in soil (Figure 39). Orders of the contents of lanthanides in plants decrease with increase in the atomic number (Table 90). Woody plants seem to have the highest ability to absorb lanthanides, and hickory trees (*Carya* sp., Juglandaceae family) are most often reported as lanthanide-accumulating plants.

Table 87
MEAN CONCENTRATIONS
OF LANTHANIDES IN
SOILS, AS GIVEN BY
VARIOUS AUTHORS (PPM
DW)

Element	a	b	c
La	33.5	40	29.5
Ce	48.5	50	29.5
Pr	7.7	3—12	6.7
Nd	33.0	35	27.9
Sm	6.1	4.5	5.1
Eu	1.9	1	1
Gd	3.0	4	4.7
Tb	0.63	0.7	0.7
Dy	3.8	5	—
Ho	0.38	0.6	1.1
Er	2.0	2	2.8
Tm	0.16	0.6	0.4
Yb	2.3	3	3.1
Lu	0.34	0.4	0.3

Note: Sources are as follows: a, 818; b, 94; c, 462.

Lanthanides are known to be toxic to cell metabolism; however, there are not many available data on their inhibitory effects on plants. Weinberg[856] reported that lanthanides (La^{3+}, Pr^{3+}, Nd^{3+}, Eu^{3+}, and Tb^{3+}) inhibit, specifically and competitively, Ca accumulation by mitochondria of microorganism cells.

X. ACTINIDES

A. Soils

Among the actinide series of metals of Group IIIb, Th and U are the only naturally occurring elements in the earth's crust. Acid rocks usually contain more Th and U than do mafic rocks, and in sediments these elements are likely to be more concentrated in argillaceous deposits than in sandstones and limestones (Table 91). However, in some kinds of alkaline magmatic rocks, Th and U are also known to be concentrated to as much as XOO ppm.

During weathering, Th and U are easily mobilized in forms of various complex inorganic cations and in organic compounds. The importance of microorganisms in the geochemical cycle of U has been emphasized recently by Wildung and Garland[873] and Trudinger and Swaine.[809]

Harmsen and de Haan[306] reviewed the behavior of Th and U in the soil environment and pointed out that the formation of hydrated cations of UO_2^{2+} and of Th^{4+} are responsible for the solubility of these metals over a broad range of soil pH. Also, several organic acids may increase the solubility of Th and U in soils. The mobility of Th and U in soils may be limited due to both the formation of slightly soluble precipitates (e.g., phosphates, oxides) and adsorption on clays and organic matter.

Sorption is a key process in the U cycle, and thus significant accumulations of U in organic deposits (coal or peat) are often reported.[299,688] The affinity of the clay soil fraction to adsorb Th and U is shown in Figure 24. Megumi and Mamuro[529] explained that the

Table 88
LANTHANIDE CONTENTS OF SURFACE SOILS (PPM DW)

Element	Podzols Range	Podzols Mean	Fluvisols Mean	Chernozems Range	Chernozems Mean	Forest soils Range	Forest soils Mean	Various soils Range	Various soils Mean	Ref.
La	34.4—49.8	40.2	20.1	19.5—35.8	30.2	17.9—72.0	39.1	13.9—56.3	34.5	558
	—							16.2—54.6	36.0	409
Ce	49.4—85.3	63.0	31.9	41.3—68.8	57.3	29.9—94.1	61.5	21.2—75.7	48.1	558
Nd	—	21.0						—	—	196
	—	7.0						—	—	196
		21.4	14.9	17.9—47.0	24.3	7.9—38.7	20.2	8.6—35.0	19.5	558
								16.5—56.0	36.1	409
Sm	5.05—8.39	6.32	3.37	4.17—6.75	5.18	2.27—10.19	6.22	2.32—7.68	5.54	558
	—							3.3—11.9	7.05	409
Eu	1.21—1.50	1.34	0.74	1.12—1.22	1.17	0.37—2.05	1.1	0.43—2.39	1.36	558
	—							0.69—3.21	1.65	409
Tb	0.89—1.7	1.01	0.66		0.74	0.60—1.61	0.92	0.11—1.34	0.79	558
	—							0.49—1.66	0.90	409
Yb	2.06—2.42	2.28	1.15		2.35	0.81—4.45	2.52	1.79—3.43	2.57	558
	—							1.57—3.66	2.46	409
Lu	0.43—0.51	0.48	0.24	0.19—0.41	0.31	0.10—0.72	0.40	0.10—0.67	0.43	558
	—							0.24—0.52	0.37	409

Table 89
LANTHANUM AND YTTERBIUM CONTENTS OF SURFACE SOILS OF THE U.S. (PPM DW)706

Soil	La Range	La Mean	Yb Range	Yb Mean
Sandy soils and lithosols on sandstones	<30—100	45	<1.0—10.0	3.0
Light loamy soils	<30—100	50	1.0—7.0	3.9
Loess and soils on silt deposits	<30—70	40	1.0—7.0	3.6
Clay and clay loamy soils	<30—100	50	1.5—10.0	4.0
Alluvial soils	<30—150	45	1.5—5.0	2.6
Soils over granites and gneisses	30—100	55	1.0—3.0	2.2
Soils over volcanic rocks	<30—150	55	1.5—7.0	3.6
Soils over limestones and calcareous rocks	<30—100	45	1.0—7.0	3.5
Soils on glacial till and drift	<30—50	40	2.0—3.0	2.3
Light desert soils	30—150	55	1.0—10.0	3.8
Silty prairie soils	<30—50	30	1.5—3.0	2.2
Chernozems and dark prairie soils	<30—100	40	1.0—7.0	3.0
Organic light soils	<30—100	45	1.0—15.0	2.5
Forest soils	<30—150	50	1.0—20.0	4.1
Various soils	30—200	55	1.0—7.0	2.9

observed conspicuous enrichment of ^{230}Th in soil particles is the result of a higher solubility and thus a higher leaching of uranyl ions (UO_2^{2+}) than of Th^{4+} ions during soil-forming processes.

Comparatively little information seems to be available with respect to Th and U in soils. Only surface soils of the U.S. have been extensively studied, and the results show relatively small variation in the contents of these elements with soil types. The world-wide mean of the Th content of soils ranged from 3.4 to 10.5 ppm, and for U, it ranged from 0.79 to 11 ppm (Table 92). Apparently, large amounts of Th and U are introduced into the biosphere from fossil-fuel power plants and also from P fertilizer works. The enrichment of these two elements in surface soils of industrial areas has not yet been very extensive.[355]

There is considerable contemporary interest in studies on the behavior of other elements of the actinides, the so-called transuranic radionuclides, which are isotopes released from the nuclear fuel power industry. This group includes various isotopes of plutonium (Pu), americium (Am), curium (Cm), and neptunium (Np), of which the long-lived and highly radiotoxic ^{239}Pu and ^{241}Am are of the greatest concern. Most of the aerosols produced and discharged from nuclear facilities, as well as in nuclear bomb fallout, are composed of PuO_2, but some of the Pu may come from the evaporation of $Pu(NO_3)_4$.

The behavior of these radionuclides, and of ^{239}Pu in particular, was extensively reviewed by Wildung and Garland[873] and Vyas and Mistry.[832] Regardless of the forms of Pu and Am entering the soil, their solubilities are controlled by various soil factors, and thus they occur mainly as complexes of fulvic and humic acids and in forms adsorbed by clay mineral particles. Unlike the tightly bound PuO_2, $Pu(NO_3)_4$ belongs to the soluble compounds and easily hydrates and hydrolysates in the soil solution, forming soluble hydroxides. Thus, the soluble, and therefore plant-available, fraction of Pu, and apparently of other transuranic elements, appears to be largely present as particulates of hydrated oxides and as organometal complexes. Jakubick[350] studied in detail the behavior of ^{239}Pu and ^{240}Pu in meadow soil around Heidelberg, West Germany, where the isotopes have been concentrated from 290 to 450 μCi km^{-2} in the 10-cm-deep upper layer, whereas the activity of these isotopes range from 1.6 to 5.8 μCi km^{-2} at a depth below 25 cm.

Table 90
LANTHANIDES IN VARIOUS TERRESTRIAL PLANT SPECIES

Element	Approximate detection (%)	Various land plants[94] (ppb DW)	Lichens and bryophytes[94] (ppb DW)	Cheatgrass (Bromus sp.)[462] (ppb DW)	Vegetables (ppb DW)[94,462,547]	Woody plants[710] (ppm AW)	Lichens and mosses[217] (ppm AW)	Horsetail (Equisetum sp.)[215] (ppm AW)
La	100	3—15,000	400—3,000	170	0.4—2,000	30—300	13—150	1—30
Ce	100	250—16,000	600—5,600	330	2—50	—	9—280	1—90
Pr	90	60—300	80—620	40	1—2	700	1.2—31	0.5—6
Nd	90	300	240—3,000	150	10	—	8—150	3—50
Sm	90	100—800	60—800	35	0.2—100	200—700	2—40	2—4
Eu	80	30—130	20—170	8	0.04—70	—	1—8.7	1—2
Gd	80	2—500	60—560	37	<2	<100—300	2—28	3—8
Tb	70	1—120	6—70	9	0.1—1	—	0.3—3.3	1—2
Dy	70	50—600	40—360	—	—	50—300	1.3—26	2—9
Ho	70	30—110	4—70	<20	0.06—0.1	150	0.2—4.5	1—2
Er	70	80—380	10—190	<500	0.5—2	<100—300	0.6—13	2—7
Tm	50	4—70	1—26	50	0.2—4	—	0.07—2.2	1
Yb	50	20—600	10—900	20	0.08—20	300	0.5—26	1—2
Lu	40	30	1—20	3	0.01—60	—	0.05—2.2	—

Table 91
**THORIUM AND URANIUM IN MAJOR ROCK
TYPES (PPM) (VALUES COMMONLY FOUND,
BASED ON VARIOUS SOURCES)**

Rock type	Values commonly found	
	Th	U
Magmatic Rocks		
Ultramafic rocks	0.004—0.005	0.003—0.010
Dunites, peridotites, pyroxenites		
Mafic rocks	1—4	0.3—1.0
Basalts, gabbros		
Intermediate rocks	7—14	1.4—3.0
Diorites, syenites		
Acid rocks	10—23	2.5—6.0
Granites, gneisses		
Acid rocks (volcanic)	15	5
Rhyolites, trachytes, dacites		
Sedimentary rocks		
Argillaceous sediments	9.6—12.0	3—4
Shales	12	3.0—4.1
Sandstones	1.7—3.8	0.45—0.59
Limestones, dolomites	1.7—2.9	2.2—2.5

Table 92
**THORIUM AND URANIUM CONTENTS OF SURFACE
SOILS OF DIFFERENT COUNTRIES (PPM DW)**

Country	Th		U		Ref.
	Range	Mean	Range	Mean	
Bulgaria	3.6—17.8	9.3	—	—	558
Canada	4.2—14.1	8.0	0.72—2.05	1.22	409
Great Britain	—	10.5	—	2.60	818
West Germany	0.4—15.0	8.0	0.42—11.02	—	235, 688
India	—	—	—	11.00	276
Poland	1.4—7.2	3.4	0.10—2.33	0.79	355
U.S.	2.2—21.0	7.6	0.30—10.70	3.70	462, 706

B. Plants

The assessment of the transfer of Th, U, and transuranic radionuclides from contaminated soils to plants is important in environmental research. However, little information is available on this subject.

The soluble fractions of these elements in soils seem to be readily absorbed by plants, and this is clearly supported by the studies conducted in the U geochemical province where plants accumulated up to 100 times more U than did plants from other areas.[423] Wildung and Garland[873] stated that plants possess the ability to effectively accumulate soluble Pu and to transport the Pu from roots to shoots. Tiffin[789] reported that a U protein complex was found in leaves of *Coprosma australis* (Rubiaceae family).

Shacklette et al.[710] found the highest U concentration in trees on mineralized ground to be 2.2 ppm (AW), and Goswani et al.[276] reported the U range from 0.5 to 4.4 ppm (AW) (average, 1.8) in xerophytic and mesophytic vegetation. Sagebrush grown near a P fertilizer works accumulated up to 8 ppm U (AW).[278] Bowen[94] gave the range in U concentrations in terrestrial plants as 5 to 60 ppb (DW), whereas Lual et al.[462] reported U concentration in corn and potatoes to be 0.8 ppb (DW). Somewhat higher and wider range in the Th content of land plants was reported by Bowen[94] to be <8 to 1300 ppb (DW), while Laul et al.[462] found Th to range from <5 to 20 ppb (DW) in vegetables.

Chapter 9

ELEMENTS OF GROUP IV

I. INTRODUCTION

The geochemical properties and terrestrial abundances of the elements in Group IV diverge widely. Their common characteristics are the weak solubility of their hydroxides and oxides in water, an affinity to bond with oxygen, and they often have the coordination number four.

Group IVa contains Si, Ge, Sn, and Pb. Si, in combination with oxygen, is the basic nonmetallic component of all rocks (Table 93) and is considered as a trace element only in respect to its biochemical role. The next elements in this group, Ge, Sn, and Pb, are trace metals, and show chalcophilic properties in the terrestrial environment.

Of the elements of Group IVb, Ti is definitely an oxyphile, is associated with silicate minerals, and is considered a trace element only because of its low concentration in plant tissues. The next two metals, Zr and Hf, are widely distributed in both the litho- and biosphere, and are included in the group of rare-earth metals.

II. SILICON

A. Soils

Si is the most abundant and, relatively, the most stable electropositive element in the earth's crust; however, under specific conditions it can be dissolved and transported, but it moves mainly in its colloidal phase. All silicate minerals are built of a fundamental structural unit, SiO_4, the so-called tetrahedron.

Quartz, SiO_2, is the most resistant mineral in soils and is also known to occur in a noncrystalline form, opal, which is believed to have had a biological origin. In soils, amorphous silicates apparently contribute to anion adsorption processes, and it has been suggested that silicate and phosphate ions compete for sites on mineral soil particles.[530] Tiller[793] has shown that the presence of monosilicic acid in solution increases the sorption of heavy metal cations, such as Co, Ni, and Zn, by clays.

In general, Si is released rapidly from minerals into the soil solution where it occurs at near-equilibrium concentrations. Carlisle et al.[121] reported that soluble Si (mainly as H_4SiO_4) in the soil solution ranges from 1 to about 200 mg ℓ^{-1}. The concentration of Si in soil solutions and drainage waters is highly dependent on several soil and climatic factors. The pH level has an especially marked effect on Si concentrations in solutions, although the mobility of Si in soils cannot be predicted accurately from the pH alone. Usually Si is more mobile in alkaline soils, but, as Carlisle et al.[121] have described, increasing the pH to about 9 decreases the Si concentration in solutions. A pH beyond approximately 9.5 results in a sharply increased Si content of the solution.

Several interferences between Si and other ions such as P, Al, Ca, and Fe may occur in soil and modify the behavior of Si. For example, in acid soils silicate and phosphate ions form insoluble precipitates which may fix several other cations, e.g., Fe and Al oxides that have a marked capacity to sorb dissolved Si as H_4SiO_4. Appreciable amounts of organic matter in flooded soil induce a higher Si mobility, apparently due to the reduction of Fe hydrous oxides, which releases adsorbed monosilicic acid.

B. Plants

Si is a common constituent of plants, and its amounts may vary by two orders of magnitude.

Table 93
SILICON, GERMANIUM, TIN, LEAD, TITANIUM, ZIRCONIUM, AND HAFNIUM IN MAJOR ROCK TYPES (VALUES COMMONLY FOUND, BASED ON VARIOUS SOURCES)

Rock type	Si (%)	Ge (ppm)	Sn (ppm)	Pb (ppm)	Ti (%)	Zr (ppm)	Hf (ppm)
Magmatic Rocks							
Ultramafic rocks Dunites, peridotites, pyroxenites	19.0—20.5	0.7—1.5	0.35—0.50	0.1—1.0	0.03—0.30	20—40	0.1—0.6
Mafic rocks Basalts, gabbros	23—24	0.8—1.6	0.9—1.5	3—8	0.90—1.38	80—200	1.0—4.8
Intermediate rocks Diorites, syenites	26.0—29.1	1.0—1.5	1.3—1.5	12—15	0.35—0.80	250—500	2—10
Acid rocks Granites, gneisses	31.4—34.2	1.0—1.4	1.5—3.6	15—24	0.12—0.34	140—240	2—5
Acid rocks (volcanic) Rhyolites, trachytes, dacites	30.8—33.6	1	2—3	10—20	0.27	150—300	4.5
Sedimentary Rocks							
Argillaceous sediments	24.5—27.5	1.0—2.4	6—10	20—40	0.38—0.46	160—200	2.8—6.0
Shales	24.0—27.5	1.3—2.0	6	18—25	0.44—0.46	150—200	2.8—4.0
Sandstones	31.6—36.8	0.8—1.2	0.5	5—10	0.15—0.35	180—220	3.0—3.9
Limestones, dolomites	2.4—4.0	0.3	0.5	3—10	0.03—0.04	20	0.3

Metson et al.[536] reported the mean Si content to range from 0.3 to 1.2% (DW) in grass, whereas this range is 0.04 to 0.13% in clover and 0.1 to 0.2% in alfalfa. Some species of plants may accumulate a much higher amount of Si (e.g., diatoms, sedges, nettles, and horsetails). Rice plants are known as accumulators of Si and can contain up to 10% (DW) in hulls and about 15% (AW) in leaves.[121,395]

Si is absorbed from the soil solution as monosilicic acid or silica, and its absorption is usually proportional to its concentration in the solution and to water flow. However, Tinker[798] reported that while Si uptake by most grasses appears to be passive, in rice it is rather an active process. There is also evidence that plants can restrict the uptake of Si, as do some clovers which either exclude H_4SiO_4 at the external surface or bind it within the root tissue, and thus can reduce the concentration of Si in the xylem sap to about 6% of that in the external solution.[121] Although Si-organic complexes have not yet been isolated, plant Si has been shown to exist in at least two forms, one of which is believed to be a hydrogen-bonded Si-organic complex.[121]

Si (possibly as amorphous silica) impregnates the walls of epidermal and vascular tissues.[387] Thus, Si strengthens plant tissues, reduces water loss, and retards fungal infection. Where large amounts of Si are accumulated, intercellular deposits can be formed as plant opal. Residues of such plants contribute to the formation of amorphous silica in soils.

Wallace[840] reported that soluble Si stimulated plant growth. This stimulation seems to be related to the observed effects of Si on increased P and Mo uptake by plants, as well as on Mn transport within plant tissues.[326]

Antagonistic effects of Si on the uptake of B, Mn, and Fe have also been observed (Figure 16). It has been suggested that Si enhances phosphorylation of sugars, which improves the energy supply for metabolism and enhances sugar synthesis that is reflected in more growth.[3] Miyake and Takahashi[546] reported that Si deficiency affected the reproductive growth of tomato plants grown in culture solution. The biochemical role of Si has not yet been clarified.

Table 94
GERMANIUM AND TIN CONTENT OF SURFACE SOILS OF THE
U.S. (PPM DW)[706]

Soil	Ge Range	Ge Mean	Sn Range	Sn Mean
Sandy soils and lithosols on sandstones	0.6—2.1	1.1	<0.1—7.7	1.1
Light loamy soils	0.6—1.6	1.2	<0.1—2.2	0.9
Loess and soils on silt deposits	0.9—1.6	1.3	0.3—1.8	1.1
Clay and clay loamy soils	0.7—2.0	1.5	0.3—3.1	1.2
Alluvial soils	0.6—2.1	1.3	0.3—4.2	1.7
Soils over granites and gneisses	1.0—1.4	1.3	0.9—1.5	1.2
Soils over volcanic rocks	1.1—1.8	1.4	0.8—1.7	1.2
Soils over limestones and calcareous rocks	0.6—1.3	1.0	<0.1—1.8	1.1
Soils on glacial till and drift	0.9—1.7	1.2	0.1—1.1	0.6
Light desert soils	0.8—1.6	1.2	0.7—1.9	1.2
Silty prairie soils	0.7—1.4	1.1	0.4—1.9	0.9
Chernozems and dark prairie soils	0.8—1.6	1.3	0.2—5.0	1.4
Organic light soils	<0.1—1.1	0.8	0.1—7.9	1.2
Forest soils	0.7—1.8	1.4	0.2—2.8	1.1
Various soils	1.5—1.8	1.6	—	—

III. GERMANIUM

A. Soils

Common amounts of Ge in the major rock types range from 0.3 to 2.4 ppm (Table 93). The distribution of Ge resembles that of Si, and thus the lowest Ge contents are in calcareous sediments and mafic magmatic rocks. Ge is associated with sulfide ores of some heavy metals.

During weathering Ge is partly mobilized, but then is readily fixed, apparently in the form of $Ge(OH)_4$, to clay minerals, Fe oxides, and organic matter. Its high concentrations in coals have often been reported.

Ge may occur as the divalent cation, but its complex anions are also known, such as $HGeO_2^-$, $HGeO_3^-$, and GeO_3^{2-}. The abundance of Ge in surface soils of the U.S. is fairly uniform and averages 1.1 ppm (Table 94). There is very little information about Ge in soils of other countries.

B. Plants

Although Ge is reported to occur in plants, it is not known to have any physiological functions. Schroeder and Balassa[695] gave the range of Ge concentrations in grains as 0.09 to 0.7 ppm (FW) and in vegetables as 0.02 to 1.07 ppm (FW). Duke[197] found Ge in food plants of a Central American region to range from <0.01 to <0.1 ppm (DW). Connor and Shacklette[145] reported the Ge content of plant ash, if detected, to average 20 ppm, but the element was rarely detected. Ge was found in only 1 of 123 samples of Spanish moss (*Tillandsia usneoides*, Bromeliaceae family) at a concentration of 15 ppm (AW), and this sample was from an area subject to air pollution from industrial operations.[708]

Plants seem to absorb Ge at a relatively high rate, possibly in the form of GeO_2.[395] Rice plants may accumulate Ge readily and concentrate this element to a level as high as 1% (AW) in the tops, although Ge is highly toxic to plants and to rice plants in particular.[516]

It is assumed, although little is actually known, that interaction between Ge and Si exists and that those plants which need Si for growth are most sensitive to Ge. Even at low concentrations, Ge has been shown to inhibit germination and plant growth.[679]

IV. TIN

A. Soils

The abundance of Sn in common rocks shows an increased concentration in argillaceous sediments (6 to 10 ppm) and lower amounts in ultramafic and calcareous rocks (0.35 to 0.5 ppm) (Table 93). Sn tends to form only a few independent minerals, of which cassiterite (SnO_2) is the most important Sn ore, and is strongly resistant to weathering. Sn is known to occur as Sn^{2+} and Sn^{4+} and to form several complex anions of oxides and hydroxides.

The mobility of Sn during weathering is highly pH dependent. Especially Sn^{2+}, a strong reducing agent, can be present only in acid and reducing environments. Soluble Sn follows the behavior of Fe and Al and remains in the weathered residue along with hydroxides of these metals. The ability of Sn to form complexes with organic substances, both soluble and insoluble, has been reported, therefore, Sn is generally enriched in bioliths.

The occurrence of Sn in soil has not receive much study; only Shacklette and Boerngen[706] have presented comprehensive data on Sn in soils (Table 94). Although Sn in soils is largely derived from Sn in the bedrock, all soil surface horizons contain fairly similar amounts of this element, averaging 1.1 ppm. Presant[629] gave the range in Sn concentrations in soils as 1.1 to 4.6 ppm, and Kick et al.[390] gave the range as 1 to 4 ppm. Ure and Bacon[818] found 4.5 ppm Sn in the standard soil samples. Chapman[131] cited the common range of Sn in soils' as 1 to 11 ppm, and Gordon (see Griffitts and Milne[286]) reported Sn concentrations in peats to range from 50 to 300 ppm (AW).

B. Plants

There is no evidence that Sn is either essential or beneficial to plants, although plants may easily take up Sn, if present in the nutrient solution, but most of the absorbed Sn remains in roots.[662] Under natural soil conditions, however, Sn apparently is less available, therefore, measurable amounts of Sn are not found in all plant species.

Gough et al.[279] reported the common range of Sn to be 20 to 30 ppm (AW). Zook et al.[906] found 5.6 to 7.9 ppm (DW) Sn in wheat grains, and Duke[197] gave the range in Sn concentrations in food plants of a Central American forest region as <0.04 to <0.1 ppm (DW), and Connor and Shacklette[145] reported Sn to average 15 ppm in plant ash, but it was not detectable in all samples. Chapman[131] reported that Sn ranges in grass from 0.2 to 1.9 ppm (DW) and that in corn grains it averages 2.9 ppm.

Plants grown in mineralized areas accumulated Sn to levels as high as 80 ppm (AW),[686] and even to 300 ppm (AW).[613] Sedges and mosses were found to be the best Sn accumulators.[286]

The Sn in plants grown in contaminated soils may be highly enriched. Pešek and Kolsky[608] found Sn concentration to be about 1000 ppm (DW) in sugar beets that were grown in the vicinity of a chemical factory, and Peterson et al.[610] reported Sn concentrations of about 2000 ppm (DW) in vegetation from the proximity of a Sn smelter. Sn is very toxic to both higher plants and fungi.

V. LEAD

A. Soils

The terrestrial abundance of Pb indicates a tendency for Pb to concentrate in the acid series of magmatic rocks and argillaceous sediments in which the common Pb concentrations range from 10 to 40 ppm, while in ultramafic rocks and calcareous sediments its range is from 0.1 to 10 ppm (Table 93).

Pb has highly chalcophilic properties and thus its primary form in the natural state is galena (PbS). Pb occurs mainly as Pb^{2+}, although its oxidation state, +4, is also known, and it forms several other minerals which are quite insoluble in natural waters.

During weathering Pb sulfides slowly oxidize and have the ability to form carbonates and also to be incorporated in clay minerals, in Fe and Mn oxides, and in organic matter. The geochemical characteristics of Pb^{2+} somewhat resemble the divalent alkaline-earth group of metals, thus Pb has the ability to replace K, Ba, Sr, and even Ca, both in minerals and in sorption sites.

The natural Pb content of soil is inherited from parent rocks. However, due to widespread Pb pollution, most soils are likely to be enriched in this metal, especially in the top horizon. There is much data available in the literature on soil Pb, but sometimes it is difficult to separate the data for background Pb levels in soils from those of anthropogenically influenced amounts in surface soils.

Values presented in Table 95 for the natural Pb occurrence in top horizons of different soils from various countries show that amounts range from 3 to 189 ppm, while mean values for soil types range from 10 to 67 ppm and average 32 ppm. High Pb levels (above 100 ppm) have been reported only for Denmark, Japan, Great Britain, and Ireland and most probably reflect the impact of pollution. Davies[165] stated that an upper limit for the Pb content of a normal soil could be established as 70 ppm.

The Pb content of surface soils of the U.S. averages 20 ppm (Table 96). Therefore, the mean Pb concentration for surface soil on the world scale could be estimated as 25 ppm.

1. Reactions with Soil Components

The natural Pb content of soils is strongly related to the composition of the bedrock, and Pb is reported to be the least mobile among the other heavy metals. The relatively low Pb concentrations in natural soil solutions (Table 12) support this statement. Although the Pb species can vary considerably from one soil type to another, it may be concluded from the results given by Norrish,[570] Riffaldi et al.,[653] Tidball,[787] and Schnitzer and Kerndorff[690] that Pb is associated mainly with clay minerals, Mn oxides, Fe and Al hydroxides, and organic matter. However, in some soil Pb may be highly concentrated in Ca carbonate particles or in phosphate concentrations.

The solubility of Pb can be greatly decreased by liming. A high soil pH may precipitate Pb as hydroxide, phosphate, or carbonate, as well as promote the formation of Pb-organic complexes.

Hildebrand and Blume[319] reported that illites show much greater affinity to sorb Pb than other clay minerals, whereas this reaction was not observed by Kabata-Pendias[377] in studies of Pb sorption by clays over a wide range of pH. Farrah and Pickering[228] emphasized that adsorption of Pb is highly dependent on kinds of ligands involved in the formation of hydroxy complexes of Pb (e.g., $PbOH^+$, $Pb_4(OH_4)^{4+}$). These authors discussed several possible mechanisms of adsorption of hydroxy species and suggested that Pb sorption on montmorillonite can be interpreted as simply cation exchange processes, while on kaolinite and illite Pb is rather competitively adsorbed. Abd-Elfattah and Wada[2] found a higher selective adsorption of Pb by Fe oxides, halloysite, and imogolite than by humus, kaolinite, and montmorillonite. In other studies, the greatest affinity to sorb Pb were reported for Mn oxides.[377,525]

The characteristic localization of Pb near the soil surface in most soil profiles is primarily related to the surficial accumulation of organic matter. The greatest Pb concentrations are also often found in the organically rich top horizons of uncultivated soils, as was reported by Fleming et al.[236] Therefore, organic matter should be considered as the important sink of Pb in polluted soils.

2. Contamination of Soils

The fate of anthropogenic Pb in soils has recently received much attention because this metal is hazardous to man and animals from two sources — the food chain and soil dust inhalation (or, with children, pica for soil).

Table 95
LEAD CONTENT OF SURFACE SOILS OF DIFFERENT
COUNTRIES (PPM DW)

Soil	Country	Range	Mean	Ref.
Podzols and sandy soils	Australia	—	57	196
	Canada	2.3—47.5	10.4	243
	Madagascar	—	37	557a
	Poland	8.5—23.5	16	378
Loess and silty soils	Poland	14—32	26	378, 665
Loamy and clay soils	Canada	1.5—50.1	16.6	243, 629
	Chad	20—45	—	39
	Madagascar	—	48	557a
	Poland	12.5—52	25	378
	U.S.S.R.	—	40	631
Soils on glacial till	Denmark	11.3—17.3	14.7	801, 802
Fluvisols	Austria	16—22	19	6
	Great Britain	24—96	63	166
	Madagascar	19—47	—	557a
	Poland	12.5—48.5	39	378
Paddy soils	Japan	6—189	29	395
Gleysols	Chad	20—50	—	39
	Poland	19.5—48.5	30	378
	U.S.S.R.	—	67	631
	Great Britain	17—63	40	874
Rendzinas	Ireland	25—45	—	236
	Madagascar	—	20	557a
	Poland	17—46	28.5	378,685
Kastanozems and brown soils	Austria	13—31	21	6
	Great Britain	20—50	35	874
Ferralsols	Chad	10—30	—	39
	Sierra Leone	3—91	47	168
Chernozems	Poland	19—29	25	378
Prairien and meadow soils	U.S.S.R.	—	61	631
	West Germany	11.5—79.5	—	46
Histosols, other organic soils	Canada	1.5—50.0	12.6	243
	Denmark	43—176	50.5	1, 801
	Great Britain	26—142	84	69, 874
	Ireland	120	—	236
	Poland	18—85	—	681
Forest soils	China	—	26	225
	U.S.S.R.	10—56	37	9, 631
Various soils	Austria	21—33	29	6
	Canada	—	20[a]	521
	Great Britain	15.5—41	29	69, 100, 818
	Japan	5—189	35	395, 403
	West Germany	15—68	—	46

[a]   Mean for whole profiles of arable soils.

The steadily increasing amounts of Pb in surface soils, both arable and uncultivated, have been reported for various terrestrial ecosystems.[331] The accumulation of Pb in surface soil exposed to various pollution sources at some sites has already reached a value as high as about 2% of dry soil material (Table 97). The levels of Pb in soils that are toxic to plants are not easy to evaluate; however, several authors have given quite similar concentrations, ranging from 100 to 500 ppm (Table 6).

Studies on Pb compounds in contaminated soils have been reviewed by Hildebrand,[318] Olson and Skogerboe,[582] Harmsen,[305] and Zimdahl and Hassett.[903] The main Pb pollutants

Table 96
LEAD CONTENT OF SURFACE SOILS OF THE
U.S. (PPM DW)

Soil	Range	Mean
Sandy soils and lithosols on sandstones	<10—70	17
Light loamy soils	<10—50	20
Loess and soils on silt deposits	10—30	19
Clay and clay loamy soils	10—70	22
Alluvial soils	10—30	18
Soils over granites and gneisses	10—50	21
Soils over volcanic rocks	10—70	20
Soils over limestones and calcareous rocks	10—50	22
Soils on glacial till and drift	10—30	17(a)
Light desert soils	10—70	23
Silty prairie soils	10—30	21(a)
Chernozems and dark prairie soils	10—70	19
Organic light soils	10—50	24
Forest soils	10—50	20(a)
Various soils	<10—70	26

Note: Sources are as follows, 706; a, 218, 219.

Table 97
LEAD CONTAMINATION OF SURFACE SOILS (PPM DW)

Site and pollution source	Mean or range of content	Country	Ref.
Old mining area	51—21,546	Great Britain	165, 166, 808, 659
Nonferric metal mining	170—4,563	Great Britain	165, 166
	>300	West Germany	390
	15—13,000	U.S.	330
	21—3,044	U.S.S.R.	567
Metal processing industry	291—12,123	Canada	363
	1,250—18,500	Greece	559
	628—1,334	Holland	305
	310—2,100	Japan	403
	104	Norway	732
	72—1,350	Poland	871, 507
	500—6,500	U.S.	538
	92—2,580	Zambia	573
	3,000	U.S.S.R.	854
Urban garden and urban vicinity	6—888	Canada	585
	270—15,240	Great Britain	48, 786a
	17—165	Poland	159
	218—10,900	U.S.	127, 208, 628
Sludged farmland	425	Great Britain	59
	60—253[a]	Japan	395
	80—254[b]	Holland	314
Roadside soil	132—397	Japan	395
	114—885	West Germany	397
	960—7,000	U.S.	582

[a] Paddy soil.
[b] 6 and 16 Tonnes dry matter sludge/ha/year, for 5 years.

emitted from smelters occur in mineral forms (e.g., PbS, PbO, $PbSO_4$, $PbO \cdot PbSO_4$), while Pb in automobile exhausts is in the form of halide salts (e.g., PbBr, PbBrCl, Ph(OH)Br, $(PbO)_2PbBr_2$). Exhaust Pb particles are unstable and readily convert into oxides, carbonates, and sulfates.

Because Pb enters the soil in various and complex compounds, its reactions may differ widely among areas. Indeed, opinion appears to differ as to whether Pb as a pollutant is a mobile or a stable soil component. Tyler[816] reported that Pb is the most stable metal in forest soil, and the time necessary for a 10% decrease of its total concentration by leaching was calculated to be 200 years for polluted soil and 90 years for "control" soil. Kitagishi and Yamane[395] calculated the period in which the amount of Pb in soil will decrease by one half to range from 740 to 5900 years, depending on the kind of soil, the water management, and the organic matter present. Stevenson and Welch[759] observed that Pb moved from the top soil treated with Pb acetate into the subsoil, even though the soil (silty clay loam) was shown to have a high capacity for binding Pb in nonexchangeable forms. This mobility was attributed to the metal leaching as soluble chelated complexes with organic matter. In general, however, several observations of the Pb balance in various ecosystems show that the input of this metal greatly exceeds its output (Table 13). For example, Hansen and Tjell[304] estimated that an annual increase of Pb in Danish agricultural soils is 3.7% of its total amount in the soil. It must be emphasized that contamination of soils with Pb is mainly an irreversible and, therefore, a cumulative process in surface soils will continue, even if inputs are low.

The accumulation of Pb in surface soils is of great ecological significance because this metal is known to greatly affect the biological activity of soils. This topic has been reviewed by Tyler[812,813] Anderson,[19] Doelman and Haanstra,[185] and Hughes et al.[331] who showed that increased levels of Pb in soil are likely to limit enzymatic activity of microbiota and as a consequence, markedly increase the accumulation of incompletely decomposed soil organic matter, particularly those materials that do not decompose readily, such as cellulose. A significant accumulation of nitrates in soils enriched in Pb was observed by Woytowicz.[889]

Niyazova and Letunova[567] reported a strong tendency of soil microflora to accumulate Pb at a very high rate that was proportional to the metal content of soils (Figure 40). Consumers such as earthworms also at times concentrate Pb from the soil substrata (Table 20), which greatly contributes to a secondary deposition of Pb in surface soils.

Recently Pb concentrations have gradually increased in soils and may seriously inhibit microbial processes. These effects should be expected principally in soils with a low CEC value. However, in the long run, they may also occur in other soils with a higher CEC.

B. Plants

Although Pb occurs naturally in all plants, it has not been shown to play any essential roles in their metabolism. Broyer et al.[108] reviewed this topic and concluded that if Pb is necessary for plants its concentration at the level of 2 to 6 ppb should be sufficient. Pb has recently received much attention as a major chemical pollutant of the environment and as an element toxic to plants.

1. Absorption and Transport

Zimdahl[902] and Hughes et al.[331] extensively reviewed the findings on Pb absorption by roots and concluded that the mode of its uptake is passive and that the rate of uptake is reduced by liming and by low temperature. Pb, although not readily soluble in soil, is absorbed mainly by root hairs and is stored to a considerable degree in cell walls.

Warren[849] and Kovalevskiy[417] described Pb as a very useful element for geochemical prospecting. The Pb content of plants grown on mineralized areas is, in general, highly correlated with the Pb concentration in soil, although this relationship differs among organs of the plant (Figure 41).

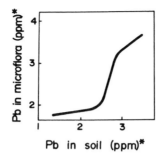

FIGURE 40. Concentration of Pb in soil microflora as a function of its content in soil. *Values given in powers of ten.[567]

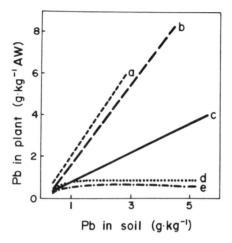

FIGURE 41. Concentration of Pb in various organs of larch (*Larix dahurica*) as a function of its content in soil overlying an ore deposit. (a) Roots; (b) bark; (c) needles; (d) twigs; (e) wood.[417]

When Pb is present in soluble forms in nutrient solutions, plant roots are able to take up great amounts of this metal, the rate increasing with increasing concentration in the solutions and with time (Figures 42 and 43). The translocation of Pb from roots to tops is greatly limited, and as Zimdahl[902] described, only 3% of the Pb in the root is translocated to the shoot.

The degree to which soil Pb is available to plants is of great environmental concern. Cannon[118] and Zimdahl and Koppe[904] reviewed this topic and showed that in spite of several statements in the literature on the slight effect of soil Pb on concentrations of Pb in plant tissues, plants are able to take up Pb from soils to a limited extent. Apparently most of the Pb in soil is unavailable to plant roots. Wilson and Cline[882] studied the absorption of ^{210}Pb by barley grown by using a modified Neubauer technique and showed that only 0.003 to 0.005% of the total Pb in soils may be taken up by plants. This uptake, however, varies significantly over the concentration ranges currently present in soils and with the various forms of Pb that occur in soils.

Zimdahl and Koeppe[904] cited an alternate hypothesis to explain the Pb uptake from soil, in which Pb is not taken up directly from soil by plant roots, but rather is sorbed from dead plant materials accumulated near the soil surface. Nevertheless, there is much evidence that Pb is taken up from soils by roots, at both low and high Pb concentrations, and that this process is strongly governed by soil and plant factors (Figures 44 and 45).

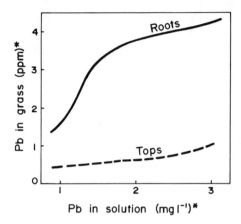

FIGURE 42.   Pb content of a grass (*Bromus uni-oloides*) as a function of its concentration in nutrient solution. *Values given in powers of ten.[915]

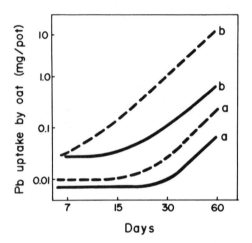

FIGURE 43.   Pb uptake by oats from sand culture as a function of time and of Pb concentration in nutrient solution. (a) 25 mg ℓ^{-1}; (b) 200 mg ℓ^{-1}. Solid lines, tops; broken lines, roots.[735]

Airborne Pb, a major source of Pb pollution, is also readily taken up by plants through foliage. Much controversy exists in the literature on the question of how much airborne Pb is fixed to hairy or waxy cuticles of leaves and how much Pb is actually taken into foliar cells. A number of studies have shown that Pb deposited on the leaf surface is absorbed by these cells. Although it has been suggested that most of the Pb pollution can be removed from the leaf surfaces by washing with detergents, there is likely to be a significant translocation of Pb into plant tissues.[340]

Zimdahl and Koeppe[904] summarized recent results of translocation and uptake studies and showed that under certain conditions Pb is mobile within the plant. It is generally agreed, however, that Pb from a soil source is not readily translocated to edible portions of plants. These authors stated that the main process responsible for Pb accumulation in root tissue is the deposition of Pb, especially as Pb pyrophosphate, along the cell walls. Malone et al.[506] identified the deposits in cell walls outside the plasmalemma as Pb precipitates and Pb

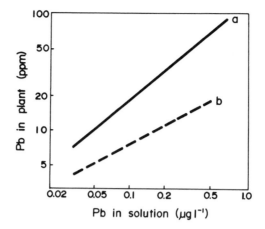

FIGURE 44.   Pb content of buckwheat (tops) as a function of its concentration in natural soil solutions obtained
from (a) sandy soil; (b) loamy soil.[915]

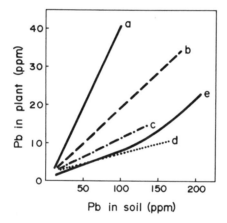

FIGURE 45.   The relation of Pb contents of various plants to EDTA
extractable soil Pb. (a) Potato weed, *Heliotropium europaeum*; (b) Ward's
weed, *Carritchera annua*; (c) *Medicago* spp.; (d) wireweed, *Polygonum
aviculare*; (e) lower part of corn, (*Zea mais*) nodes.[448,533]

crystals. Similar Pb deposits observed in roots, stems, and leaves suggest that Pb is transported and deposited in a similar manner in all tissues of the plant.

2. Biochemical Roles

Although there is no evidence that Pb is essential for the growth of any plant species,
there are many reports on the stimulating effects on plant growth of some Pb salts (mainly
$Pb(NO_3)_2$ at low concentrations. Moreover, other reports have described inhibitory effects
of low Pb levels on plant metabolism. Due to interactions of Pb with other elements and
with many environmental factors, it has not been simple to establish Pb concentrations that
are toxic to vital plant processes (Table 25). Several reports describe the toxic effects of Pb
on processes such as photosynthesis, mitosis, and water absorption; however, the toxic
symptoms in plants are not very specific (Table 29).

Subcellular effects of Pb on plant tissues are related to the inhibition of respiration and
photosynthesis due to the disturbance of electron transfer reactions. These reactions have

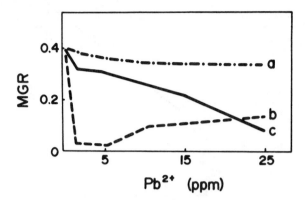

FIGURE 46.   The influence of PbCl₂ in culture solution on the
maximal specific growth rate (MGR) of (a) Pb-tolerant bacteria
strain; (b, c) Pb-sensitive bacteria strains.[184]

been found to be inhibited by Pb concentrations as low as 1 ppm in corn mitochondria.[902]
Photosynthesis processes in sunflower leaves were also reduced by half at Pb concentration
of about 1 μM g^{-1}.[661]

Several plant species, ecotypes, and bacterial strains are able to develop Pb-tolerance
mechanisms. This tolerance seems to be associated with the properties of membranes. Lane
et al.[460] stated that Pb becomes strongly bound to cell walls and that pectic acid is most
active in Pb sorption; thus Pb has a marked influence on the elasticity and plasticity of cell
walls, resulting in an increase in tissue wall rigidity.

Sensitive plant species or bacterial strains attract more Pb to their cell walls than do Pb-
tolerant plants. A large deposition of Pb on membranes apparently impairs their functions.
On the other hand, Pb-tolerant ecotypes may accumulate Pb in inactive forms such Pb pyro-
or orthophosphates which have been identified in roots of plants growing under Pb stress.[902]
The behavioral differences between Pb-sensitive and Pb-tolerant bacteria strains are manifested mainly by their growth rates, as is demonstrated with bacterial cultures (Figure 46).

Although even a very low Pb concentration may inhibit some vital plant processes, Pb
poisoning has seldom been observed in plants growing under field conditions. Zimdahl[902]
discussed that problem and suggested two possible explanations: (1) the relatively low Pb
concentration in soil and (2) a low Pb availability even under contaminated soil conditions.

3. Interactions with Other Elements

The interference of Pb with trace elements has been reported only for Zn and Cd (Figure
16). The stimulating effect of Pb on Cd uptake by plant roots may be a secondary effect of
the disturbance of the transmembrane transport of ions. The Zn-Pb antagonism adversely
affects the translocation of each element from roots to tops.

The interference of Pb with Ca is of metabolic importance since Pb can mimic the
physiological behavior of Ca and thus can inhibit some enzymes. The beneficial effect of
liming on reducing Pb absorption by roots is governed by soil type. In soils where Pb-
organic complexes are formed, Ca^{2+} cations do not significantly limit the availability of
Pb.

A favorable P regime is known to reduce the effects of Pb toxicity. This interference is
due to the ability of Pb to form insoluble phosphates in plant tissues, as well as in soils. S
is known to inhibit the transport of Pb from roots to shoots. Jones et al.[370] concluded that
S-deficiency increases markedly the Pb movement into the tops.

Table 98
MEAN LEAD CONTENT OF PLANT FOODSTUFFS (PPM)

Plant	Tissue sample	FW basis (a)	DW basis	AW basis
Sweet corn	Grains	—	0.88(b), 3(c), <0.3(d)	34(b), 94(c)
Bean	Pods	0.08	2(c), <1.5(d)	37(c)
Beet (red)	Roots	—	2(c), 0.7(e)	28(c)
Carrot	Roots	0.012	3(c) <1.5(d), 0.5(e)	38(c)
Lettuce	Leaves	0.001	0.7(b), 2(c), 3.3(d), 3.6(e)	5(b), 13(c)
Cabbage	Leaves	0.016	1.7(c), 2.3(d)	17(c)
Potato	Tubers	—	3(c), 0.5(d)	90(c)
Onion	Bulbs	—	2(c), 1.3(d), 1.1(e)	35(c)
Cucumber	Fruits, unpeeled	0.024	—	—
Tomato	Fruits	—	3(c), 1(d), 1.2(e)	44(c)
Apple	Fruits	0.001	0.05(b), 0.2(d)	2.7(b)
Orange	Fruits	0.002	—	—

Note: Sources are as follows: a, 547; b, 705; c, 852; d, 188; and e, 131.

4. Concentrations in Plants

Pb is a major chemical pollutant of the environment, therefore, its concentration in vegetation in several countries has increased in recent decades owing to man's activities. This fact has been clearly demonstrated by Rühling and Tyler,[668] who have reported a severalfold increase of Pb concentration in the moss *Hypnum cupressiforme* during the last century (1860 to 1970). For this reason it is important, as Cannon[118] stated, that information on the Pb content of plants be documented as to both date of collection and location of the samples. There is an urgent need to collect and preserve samples and data from unpolluted areas throughout the world.

The great variation of Pb contents of plants is influenced by several environmental factors, such as the presence of geochemical anomalies, pollution, seasonal variation, and genotype ability to accumulate Pb. Nevertheless, natural Pb in plants growing in uncontaminated and unmineralized areas appears to be quite constant, ranging from 0.1 to 10 ppm (DW) and averaging 2 ppm (DW).[13,118]

The Pb contents of edible portions of plants grown in uncontaminated areas, as reported by various authors for the decade 1970 to 1980, range from 0.001 to 0.08 ppm (FW), from 0.05 to 3.0 ppm (DW), and from 2.7 to 94 ppm (AW) (Table 98). Although mean Pb contents calculated for cereal grains of various countries seem to vary considerably (from 0.01 to 2.28 ppm DW), the grand mean, when calculated with the exclusion of two extreme values, is 0.47 and is quite close to the commonly reported concentrations (Table 99).

The background levels of Pb in forage plants average 2.1 ppm (DW) for grasses and 2.5 ppm (DW) for clovers, when two extreme values (0.36 ppm Pb in grass in Finland, and 8.0 ppm Pb in clover in Sweden) are excluded from the calculation (Table 100). The commonly observed increase in Pb amounts in forage plants during the fall and winter seasons is not yet completely understood.

Of great environmental significance is the ability of plants to absorb Pb from two sources, soil and air, even though Pb is believed to be the metal of least bioavailability and the most highly accumulated metal in root tissues. Several plant species and genotypes are adapted to grow in high Pb concentrations in the growth media; this is reflected by anomalous amounts of this metal in the plants (Table 101). The highest bioaccumulation of Pb generally is reported for leafy vegetables (mainly lettuce) grown in surroundings of nonferrous metal smelters where plants are exposed to Pb sources of both soil and air. In these locations, highly contaminated lettuce may contain as much as 0.15% Pb (DW).[658]

Table 99
LEAD CONTENT OF CEREAL GRAINS FROM DIFFERENT COUNTRIES (PPM DW)

Country	Cereal	Range	Mean	Ref.
Austria	Rye	—	0.64	323
	Wheat	—	0.59	323
Great Britain	Barley	<1.25—1.50	—	783
Canada	Barley	0.1—0.2	—	195
	Oats	—	2.28	515
Egypt	Wheat	0.10—0.92	0.51	212
Finland	Barley	0.29—0.56	0.40[a]	829
	Oats	0.33—1.08	—	829
	Wheat	0.13—0.28	0.18[a]	829
Japan	rown rice	—	0.19	395
	Buckwheat (flour)	—	0.36[b]	395
Poland	Wheat	0.2—0.8	0.32	267, 915
	Oats	<0.4—0.8	0.49	915
	Rye	<0.4—0.8	0.49	915
Sweden	Wheat	0.4—0.7	0.57	21
U.S.	Wheat	0.42—1.0	0.64	906
U.S.S.R.	Oats	—	0.01[c]	735
	Wheat	0.4—0.6	0.5	338

[a] After $PbNO_3$ addition to soil.
[b] FW basis.
[c] After addition of 25 to 100 mg Pb per liter in sand culture.

Table 100
MEAN LEVELS AND RANGES OF LEAD IN GRASSES AND LEGUMES AT IMMATURE GROWTH FROM DIFFERENT COUNTRIES (PPM DW)

Country	Grasses		Clovers		Ref.
	Range	Mean	Range	Mean	
Canada	<1.2—3.6	1.8	—	—	237
Great Britain	1—9	2.1	1—3	1.3	369a, 874
West Germany	2.4—7.8	3.3	3.3—4.7	4.2	576, 577
Finland	0.19—0.88	0.36	—	—	590
Poland	0.7—3.2	1.8	1.6—4.0	2.7	915
Sweden	5—6	—	—	8.0	668
U.S.	<0.8—5.6	1.6	<2—15	2.0[a]	118, 710

[a] Alfalfa, calculated from AW basis.

A relatively minor effect on the Pb concentrations in plants has been reported for contamination of soil due to agricultural activities. However, the utilization of Pb-enriched sludges has not been practiced for a long period of time, therefore, it is too early to assess the general environmental impact of built-up Pb levels in soils due to repeated sludge applications. Elevated Pb contents of vegetables grown in urban and industrial areas present a health risk to man.

VI. TITANIUM

A. Soils

Ti is a common constituent of rocks and commonly ranges in concentration from 0.03 to

Table 101
EXCESSIVE LEVELS OF LEAD IN PLANTS GROWN IN CONTAMINATED SITES (PPM DW)

Site and pollution source	Plant and part	Mean or range in content	Country	Ref.
Mining, or mineralized area	Grass, tops	63—232	Great Britain	165, 658, 659
	Maple, stems	135[a]	U.S.	330
	Birch, twigs	277—570	U.S.S.R.	417
Metal-processing industry	Blueberry, stems	150	Canada	715
	Lettuce, leaves	596—1506[b]	Canada	658, 659
	Potato, tubers	350—425[a]	Canada	848
	Grass, shoots	229—2714	Canada	410
	Chinese cabbage	45	Japan	403
	Lettuce, leaves	45—69	Poland	224
	Carrot, roots	27—57	Poland	224
	Blueberry, leaves	141—874	U.S.	357
	Spinach, leaves	322	Zambia	573
Battery manufacturer	Tree foliage	34—459	Canada	480
Urban garden and urban vicinity	Spinach, leaves	66	Zambia	573
	Potato, tubers	100—200[a]	Canada	848
Roadside	Corn, leaves	56	Poland	157
	Corn, leaves	16—24[b]	U.S.	594
	Grass, shoots	111—186	West Germany	397
	Grass, young shoots	67—950	Sweden	669

[a]   AW basis.
[b]   Washed and unwashed, respectively.

1.4% (Table 93). In minerals, Ti occurs predominantly in the tetravalent oxidation state, mainly as a major component of oxides, titanates, or silicates. The dioxide of Ti occurs in nature in different modifications — rutile, brookite, anatase, and leucoxene.

Minerals of Ti (mainly oxides and ilmenite) are very resistant to weathering, therefore, they occur practically undecomposed in soils. When Ti-bearing silicates are dissolved, the element is soon transformed into Ti oxide-aquate, which is transformed to anatase or rutile. Bain[41a] reported that Ti in podzol developed on glacial drift is almost entirely in the form of cryptocrystalline anatase. Hutton,[334] however, stated that Ti can enter into the structure of some layered silicates.

Although Ti minerals are known to be the most stable minerals in the soil environment, no mineral is completely insoluble, and the Ti level reported in soil solutions is 0.03 mg ℓ^{-1}.[334] The solubility of Ti in soils is, however, very limited and this effects an increase in the absolute amounts of Ti in the top horizon of soils after the loss of some clay-size layered silicates due to weathering. The total Ti (also Zr) contents of soils have been used in several studies of soil genesis and of the continuity of soil profiles.

The Ti content of surface soils generally ranges from 0.1 to 0.9% (mean, 0.35%) except for organic soils in the Soviet Union (Tables 102 and 103). Greater Ti contents are normally associated with highly weathered soils, tropical soils, and those derived from Ti-rich parent rocks. Light organic soils contain the smallest amounts of Ti. Soil exposed to effluents or emissions from certain industries (Ti alloys, Ti paint production) may become contaminated by Ti; however, this element does not create any environmental problems.

B. Plants

No clear evidence of a biochemical role of Ti has appeared, although Chapman[131] and

Table 102
TITANIUM CONTENT OF SURFACE SOILS OF DIFFERENT
COUNTRIES (% DW)

Soil	Country	Range	Mean	Ref.
Podzols and sandy soils	Ireland	0.10—0.17	—	236
	Madagascar	—	0.60	557a
	New Zealand[a]	0.37—1.70	—	861
	Poland	0.02—0.24	0.17	382
	U.S.S.R.	0.09—0.11	0.10	493
Loess and silty soils	New Zealand[a]	0.57—1.00	—	861
	Poland	0.11—0.67	0.32	382
	U.S.S.R.	—	0.20	493
Loamy and clay soils	Madagascar	—	0.50	557a
	New Zealand[a]	0.54—2.40	—	861
	Poland	0.09—0.51	0.35	382
	U.S.S.R.	—	0.14	493
Fluvisols	India	—	0.21	455
	Madagascar	—	0.60	557a
	U.S.S.R.	—	0.11	493
Rendzinas	Australia	—	0.49	334
	India[b]	0.17—0.22	—	455
	Ireland	0.2—0.3	—	236
	Madagascar	—	0.94	557a
	Poland	0.04—0.8	0.47	382
Ferralsols	Madagascar	0.05—2.10	—	557a
	India	0.30—0.32	—	455
Solonchaks	Madagascar	0.50—1.00	—	557a
Chernozems	U.S.S.R.	0.40—0.48	0.45	4
Histosols	India	0.60—0.67	0.64	455
	Poland	0.008—0.39	0.15	382
	U.S.S.R.	—	0.011	493
Forest soils	U.S.S.R.	—	0.57	4
Various soils	Great Britain	0.05—0.60	0.56	34, 818
	Madagascar	0.065—1.00	—	557a
	New Caledonia	0.20—0.50	—	39

[a]   Soils derived from basalts and andesites.
[b]   Desert calcareous soils.

Shkolnik[718] described its possible catalytic function in N fixation by symbiotic microorganisms and in photooxidation of N compounds by higher plants, as well as in some processes of photosynthesis. Pais et al.[597] observed an increase of chlorophyll in tomato plants grown in culture solution after spraying with a Ti-chelate solution. Although little attention has been given to Ti absorption by plants, this element is considered to be relatively unavailable to plants and not readily mobile in them.

Levels of Ti in plants vary rather considerably within the range of 0.15 to 80 ppm (DW) (Table 104). Some weeds, especially horsetail and nettle, are known to accumulate much more Ti, and diatoms are reported to contain Ti concentrations ranging from 15 to 1500 ppm (DW).[855] Only one report, Wallace et al.,[841] described Ti toxicity symptoms, and this report cites the chlorotic and necrotic spots that were observed on leaves of bush bean that contained Ti at a concentration of about 200 ppm (DW).

VII. ZIRCONIUM

A. Soils
The crustal abundance of Zr generally varies from 20 to 500 ppm, being lowest in

Table 103
TITANIUM AND ZIRCONIUM CONTENTS OF SURFACE SOILS OF THE U.S.[706]

Soil	Ti (%) Range	Ti (%) Mean	Zr (ppm) Range	Zr (ppm) Mean
Sandy soils and lithosols on sandstones	0.02—1.00	0.28	70—2000	305
Light loamy soils	0.07—1.00	0.27	70—2000	270
Loess and soils on silt deposits	0.05—1.00	0.41	30—500	255
Clay and clay loamy soils	0.10—1.00	0.36	100—700	255
Alluvial soils	0.07—0.50	0.20	30—300	140
Soils over granites and gneisses	0.10—0.70	0.30	50—700	170
Soils over volcanic rocks	0.15—1.00	0.53	70—500	195
Soils over limestones and calcareous rocks	0.07—1.00	0.26	50—500	240
Soils on glacial till and drift	0.10—0.30	0.21	70—200	140
Light desert soils	0.07—0.70	0.29	70—1500	330
Silty prairie soils	0.15—0.70	0.26	70—500	185
Chernozems and dark prairie soils	0.07—0.70	0.26	70—500	205
Organic light soils	0.03—0.50	0.14	50—1500	200
Forest soils	0.15—0.50	0.36	70—700	240
Various soils	0.05—1.00	0.30	50—890	230

Table 104
MEAN TITANIUM CONTENT OF PLANT FOODSTUFFS (PPM)

Plant	Tissue sample	FW basis(a)	DW basis	AW basis
Wheat	Grain	—	0.9(b)	—
Corn	Grain	—	2.0(c)	<5—20(d)
Asparagus	Stem	—	—	180(e)
Snap bean	Pods	—	3.2(d)	45(d)
Bean	Pods	<0.2	—	—
Lettuce	Leaves	<0.3	—	—
Cabbage	Leaves	<0.7	—	—
Carrot	Roots	<0.5	—	<5(e)
Potato	Tubers	—	—	18(e)
Onion	Bulbs	—	1.6(d)	37(d)
Cucumber	Fruit	<0.5	—	19(e)
Apple	Fruit	<0.004	0.18(d)	10(d)
Orange	Fruit	<0.1	0.15(d)	4(d)
Food plants	Edible parts	—	0.2—80(c)	—

Note: Sources are as follows: a, 574; b, 267; c, 197; d, 705; and e, 145.

ultramafic rocks and calcareous sediments (Table 93). The predominant stable valency of Zr is +4, and the prevailing reaction is bonding with oxygen. The widely distributed Zr mineral, zircon, is highly resistant to weathering, therefore, Zr is considered to be only slightly mobile in soils. However, organic acids seem to be the transporting agent for the migration of Zr in soils. Smith and Carson[741] reported that small dissolutions of Zr have been noted in both acidic podzolic soils and in alkaline laterites.

Like Ti, Zr has been used as an index element in soil studies. Hutton[334] extensively reviewed this topic and concluded that reliable results for soil genesis are obtained when Ti to Zr ratios are considered.

The Zr content of soils generally is inherited from parent rocks, therefore, no significant variation in the Zr content is observed among the soil types. However, lower amounts of

Table 105
ZIRCONIUM CONTENT OF PLANT FOODSTUFFS (PPM)

Plant	Tissue sample	FW basis	DW basis	AW basis(a)
Cereal	Grains	0.08—10(b)	0.02—1(b)	8—1,033(b)
Corn	Grains	—	—	<20
Bean	Pods	<0.13(c)	2.6(a)	—
Lettuce	Leaves	0.41—0.62(b)	0.56(a)	4
Cabbage	Leaves	<0.4(c)	—	<20
Carrot	Roots	<0.32(c)	—	<20
Potato	Tubers	—	0.5(a)	12
Onion	Bulbs	0.45—0.84(b)	—	<20
Tomato	Fruits	0—1.79(b)	—	<20
Apple	Fruits	0.31(b)	—	<20
Orange	Fruits	0.05(b)	—	<20
Grape	Raisins	—	1.5(b)	—
Peanuts	Seed	—	2.3(b)	—
Food plants	Edible parts	—	0.005—0.2(d)	—

Note: Sources are as follows: a, 705; b, 741; c, 574; and d, 197.

Zr are in soils on glacial drift and higher amounts are in residual soils derived from Zn-rich rocks (Table 103). The average Zr content calculated for soils of the U.S. is 224 ppm, while Hutton[334] gave the average of about 350 ppm Zr for Australian soils. Wells[861] found Zr in soils derived from basalts and andesites to range from 330 to 850 ppm, whereas Lukashev and Pietukhova[493] gave the range in Zr for sandy soils as 90 to 160 ppm and for peat soil as little as 32 ppm. Chattopadhyay and Jervis[132] reported the range of Zr in garden soils to be 200 to 278 ppm. Smith and Carson[741] reviewed the current and potential uses of Zr in industry and concluded that the only potentially harmful effect on the soil environment would be from production-waste streams containing soluble Zr salts.

B. Plants

Although most soils contain significant amounts of Zr, its availability to plants is greatly limited. The Zr content seems to be higher in roots, especially in nodules and roots of legumes, than in tops, which apparently indicates a low mobility of this metal in plants.[283]

Smith and Carson[741] reviewed the history of studies of stable Zr and ^{95}Zr in plants and showed that concentration factors for Zr derived from soil are low in plants, while those derived from rainwater are much higher for both soil-rooted plants and epiphytes. Aquatic plants are also likely to rapidly take up soluble species of Zr.

There are not many data available on the Zr status of plants. The Zr levels in food plants vary from 0.005 to 2.6 ppm (DW) and do not show any trend in their distribution within plant tissues (Table 105). Some herbage, especially legumes, as well as shrubs and mosses are likely to concentrate more Zr than are other plants, and thus the highest Zr level reported for leaves and stems of deciduous trees is as much as 500 ppm (AW). However, as Shacklette et al.[710] reported, Zr was detected only in about 30% of the plant species analyzed. Bowen[94] gave the range in Zr concentrations in lichens and bryophytes as 10 to 20 ppm (DW). It should be noted that, due to the low reliability of analytical results, the data presented may not be adequate for a realistic estimate of the typical Zr concentration in plants.

Although there is not much agreement among the reported values for Zr in plants, there is, apparently, no bioaccumulation of this metal in food plants. However, the ready uptake of atmospheric ^{95}Zr by terrestrial plants due to deposition of fallout particulates is reported to contribute a great proportion of this radionuclide in the Japanese diet (Yamagata and Iwashima, cited in Smith and Carson[741]).

Although toxic effects of Zr on plants, especially on root growth, are commonly reported, its stimulating effect on the growth of yeasts and on metabolism of other microorganisms has also been observed.[741] Davis et al.[171] reported Zr to be the least toxic element, among the heavy metals, to barley seedlings.

VIII. HAFNIUM

Crystalochemical properties of Hf are similar to those of Zr; however, its terrestrial abundance is much lower. The common concentration of Hf in rocks ranges from 0.1 to 10 ppm (Table 9).

The Hf content of soils is reported to range from 1.8 to 18.7 ppm in Bulgarian soils,[558] to be 1.8 to 10 ppm in Canadian soils,[409] and to average 5 pm in standard soils of Great Britain.[818]

As reported by Furr et al.,[250] Hf is taken up by plants grown on soil amended with sewage sludges which contain, on the average, about 3 ppm Hf. The common range of Hf in plants was given as 0.01 to 0.4 ppm (DW).[381] Oakes et al.[574] reported Hf in food plants to range from 0.6 to 1.1 ppb (FW), but it was not always detectable.

Chapter 10

ELEMENTS OF GROUP V

I. INTRODUCTION

Geochemical characteristics of the trace elements of Group V are widely diverging. Especially, variation in the electrical charge and in the valence states are common features of these elements. The metalloids As and Sb and the metal Bi of the Group Va are highly chalcophilic. While As follows P in biogeochemical behavior, Sb rather resembles Bi.

Of the Group Vb elements, V is the least basic and may form both cationic and anionic compounds. Two other elements, Nb and Ta, are rare metals and resemble each other in biogeochemical properties. All metals of this group reveal an affinity for oxygen bonds.

II. ARSENIC

A. Soils

As is distributed rather uniformly in major types of rocks and its common concentrations in most rocks range from 0.5 to 2.5 ppm (Table 106). Only in argillaceous sediments is As, on the average, concentrated as high as 13 ppm.

As has a great affinity to form or to occur in many minerals, and of over 200 As-containing minerals approximately 60% are arsenates. Arsenite compounds are not common in hypergenic environments. This element is highly associated with deposits of many metals and therefore is known as a good indicator in geochemical prospecting surveys.

Although As minerals and compounds are readily soluble, As migration is greatly limited due to the strong sorption by clays, hydroxides, and organic matter. The As enrichment in argillaceous sediments as well as in surface soils, as compared to concentrations in igneous rocks, apparently reflects also some external As sources, such as volcanic exhalations and pollution.

The oxidation states of As are -3, 0, $+3$, and $+5$, of which As^0 and As^{3+} are characteristic of reducing environments. The complex anions AsO_2^-, AsO_4^{3-}, $HAsO_4^{2-}$, and $H_2AsO_3^-$ are the most common mobile forms of As. The behavior of arsenate (AsO_4^{3-}) resembles that of phosphates and vanadates.

The reactions of As in soils are highly governed by its oxidized state. However, arsenate ions are known to be readily fixed by such soil components as clays, phosphatic gels, humus, and calcium, and the most active in As retention are hydrated Fe and Al oxides. Hydroxy-Al on the external surfaces of micaceous minerals is reported to be especially significant in the retention of As.[329] The strong association of As with Fe (mainly goethite) in soils for both natural and added As has been reported by Norrish.[570]

El-Bassam et al.[204] reported that strongly adsorbed As in soil is hardly likely to be desorbed again and that generally the retention of As by soil progressed with the years. However, As combined with Fe and Al oxides may be liberated upon hydrolysis with the reduction of soil potential. Little is known about the As compounds in soil, but their formation is good evidence that arsenate in soils behaves like phosphate, and in acid soils Fe or Al arsenates are likely to be most common.[570,887]

Several strains of bacteria accelerate the oxidation of arsenites to arsenates and are also involved in methylation and alkylation of As. Thus, microbiota may highly govern the processes of As migration, precipitation, and volatilization as cited by Boyle and Jonasson,[98] Jernelöv,[359] and Weinberg.[856] Methylation of As conducted by certain yeasts under oxic conditions, and by methanogenic bacteria under anoxic conditions, plays significant roles in the release of volatile As from the soil to the atmosphere.

Table 106

**ARSENIC, ANTIMONY, BISMUTH, VANADIUM, NIOBIUM, AND TANTALUM
IN MAJOR ROCK TYPES (PPM) (VALUES COMMONLY FOUND, BASED ON
VARIOUS SOURCES)**

Rock type	As	Sb	Bi	V	Nb	Ta
Magmatic Rocks						
Ultramafic rocks Dunites, peridotites, pyroxenites	0.5—1.0	0.1	0.001—0.02	40—100	1—15	0.02—1.0
Mafic rocks Basalts, gabbros	0.6—2.0	0.2—1.0	0.01—0.15	200—250	10—20	0.5—1.0
Intermediate rocks Diorites, syenites	1.0—2.5	0.X	0.01—0.10	30—100	20—35	0.7—2.1
Acid rocks Granites, gneisses	1.0—2.6	0.2	0.01—0.12	40—90	15—25	2.0—4.0
Acid rocks (volcanic) Rhyolites, trachytes, dacites	1.5—2.5	0.2	0.01—0.12	70	20—60	3
Sedimentary Rocks						
Argillaceous sediments	13	1.2—2.0	0.05—0.40	80—130	15—20	0.8—1.5
Shales	5—13	0.8—1.5	0.05—0.50	100—130	15—20	1—2
Sandstones	1.0—1.2	0.05	0.10—0.20	10—60	0.05	0.05
Limestones, dolomites	1.0—2.4	0.3	0.10—0.20	10—45	0.05	0.05

Table 107

**ARSENIC, VANADIUM, AND NIOBIUM CONTENTS OF SURFACE SOILS OF
THE U.S. (PPM DW)**

Soil	As Range	As Mean	V Range	V Mean	Nb Range	Nb Mean
Sandy soils and lithosols on sandstones	<0.1—30.0	5.1	7—150	47	5—30	10
Light loamy soils	0.4—31.0	7.3	20—150	77	5—30	12
Loess and soils on silt deposits	1.9—16.0	6.6	20—300	102	5—20	12
Clay and clay loamy soils	1.7—27.0	7.7	20—150	87	5—30	13
Alluvial soils	2.1—22.0	8.2	30—150	79	5—20	10
Soils over granites and gneisses	0.7—15.0	3.6	50—200	100	5—30	13
Soils over volcanic rocks	2.1—11.0	5.9	30—300	136	5—30	14
Soils over limestones and calcareous rocks	1.5—21.0	7.8	10—150	72	5—20	12
Soils on glacial till and drift	2.1—12.0	6.7	30—200	95	5—15	7
Light desert soils	1.2—18.0	6.4	30—150	93	5—30	15
Silty prairie soils	2.0—12.0	5.6	50—150	87	5—15	7
Chernozems and dark prairie soils	1.9—23.0	8.8	30—150	92	5—30	10
Organic light soils	<0.1—48.0	5.0	<7—150	38	5—50	11
Forest soils	1.5—16.0	6.5	15—200	85	5—100	14
Various soils	<1.0—93.2	7.0	0.7—98(a)	—	5—70	13

Note: Sources are as follows: a, 600; 706, 707.

The biotransformation of organic arsenical pesticides recently has received much attention because their inorganic derivatives may be quite toxic. The background As levels in topsoils are generally low, although they exceed those in rocks several times. The range in As in soils of the U.S. is broad, from <0.1 to 69 ppm (Table 107) and, similarly in worldwide uncontaminated soils, the As content ranges from <1 to 95 ppm (Table 108). The grand

Table 108
ARSENIC CONTENT OF SURFACE SOILS OF DIFFERENT COUNTRIES (PPM DW)

Soil	Country	Range	Mean	Ref.
Podzols and sandy soils	Canada	1.1—28.9	5.8	243
	Great Britain	5.1—6.8	—	819
	Japan	1.2—6.8	4.0	395
	Korea	2.4—6.8	4.6	395
	Thailand	—	2.4	395
Loamy and clay soils	Canada	1.3—16.7	4.8	243, 629
	Thailand	7.2—18.4	12.8	395
Fluvisols	Bulgaria	3.4	—	612
	Great Britain	20—30	25	786
Soils on mafic rocks	Great Britain	5.0—8.2	—	819
Chernozems	Bulgaria	8.2—11.2	8.2	612
Histosols	Canada	1.8—66.5	13.6	98, 243
Forest soils	Norway	0.6—5.0	2.2	442
Various soils	Bulgaria	2—10.4	5.6	612
	Canada	<1—30	5.8	98, 409, 480, 629
	Great Britain	4—95	16.3	786,818
	Japan	0.4—70	11	395
	Norway	0.7—8.8	2.5	754
	Taiwan	—	7.9	472
	U.S.S.R.[a]	5—30	—	656

[a] Biogeochemical province in Usbekistan.

Table 109
ARSENIC CONTAMINATION OF SURFACE SOILS (PPM DW)

Site and pollution source	Content	Country	Ref.
Nonferric metal mining	90—900	Great Britain	144
Metal-processing industry	33—2000	Canada	480, 777
	38—2470	Japan	336, 767
	69[a]	Hungary	328
	72—340[b]	Norway	447
	10 — 380	U.S.	153, 538
Chemical works	10—2000	Hungary	328
Application of arsenical pesticides	10—290	Canada	98, 585
	38—400	Japan	303
	31—625	U.S.	208, 279, 887

[a] EDTA soluble.
[b] HNO$_3$ soluble.

mean of soil As is calculated to be 6.7 and 8.7 ppm, respectively, for the U.S. and the world. The geometric mean for As in surficial materials of the U.S. is reported to be 5.8 ppm.[707]

The lowest As levels are found in sandy soils and, in particular, in those derived from granites, whereas higher As concentrations are related most often to alluvial soils and soils rich in organic matter. Due to common As pollution, the levels of this element are likely to be increased in topsoils, therefore, the As ranges reported as background values are also given for contaminated soils.

The As content of certain contaminated soils has already been built up to as high as 0.2% (Table 109). Significant anthropogenic sources of As are related to industrial activities (metal

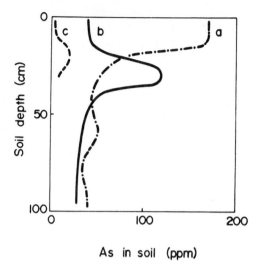

As in soil (ppm)

FIGURE 47.   Patterns of vertical distribution of total
As in polluted soils from a nonferrous metal mining
area in Japan. (a) Heavy gley soil; (b) light gley soil;
(c) unpolluted paddy soil.[395]

processing, chemical works based on S and P minerals, coal combustion, and geothermal power plants) and to the use of arsenical sprays, particularly in orchards.

As pollution is reported most frequently for Japanese soils, Kitagishi and Yamane[395] have reviewed several studies on As behavior in contaminated soils and have shown that paddy soils accumulated most often considerable amounts of As and that this fact is due to the high sorption capacity of these soils and also to the As transportation by irrigation water. Since As behavior is dependent on the soil oxidation state, this element varies in its vertical distribution. In heavy gley soil As is accumulated only in the top horizon and is readily leached from the subsurface layer with a high reduction potential. The As pattern in light well-drained soils is almost the opposite (Figure 47). Also in uncontaminated soils the patterns of As profile distribution may diverge highly.

The mobility of As in soil was shown to be proportional to the As added and inversely proportional to time and to Fe and Al contents. The toxicity of As depends on the concentration of soluble As, therefore, sodium arsenate and arsenic trioxide, formerly used as herbicides, are the most toxic. However, plant growth response may also be related to total soil As (Figure 48). Phytotoxicity of As is highly dependent on soil properties, and while in heavy soil about 90% growth reduction appears at 1000 ppm As addition, in light soil 100 ppm As is equally toxic.[887] The maximum allowable limit of As in paddy soils is proposed as 15 ppm.[395]

The toxicity of As in soils may be overcome in several ways, depending on As pollution sources and on soil properties. Increasing the oxidation state of flooded (paddy) soils limits As bioavailability.[395] Application of materials that produce precipitates with As in soil (e.g., ferrous sulfate, calcium carbonate) is reported to be effective when added to soils having less than 10 ppm of soluble As (in 0.05 N HCl).[303] Also fertilizing, mainly the application of phosphate, decreases As bioavailability. However, the literature on phosphate retention of As in soils is confused, for phosphate could also displace adsorbed or fixed As from sorbing complexes and thereby initially increase the amount of soluble As in soils. The alleviating effect of S application on As toxicity has also been reported. Kitagishi and Yamane[395] have reviewed this topic and presented several results that require further study.

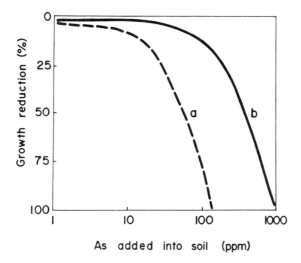

FIGURE 48.   Corn growth response to As added to two soils.
(a) Light soil with low organic matter content and with ka-
olinitic clay predominating; (b) heavy soil with high organic
matter content and with vermiculitic clay predominating.[887]

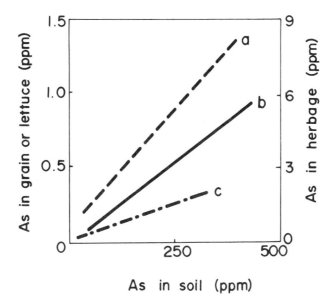

FIGURE 49.   Concentration of As in plants as a function of total soil
As. (a) Lettuce; (b) pasture herbage; (c) barley grains.[783]

B. Plants

As is a constituent of most plants, but little is known about its biochemical role. Several
reports on the linear relationship between As content of vegetation and concentrations in
soil of both total and soluble As suggest that plants take up As passively with the water
flow. Some plants, particularly Douglas fir, show a remarkable ability to take up As far
more than many of the associated plants. Such plants are useful guides to the recognition
of subsurface mineralization with certain metals.[624,851]

Thoresby and Thornton[783] described the ready uptake of As by various plant species
(Figure 49). Apparently, As is translocated in plants since its concentration in the grain also

Table 110
ARSENIC CONTENT OF FOOD AND FORAGE
PLANTS (PPB)

Plant	Tissue sample	FW basis	DW basis
Barley	Grains	—	3—18(a)
Oats	Grains	—	10(b)
Wheat	Grains	—	50(b), 3—10(a)
Brown rice	Grains	—	110—200(c)
Sweet corn	Grains	25	30—400, 30(d)
Snap beans	Pods	0.74—<6.7	7—100
Cabbage	Leaves	1.2—<16.0	20—50
Spinach	Leaves	—	200—1,500
Lettuce	Leaves	<5.3	20—250
Carrot	Roots	4.8—<13.0	40—80
Onion	Bulbs	4.5	50—200
Potato	Tubers	—	30—200
Tomato	Fruits	0.46	9—120
Apple	Fruits	<0.21	50—200
Orange	Fruits	1.4—<5.2	11—50
Edible mushroom	Whole	—	280(e)
Clover	Tops	—	20—160(b)
Grass	Tops	—	280—330(f)

Note: Sources are as follows: 381, 574, 705; a, 447; b, 915; c, 395; d, 462; e, 414; and f, 624; i, 385.

has been reported. With increasing soil As, however, the highest As concentrations were always recorded in old leaves and in roots.

Concentrations of As in plants grown on uncontaminated soils vary from 0.009 to 1.5 ppm DW, with leafy vegetables being in the upper range, and fruits in the lower range (Table 110). Mushrooms are found to be relatively high As accumulators.

Several plant species are known to tolerate a high level of As in tissues. As toxicity has commonly been noted in plants growing on mine waste, on soils treated with arsenical pesticides, and on soils with As added by sewage sludge treatment. The symptoms of As toxicity are variously described as leaf wilting, violet coloration (increased anthocyanin), root discoloration, and cell plasmolysis. The most common symptom, however, is growth reduction. Kitagishi and Yamane[395] reported that rice grown on apple orchard soil containing 77 ppm As produced almost no yield the first year. The toxic effect of As was partly reduced after 3 years of cultivation without any special treatment, but an application of S greatly limited the phytotoxicity of As.

Wallace et al.[845] have shown that bush bean plants grown in solution culture with 10^{-4} M arsenate accumulated approximately 4, 19, and 42 ppm As, respectively, in leaves, stems, and roots, causing considerable damage. Decreased Mn, P, and Ca concentrations in all plant parts, and K in roots, was reported.

Gough et al.[279] reviewed recent findings on As phytotoxicity and reported that the As content of injured leaves of fruit trees ranged from 2.1 to 8.2 ppm DW. In general, the residue tolerance for As in plants is established as 2 ppm DW (Table 25). However, the critical value in rice plants is as high as 100 ppm DW in tops and 1000 ppm DW in roots.[395] Davis et al.[171] gave the critical value of 20 ppm DW for barley seedlings.

Although there are some reports of the stimulating effects of As on the activity of soil microorganisms, As is known as a metabolic inhibitor, therefore, yield reduction of vegetation under a high level of bioavailable As should be expected. Apparently, As is less toxic when the plant is well supplied with P.

Table 111
EXCESSIVE LEVELS OF ARSENIC IN PLANTS GROWN IN
CONTAMINATED SITES (PPM DW)

Site and pollution source	Plant and part	Mean or range of content	Country	Ref.
Mining or mineralized area	Douglas fir, stems	140—8200[a]	Canada	851
	Grass, tops	460—6640	Great Britain	624
Metal-processing industry	Grass	0.5—62	Canada	777
	Tree foliage	27—2740[b]	Canada	480
	Rice, leaves	7—18	Japan	395
	Hay	0.3—2.6	Norway	441
Battery manufacturer	Tree foliage	16—387[b]	Canada	480
Sludged or irrigated field	Brown rice	1.2 (max.)	Japan	336
Application of arsenical pesticides	Turnip, roots	1.08	Canada	138
	Potato, tuber peels	1.10	Canada	138
	Carrot, roots	0.26	Canada	138

[a] AW basis.
[b] Washed leaves.

Depending on the location and pollution source, plants may accumulate extremely large amounts of As — above 6000 ppm DW and above 8000 ppm AW (Table 111). Although As poisoning from plants to animals is believed to be very uncommon, unfavorable health effects of such a high As concentration in vegetables and in forage plants can not be precluded.

III. ANTIMONY

A. Soils

The crustal abundance of Sb is low and, with the exception of argillaceous sediments which contain up to 2 ppm Sb, does not exceed 1 ppm (Table 106). The geochemical characteristics of Sb are closely related to those of As and, in part, to those of Bi. Sb usually occurs with the valence of $+3$ and occasionally of $+5$ and shows amphoteric behavior.

The reactions of Sb during weathering are not yet well known. However, the common occurrence of Sb in waters, its concentrations in coals, and its association with Fe hydroxides indicate a relatively high mobility in the environment. In surface soil Sb ranges from 0.05 to 4.0 ppm (Table 112) and, if the above results are representative, Sb is concentrated in soils as compared to rocks. The average for soil Sb is given by Wedepohl[855] as 1 ppm, and the grand mean calculated from the data presented in Table 112 is 0.9 ppm.

Like As, Sb may be associated with nonferrous ore deposits and is likely to be a pollutant in industrial environments. Thus, the increased level of Sb to about 200 ppm was found in soils near a Cu smelter,[153] and elevated air concentrations of Sb have been reported for different smelter operations and urban areas.[210] The trend of variation in the Sb content of surface soil in Norway clearly indicates pollution by long-range transport.[754]

B. Plants

Sb is considered a nonessential metal and is known to be easily taken up by plants if present in soluble forms. Shacklette et al.[710] found the Sb content of trees and shrubs growing in mineralized areas to range from 7 to 50 ppm DW, whereas Bowen[94] reported an Sb mean of 0.06 ppm DW in land plants. Oakes et al.[574] gave the range of Sb concentrations in edible plants as 0.02 to 4.3 ppb FW, with cabbage being in the upper range and apple fruits in the lower range. Laul et al.[462] reported Sb to be <2 ppb DW in corn grains and potato tubers and 29 ppb in grass. Ozoliniya and Kiunke[588] showed that the Sb content of barley and flax roots was, respectively, 122 and 167 ppb DW and exceeded that of leaves having

Table 112
**ANTIMONY CONTENT OF SURFACE SOILS OF DIFFERENT
COUNTRIES (PPM DW)**

Soil	Country	Range	Mean	Ref.
Podzols and sandy soils	Canada	0.05—1.33	0.19	244
	Great Britain	0.34—0.44	—	819
Loamy and clay soils	Canada	0.05—2.0	0.76	244, 409, 629
Soils on mafic rocks	Great Britain	0.29—0.62	—	819
Fluvisols	Bulgaria	—	0.82	558
Chernozems	Bulgaria	—	0.99	558
Histosols	Canada	0.08—0.61	0.28	244
Forest soils	Bulgaria	1.25—2.32	1.77	558
Various soils	Nigeria	1—2[a]	—	855
	Canada	0.29 — 4	1.67	132, 409, 629
	Great Britain	0.56—1.3	0.81	818, 819
	Norway	0.17—2.2	0.61	754
	U.S.	0.25—0.6	—	462

[a]   Possible influence of Pb-Zn mineralization.

10 and 27 ppb DW. The same plants grown in peat soil contained several times less Sb in roots, while the concentration in leaves was about the same as that given above. There are no reports of plant toxicity caused by Sb; however, Sb levels may be expected to increase in plants growing in soils contaminated by industrial emissions or sewage sludge applications.

IV. BISMUTH

Bi is considered a rare metal in the earth's crust, and its higher concentrations in argillaeous sediments do not exceed 0.5 ppm (Table 106). However, Bi accumulation in coals and in graphite shales to about 5 ppm has been reported.

Bi reveals chalcophilic properties, but during weathering is readily oxidized, and when it becomes carbonated (e.g., $Bi_2O_2CO_3$) it is very stable. Therefore the Bi content of most surface soils is directly inherited from parent rocks.

There is a paucity of information on the Bi content of soils. Ure et al.[819] reported the mean Bi content of arable Scottish soils derived from different rocks to be 0.25 ppm (range, 0.13 to 0.42). Chattopadhyay and Jervis[132] reported the range of Bi in garden soils of Canada to be from 1.33 to 1.52 ppm. Bowen[94] gave the approximate mean of soil Bi as 0.2 ppm. Aubert and Pinta[39] reported Bi in ferralitic calcareous soils of Madagascar to be 10 ppm. An increase of Bi in soil horizons rich in Fe oxides and organic matter should be expected.

The Bi content of plants has not been studied extensively. Shacklette et al.[710] reported that Bi was found only in about 15% of a number of samples of Rocky Mountain trees and that the Bi range was from 1 to 15 ppm AW. Bowen[94] gave the mean Bi as <0.02 ppm DW in land plants and 0.06 ppm DW in the edible parts of vegetables. Erämetsä et al.[216] found a range of Bi in *Lycopodium* sp. from <1 to 11 ppm DW, with about 60% detectability. Bi is likely to be concentrated at polluted sites due to its high concentration in some coals and sewage sludges.

V. VANADIUM

A. Soils
The general abundance pattern of V in common rocks compiled in Table 106 shows that although there is a large variation in the content of rocks, this metal is concentrated mainly

in mafic rocks and in shales (within the common range of 100 to 250 ppm). The geochemical characteristics of V are strongly dependent on its oxidation state ($+2$, $+3$, $+4$, and $+5$) and on the acidity of the media. V is known to form various complexes of cationic and anionic oxides and hydroxy oxides,[256] therefore, V displays various behaviors. It usually does not form its own minerals, but rather replaces other metals (Fe, Ti, Al) in crystal structures.

During weathering the mobility of V is dependent on the host minerals, and finally V remains in the residual rock-forming minerals or is adsorbed or incorporated in mineral structures of clays or Fe oxides. A high degree of association of V with Mn and with the K content of soil has also been reported by Norrish.[570] V tends to be associated with organic matter, and therefore its elevated concentrations in organic shales and bioliths is common. An especially great range of V in some coals and crude oil has often been observed. Wedepohl[855] explained these phenomena, following the assumption of Bertrand, by postulating that in former geologic periods there might have existed some plants with much higher V contents than at present are found in normal plants. Yen[894] described high concentrations of V in organic sediments as the results of V^{3+} sorption by lipids and cholines, the basic compounds in further formation of porphyrins.

The behavior of V in soil has received little attention. Norrish[570] reported that Fe oxides hold a reasonable fraction of the soil V that is more mobile and could supply V to plants. Berrow et al.[71] emphasized that in certain horizons of podzols the role of the clay minerals as well as organic acids might be more significant than the V fraction adsorbed by Fe oxides. Apparently, the vanadyl cation (VO^{2+}) may be an important form of V in many soils and may result from reduction of the metavanate anion (VO_3^-). Goodman and Cheshire[272] and Bloomfield[81] stated that much of the soil V, mainly the vanadyl cation, is mobilized as complexes with humic acids. Also anionic forms of V (VO_4^{3-}, VO_3^-) are known to be mobile in soils and to be relatively more toxic to soil microbiota.

Surface horizons of some podzols are reported to contain less V as a result of extensive leaching into lower horizons.[71] In general, V is distributed in soil profiles rather uniformly and the variation in V content of soil is inherited from the parent materials. Thus, the highest concentrations of V (150 to 460 ppm) are reported for soils derived from mafic rocks, while the lowest (5 to 22 ppm) were found in peat soils (Table 107 and 113). Loamy and silty soils, as well as some ferralitic soils, also contain large amounts of V which exceed that of the parent material.

The average V content of soils world-wide has been calculated to be 90 ppm and to be 84 ppm for U.S. soils (Tables 107 and 113). Shacklette and Boerngen[706] gave a geometric mean of 58 ppm for soil samples of the U.S., while Cannon et al.[119] reported the average soil V to be approximately 100 ppm.

Although there are not many reports of V pollution of soils, it is likely that industrial processing of certain mineral ores (ore smelters, cement, and phosphate rock plants) and burning of coals and oils will increase the deposition of V residues in soils. Combustion of fuel oils is an especially serious source of V in soil. Tyler[814] reported that forest mor around a densely inhabited area accumulated up to 100 ppm V, and Pawlak[605] found increased V content of soil up to 110 ppm in the vicinity of a crude oil refinery. Jacks[347] calculated that the V aerial input to soil in the vicinity of Stockholm is approximately 20 mg m^{-2} year^{-1}.

B. Plants

The evidence that V is essential for the growth of higher plants is not yet conclusive, while the essentiality of this element for green alga species is unquestionable and V is known to stimulate photosynthesis in these organisms. There are evidences that V is a specific catalyst of N_2 fixation and may partially substitute for Mo in this function as carried on by rhizobium bacteria in particular. Dobritskaya[180] reported a high accumulation of V in nodules

Table 113
VANADIUM CONTENT OF SURFACE SOILS OF DIFFERENT COUNTRIES (PPM DW)

Soil	Country	Range	Mean	Ref.
Podzols and sandy soils	Great Britain	58—91	—	819
	Madagascar	—	75	557a
	New Zealand[a]	160—220	—	861
	Poland	10—27	—	382
	U.S.S.R.	10—62	—	180, 493
Loess and silty soils	New Zealand[a]	—	185	861
	Poland	27—110	—	382
	U.S.S.R.	—	57	493
Loamy and clay soils	Chad	15—50	—	39
	Madagascar	—	95	557a
	New Zealand[a]	150—330	—	861
	U.S.S.R.	34—210	—	180, 493, 631, 714
Soil on mafic rocks	Great Britain	340—460	—	819
Fluvisols	Madagascar	48—180	103	557a
	U.S.S.R.	—	18	493
Gleysols	Chad	20—100	—	39
	U.S.S.R.	—	118	631
Rendzinas	Ireland	38—85	—	236
Kastanozems and brown soils	Ireland	25—55	—	236
	Madagascar	—	125	557a
	U.S.S.R.	—	56	714
Ferralsols	Chad	20—250	—	39
	Madagascar	28—530	137	557a
	U.S.S.R.	42—360	—	180
Solonchaks and solonetz	Chad	55—260	—	39
	U.S.S.R.	78—99	88	351
Chernozems	U.S.S.R.	37—125	—	4, 180, 351, 714
Meadow soils	U.S.S.R.	85—380	190	631, 714
Histosols and other organic soils	Denmark	6.3—10	8.1	1
	Sweden	19—22	—	814
	U.S.S.R.	—	5	493
Forest soils	China	—	92	225
	U.S.S.R.	—	145	4, 631, 714
Various soils	Great Britain	15—200	—	34, 818
	Madagascar	19—320	75	557a
	U.S.S.R.	50—87	69	283

[a] Soils derived from basalts and andesites.

of several legumes (3 to 12 ppm DW), which suggests the V association with N_2 biofixation. However, no evidence of V deficiency in higher plants was observed, and Welch and Cary claimed that if V is essential for plants, adequate levels in their tissues are less than 2 ppb DW.[860]

Soluble soil V appears to be easily taken up by roots, and some plant species show a great ability to accumulate this metal. Petrunina reviewed this topic and showed that some bryophytes and fungi,[613] especially *Amanita muscaria*, may contain as much as about 180 ppm DW when grown in mineralized areas. Other accumulator plant species are also known.[119]

Welch[859] studied the uptake of V by roots from labeled V solution, and concluded that V is passively absorbed by barley roots. The uptake was a linear function of V concentration and was highly dependent on pH (Figure 50). These results indicate that VO^{2+} species occurring under acid conditions are more rapidly absorbed by roots than are VO_3^- and

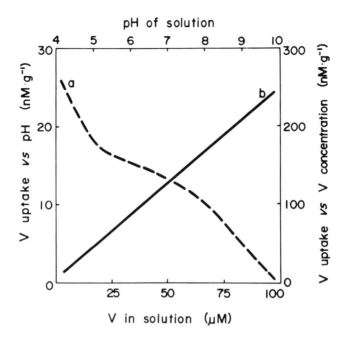

FIGURE 50. The influence of (a) pH on the rate of V absorption by barley roots from a solution of 5 μM NH_4VO_3; (b) V concentration in the solution on its content in barley roots.[858]

HVO_4^{2-} species that predominate in neutral and alkaline solutions. However, both cationic and anionic species are capable of being chelated and thus could contribute substantially to V uptake by plants from soils, but no significant relationship between vegetation and soil V has been reported.[600]

The mean concentration of V in higher plants has been calculated by Dobritskaya to be 1.0 ppm DW,[180] and the range of V in ash of most vegetables is given by Shacklette et al.[710] to be from <5 to 50 ppm. Several other authors have reported a pronounced difference in V content among most plants (Table 114).

The reported variation in V content of plants may be an effect of both analytical difficulties and pollution. The values given by Shacklette et al.[710] from the Report of the Committee on Biological Effects of Atmospheric Pollutants for V in some food plants follow (ppb FW): lettuce, 1080; apple, 330; potato, 1490; carrot, 990; beet, 880; and pea, 460. These values are much higher than those presented in Table 114 and apparently reflect a pollution source of V.

Bryophytes appear to be most sensitive to aerial sources of V. Bowen[94] gave the mean V content of mosses as 11 ppm DW. Shacklette and Connor (see Cannon et al.,[119]) found the range of V in Spanish moss (a flowering plant) to be 50 to 180 ppm AW, being higher in areas affected by emissions from crude oil refining. Gough and Severson[278] found V to be as high as 700 ppm AW in sagebrush near a P fertilizer plant. Pawlak[605] reported the mean V content of clover and grass grown in the vicinity of a crude oil refinery to be 13 and 8 ppm DW, respectively. Folkeson[239] found the range in V concentration in mosses near a peat-fired plant to be 6 to 8 ppm DW and in those near the ash heap to be 14 to 25 ppm DW.

Gough et al.[279] reviewed the topic of V phytotoxicity and stated that there are no reports indicating V toxicity under field conditions. However, under man-induced conditions, V concentrations as high as 0.5 ppm in the nutrient solution, and 140 ppm in the soil solution, may be toxic to plants. Phytotoxicity of V (chlorosis and dwarfing) may appear at about 2

Table 114
VANADIUM CONTENT OF FOOD AND FORAGE PLANTS (PPB)

Plant	Tissue sample	FW basis (a)	DW basis (b)	AW basis (b)
Wheat	Grains	—	7—10(c)	—
Oats	Grains	—	60(c)	—
Beans	Pods	3.4	—	—
Cabbage	Leaves	8	—	—
Lettuce	Leaves	5.3	280	2,800
Carrot	Roots	8.8	—	—
Potato	Tubers	—	6.4	—
Cucumber	Fruits	5.8	56	380
Tomato	Fruits	—	0.5	41
Apple	Fruits	0.01—0.1	8.6	334
Strawberries	Fruits	—	—	660
Kidney bean	Aerial parts	140	760	5,600(d)
Clover	Aerial parts	—	2,700; 380(i)	27,000(e)
Grass	Shoots	160—230(f)	180—420(i)	—
	Shoots	100—2,600(g)	—	—
	Shoots	<1—9,800(h)	—	—

Note: Sources are as follows: a, 574; b, 750; c, 860; d, 855; e, 283; f, 235; g, 119; h, 600;
i, 385.

ppm DW V in some plants as cited by Davis et al.[171] Wallace et al.[841] reported that bush
beans absorbed from culture solution as much as 13, 8, and 880 ppm V (DW) in leaf, stem,
and root, respectively, and that this resulted in smaller growth but not chlorosis.

VI. NIOBIUM

Data on the abundance of Nb in various rocks are listed in Table 106. The amounts of
Nb increase on the average in intermediate and acid magmatic rocks (15 to 60 ppm) and in
argillaceous sediments (15 to 20 ppm).

Nb has strong geochemical relations to Ta, and its association with Fe, Ti, and Zr has
been recognized. Concentrations of Nb in Mn nodules have also been reported.[855]

The +5 valence state of Nb is the most stable in the earth's crust. Most of the Nb
compounds are slightly soluble in both acid and alkaline media. However, the presence of
organic complexing agents mobilizes Nb. The behavior of Nb during weathering is highly
dependent on host minerals, therefore, Nb may be released (e.g., from biotite, amphibolite)
or may remain within resistant minerals (e.g., sphene, zircon). The accumulation of Nb in
certain residual sediments has often been reported.

There are not many reports on the Nb status of soils. The geometric mean of 12 ppm Nb
has been calculated for surface soil samples of the U.S., and the distribution of Nb does
not show any significant variation between soil types (Table 107). Ure and Bacon[818] reported
a mean Nb content of 24 ppm for standard soil samples, while Ure et al.[819] gave the range
in Nb concentrations in arable soils derived from different rocks as 31 to 300 ppm, with
the highest value for soil on trachyte enriched in this metal. Wedepohl[855] stated that Nb
averaged 24 ppm in lateritic soils from West Africa.

Nb is reported to be relatively mobile under humid conditions and therefore may be
available to plants. However, Shacklett et al.[710] found that of more than 1000 samples of a
variety of plant species from throughout the U.S. only one contained Nb (30 ppm AW).
Some mosses and lichens were reported to contain Nb ranging from 0.02 to 0.45 ppm DW
and 15 to 20 ppm AW.[94,710] Tiutina et al.[800] found that the common value of Nb in plants
is about 1 ppm DW; however, several native plants have a great capacity for extracting Nb

from soil that is enriched in this metal. Concentrations of Nb up to about 10 ppm DW, found in selected plant species (mainly *Rubus arcticus* L.) from a Nb mineralized area of the Komi, USSR, served as an exploration indicator for a Nb deposit.

VII. TANTALUM

Ta closely resembles Nb in geochemical behavior, thus Ta distribution in rocks follows that of Nb and reaches the highest concentrations in acid magmatic rocks (2 to 4 ppm) and in argillaceous sediments (1 to 2 ppm). Ta is believed to be less mobile than Nb during weathering because of its lower solubility and the slight stability of organic complexes. Thus, the Nb to Ta ratio varies, depending on environmental conditions.

There is a paucity of information on Ta occurrence in soils. Naidenov and Travesi[558] found Ta in various Bulgarian soils to range from 0.42 to 3.87 ppm (mean, 0.65 ppm), and Laul et al.[462] gave the range in Ta contents of U.S. soils as 1.1 to 2.7 ppm. The range of Ta in garden soils of Canada was reported to be from 0.17 to 0.22 ppm.[132] Oakes et al.[574] found the range of Ta in food plants to be from 0.013 to 0.48 ppb FW, whereas Bowen[94] gave the mean Ta in edible parts of vegetables as <1 ppb DW. Laul et al.[462] reported Ta in vegetation from <1 to <6 ppb DW.

Chapter 11

ELEMENTS OF GROUP VI

I. INTRODUCTION

Geochemical and biochemical behavior of trace elements of this group is complex and diverse. Se and Te resemble S in geochemical reactions. Po, also belonging to the subgroup VIa, is a natural isotope of the U-Ra transformation chain.

Trace elements of the Group VIb, Cr, Mo, and W, have strong lithophile tendencies and though they have variable oxidation states, they are preferably hexavalent in their oxygen compounds. The geochemical behavior of Mo and W is very similar, and in biochemistry some substitution by W for Mo has been observed.

II. SELENIUM

A. Soils

Se occurs in nearly all materials of the earth's crust and is present in magmatic rocks in concentrations rarely exceeding 0.05 ppm. In sedimentary rocks Se is associated with the clay fraction and thus the smallest quantities of Se are in sandstones and limestones (Table 115).

Much of the Se occurs in S and sulfide minerals where it may be concentrated up to 200 ppm; however, in S deposits of sedimentary origin the Se level is usually below 1 ppm. During chemical weathering of rocks Se is easily oxidized and the state of its oxidation, as well as its solubility, are controlled by the oxidation-reduction regime and by pH of the environment. Also the biological methylation of Se, yielding the volatile Se compounds, is common and plays a significant role in the geochemical cycle of Se. Selenite ions resulting from oxidation processes are stable and able to migrate until they are adsorbed on mineral or organic particles. In consequence, the Se level is increased in several coals, as well as in clay sediments. Apparently, selenites are the preferable species of Se being adsorbed by clay minerals, particularly by montmorillonite and Fe oxides.[246] Adsorption of SeO_3^{2-} by goethite is highly pH dependent.[93]

On the basis of the geochemistry of Se it seems to be possible to predict trends of Se behavior in a particular soil environment. A generalized summary of Se species that may occur in soils is presented in Figure 51.

Lakin and Dawidson,[454] Allaway,[13] and Paasikallio[551] have extensively reviewed Se behavior in soil and have emphasized its complex character. It can be generalized that:

1. In acid, gley soils, and soils with high organic matter content, selenides and Se-sulfides dominate — they are only slightly mobile and therefore hardly available to plants.
2. In well-drained mineral soils with pH close to neutral, selenites exist exclusively. Their alkaline metal compounds are soluble, but Fe selenites are not; moreover, selenites are rapidly and nearly completely fixed by Fe hydroxides and oxides and thus are very slightly available to plants.
3. In alkaline and well-oxidized soil selenates are likely to occur. They are easily soluble and are unlikely to be fixed by Fe oxides and may be highly mobile and readily taken up by plants.

However, several complex anions of Se as well as organic chelates greatly modify the behavior of Se in each particular soil. This has been nicely illustrated by various trends in Se distribution along soil profiles (Figure 52).

The Se content of soils has received much attention in certain countries, mainly those

Table 115
SELENIUM, TELLURIUM, CHROMIUM, MOLYBDENUM, AND TUNGSTEN IN MAJOR ROCK TYPES (PPM) (VALUES COMMONLY FOUND, BASED ON VARIOUS SOURCES)

Rock type	Se	Te	Cr	Mo	W
Magmatic rocks					
Ultramafic rocks Dunites, peridotites, pyroxenites	0.02—0.05	0.001	1600—3400	0.2—0.3	0.10—0.77
Mafic rocks Basalts, gabbros	0.01—0.05	0.001	170—200	1.0—1.5	0.36—1.10
Intermediate rocks Diorites, syenites	0.02—0.05	0.001	15—50	0.6—1.0	1.0—1.9
Acid rocks Granites, gneisses	0.01—0.05	0.005	4—25	1—2	1.3—2.4
Acid rocks (volcanic) Rhyolites, trachytes, dacites	0.02—0.05	—	4—16	2	2
Sedimentary rocks					
Argillaceous sediments	0.4—0.6	—	80—120	2.0—2.6	1.8—2.0
Shales	0.6	<0.01	60—100	0.7—2.6	1.8—2.0
Sandstones	0.05—0.08	—	20—40	0.2—0.8	1—2
Limestones, dolomites	0.03—0.10	—	5—16	0.16—0.40	0.4—0.6

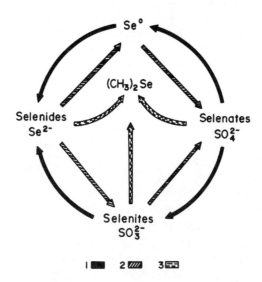

FIGURE 51.   Ionic species and transformation of Se compounds in soils. (1) Reduction; (2) oxidation; (3) methylation.

where the role of Se in animal health has been widely recognized. Surface soil on a world-wide scale contains an average of 0.40 ppm Se (Table 116 and 117). Elevated concentrations of Se are observed in some ferralsols, organic soils, and other soils derived from Se-rich parent materials (Table 118). Also in salt-affected soil the total as well as the water-soluble Se is likely to be elevated.[729] There is great concern about enriched water-soluble Se levels in soils because increased bioavailability of Se is a direct health risk to livestock.

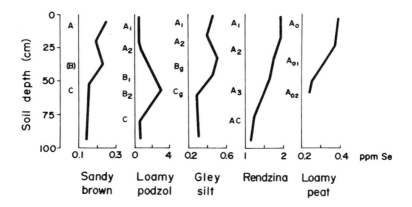

FIGURE 52. Distribution of Se in some profiles of soils in New Zealand. (Letters indicate genetic soil horizons.)[862]

Table 116
SELENIUM CONTENT OF SURFACE SOILS OF DIFFERENT COUNTRIES (PPM DW)

Soil	Country	Range	Mean	Ref.
Podzols and sandy soils	Great Britain	0.15—0.24	—	819
	Canada	0.10—1.32	0.27	244, 294
	Poland	0.06—0.38	0.14	618a
	U.S.S.R.	0.05—0.32	0.18	358
Loess and silty soils	Poland	0.17—0.34	0.23	618a
Loamy and clay soils	Canada	0.13—1.67	0.43	244, 294
	Poland	0.18—0.60	0.30	618a
Soils on mafic rocks	Great Britain	0.02—0.36	0.20	819
Fluvisols	Egypt	0.15—0.85	0.45[a]	212
	Poland	0.12—0.34	0.22	618a
Rendzinas	Poland	0.24—0.64	0.44	618a
Ferralsols	India	—	0.55	729
Chernozems	Poland	0.14—0.24	0.17	618a
	U.S.S.R.	0.32—0.37	0.34	358
Histosols and other organic soils	Canada	0.10—0.75	0.34	244
	Finland	0.08—0.18	0.13[a]	408
	U.S.S.R.	—	0.34	358
Various soils	Canada	0.41—2.09	0.94	468
	Canada	0.03—2.00	0.30[a]	521
	Great Britain	—	0.21	818
	Finland	0.005—1.24	—	734
	Ireland	0.02—0.07	0.05	408
	Ireland	—	1.27	454
	India	0.14—0.68	0.39[a]	729
	New Zealand	—	0.6	862
	Norway	0.15—2.32	0.78	442, 754
	Sweden	0.17—0.98	0.39	473

[a] Data for whole soil profile.

The solubility of Se in most soils is rather low, therefore, many agricultural areas produce crop plants and forage with a low Se content. However, in naturally Se-enriched soils, in poorly drained or calcareous soils, in soils of aridic zones, and also in soils heavily amended with sewage sludges or fly ashes, Se may be accumulated by plants in concentrations high enough to be toxic to grazing livestock.

Table 117
SELENIUM AND CHROMIUM CONTENTS OF SURFACE SOILS OF
THE U.S. (PPM DW)

Soil	Se Range	Se Mean	Cr Range	Cr Mean
Sandy soils and lithosols on sandstones	0.005—3.5	0.5(a)	3—200	40
Light loamy soils	0.02—1.2	0.33(a)	10—100	55
Loess and soils on silt deposits	0.02—0.7	0.26(a)	10—100	55
Clay and clay loamy soils	<0.1—1.9	0.5	20—100	55
Alluvial soils	<0.1—2.0	0.5	15—100	55
Soils over granites and gneisses	<0.1—1.2	0.4	10—100	45
Soils over volcanic rocks	0.1—0.5	0.2	20—700	85
Soils over limestones and calcareous rocks	0.1—1.4	0.19(a)	5—150	50
Soils on glacial till and drift	0.2—0.8	0.4	30—150	80
Lateritic soils	0.02—2.5	1.05(b)	—	—
Light desert soils	<0.1—1.1	0.5	10—200	60
Silty prairie soils	<0.1—1.0	0.3	20—100	50
Chernozems and dark prairie soils	<0.1—1.2	0.4	15—150	55
Organic light soils	<0.1—1.5	0.3	1—100	20
Forest soils	<0.1—1.6	0.4	15—150	55
Various soils	<0.1—4.0	0.31	7—1500	50

Note: Sources are as follows: 706, 707; a, 454; and b, 364.

Table 118
SELENIUM CONTENT OF SURFACE OR
SUBSURFACE SOILS FROM AREAS HAVING
Se-TOXICITY SYMPTOMS IN LIVESTOCK
(PPM DW)

Country	Soil	Range	Mean	Ref.
Great Britain	Peat	92—230	138	880
Ireland	Peat	3—360	54	234
U.S. (South Dakota)	Mineral	6—28	17	454
U.S.S.R. (Armenia)	Mineral	0.8—2.2	1.3	358

The water-soluble fraction of soil Se is considered to be the fraction that is available to plants. van Dorst and Peterson[190] reported the close positive correlation between the Se content of plants and the selenate ion concentration in the soil solution. However, other Se fractions may also be soluble in soils, and thus Elsokkary[212] concluded that, on the average, about 45% of the total soil Se could be available to plants and that plant-available fractions could be extracted with K_2SO_4 or NH_4OH solutions. Soltanpour and Workman[745] found a good correlation between Se uptake by alfalfa and the Se that was soluble in DTPA with ammonium acid carbonate.

The behavior of Se in highly calcareous soil is of special concern because when soils are low in sesquioxides the Se becomes easily water soluble. Singh[729] stated that the best effects in correcting Se toxicity to plants in such soils are obtained by the application of S, P, and even N onto the soils.

Due to several microbiological processes, methylated derivatives of Se are often volatile from soils. On the other hand, a considerable input of Se to the soils takes place through precipitation. Kubota et al.[436a] calculated that the amounts of Se in rainwater that fall on land surfaces reflect also other sources of Se such as volcanic exhalations and industrial emissions, in particular, the combustion of coals.

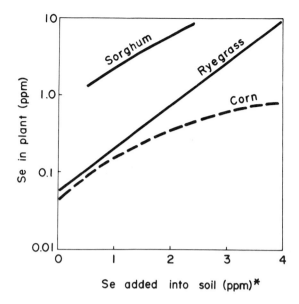

FIGURE 53. Influence of Se added as Na_2SeO_3 into the soil on Se content of plants. *Se level in soil in the experiment with sorghum is given in ten-times lower concentration (e.g., at the range of 0.05 to 0.25 ppm).[551,729]

B. Plants

Although Se in plants has been investigated in many studies, its physiological role is still unknown. There are some opinions stating that Se may be involved in certain metabolic processes, especially in plants that are Se accumulators,[555] but the essentiality of Se for plant growth has not been conclusively established. The great interest in plant Se is due to the importance of this element in nutrition, particularly for domestic animals, since Se can act both as a micronutrient and a toxin. Furthermore, the margin of safety of Se concentrations is rather narrow.

When present in soluble forms Se is readily absorbed by plants, though differences between plant species are commonly observed (Figure 53). The availability of soil Se is also controlled by several soil factors, among which pH is believed to be the most pronounced (Figure 54).

The uptake of Se by plants is also a temperature-dependent process; on a soil low in Se plants absorb a much higher amount when the temperature is $>20°C$ than during cooler seasons with a temperature of $<15°C$.[474] Rainfall may also highly influence the Se concentration of herbage. As Reuter[649] reported, low Se in plants frequently occurred in high rainfall areas.

There is a positive linear correlation between Se in plant tissues and Se content of soils, and Sippola[734] stated that total soil Se gives a better measure of plant response than does its soluble fractions. In general, soil Se uptake by plants depends on climatic condition, water regime of soil, oxidation-redox potential, pH, and sesquioxide content of the soil.

The distribution of Se within plants differs, depending on several nutritional factors, but its concentration in growing points, in seeds, and also in roots was most often observed. Arvy et al.,[37] however, studied Se distribution between various organs of corn and did not find any significant differences.

In plants Se partly resembles S in its biochemical properties and is able to replace S in amino acids as well as in several biological processes. About eight Se-organic compounds have been identified in plants.[254] Amino acids, especially cysteine, are likely to bind Se, which may have several metabolic implications. Volatile Se compounds (i.e., dimethyl

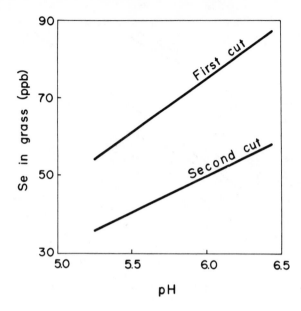

FIGURE 54.   Se content of ryegrass as a function of the soil pH.[551]

diselenide) isolated from plants, and from Se-accumulating species in particular, are known to be responsible for Se release from plants and also for the unpleasant odor of plants such as *Astragalus*.[222]

The ability of a plant to accumulate and tolerate Se is apparently related to different Se metabolisms. Opinions vary, however, as to whether Se is incorporated in proteins or in nonprotein amino acids.[531]

Toxicity of Se to plants growing under natural conditions has not been reported, but symptoms of Se toxicity to crop plants have been described (Table 29). In culture solutions, toxic effects of Se on onion roots were observed by Fiskesjö[232] at as low as 1 mg ℓ^{-1} concentration. Also a decrease in yield of several crops was effected by the application of about 2 ppm Se onto the soil.[729]

Increased Se levels in plants suppresses concentrations of N, P, and S as well as of several amino acids. Also the absorption of heavy metals, mainly Mn, Zn, Cu, Fe, and Cd, is inhibited by increasing Se. This relationship is dependent on the ratio between the elements, and thus some stimulating effects of high Se concentrations on uptake of heavy metals may also be expected.

The application of P, S, and N is known to help in detoxifying Se, which may be a result either of depressing the Se uptake by roots or of establishing a beneficial ratio of Se to these elements, even when the Se content of plants is elevated. There is also a report that the addition of lime, S, B, and Mo to the soil under field conditions did not affect the Se concentration in plant tissues.[294] In practice, the application of S is an important remedial treatment on Se-toxic soils.[729] However, as Johnson[364] reported, S fertilizers are effective in preventing Se toxicity only to plants growing on low Se soils.

Ehlig et al.[202] reported that differences among plant species in Se accumulation from soils low in Se were small, with the exception of seleniferous indicator plants such as *Astragalus* sp. or other legumes which concentrate Se to extremely high levels (up to about 1000 ppm DW). The available systematic data concerning Se content of forage plants show that, for most countries, Se ranges in grasses from 2 to 174 ppb (DW, mean 33), for clover and alfalfa ranges from 5 to 880 ppb (DW, mean 99), and for other forage plants ranges from 4 to 870 ppb (DW, mean 67). The greatest amounts of Se were reported for India, Japan,

Table 119
SELENIUM IN FORAGE PLANTS FROM DIFFERENT COUNTRIES
(PPB DW)

Country	Grasses		Clovers or alfalfa		Hay or fodder Plants		Ref.
	Range	Mean	Range	Mean	Range	Mean	
Canada	5—23	13	5—31	15	—	—	294
Denmark	—	32	—	35	—	44	265
West Germany	30—210	110	50—130	90	—	—	578
Finland	1—54	11	—	—	2—48	14	221, 371, 734
France	19—134	47	36—39	38	29—35	31	37, 551
	—	—	—	—	40—150	130	487
India	—	—	—	—	200—870	—	729
Japan	5—174	43	6—287	33	—	—	395, 765
Norway	—	—	—	—	4—28	13	371
Sweden	11—64	30	18—40	—	4—46	10[a]	473
	—	—	—	—	13—34	23	371
U.S.	10—40	32	30—880	320	28—360	98	14, 265

[a]   Cereal plants.

and West Germany (Table 119). Similarly, the highest Se levels in cereal grains were found in Egypt, West Germany, and the U.S. (Table 120). The mean Se contents of food plants do not exceed 100 ppb (DW) (Table 121). The levels of Se in crops that are adequate and safe for animals are believed to be about 100 ppb (DW), although the requirements for Se vary widely with the form of the Se ingested and with other dietary factors. Disorders in livestock may be expected when Se concentrations in forage plants are within, or below, the order of 10 ppb (DW). Other threshold values, such as a lower toxic level of 3000 ppb Se and a minimum requirement of 100 ppb Se, are also proposed for grasslands.[395]

In general, the trend in variation of Se concentrations in plants indicates higher Se levels in plants from aridic zones than in those from humidic zones. It may be summed up that Se in plants is positively correlated with pH, salinity, and $CaCO_3$ in soils. High levels of Se in plants may also occur near seacoasts where the return of Se to land surfaces may be higher than in other regions because of sea spray.

Pollution with Se is observed in industrial areas where Se is released into the atmosphere due to some metal-processing operations and by the combustion of coal. Some legumes (e.g., sweetclover) grown on coal ashes are known to contain as much as 200 ppm (DW) Se.[296] Grass grown in the vicinity of a P fertilizer plant contained up to 1.2 ppm (DW) Se,[278] the edible mushroom *Agaricus bitorquis* from an urban area accumulated 11.2 ppm (DW) Se,[636] and tree leaves from a nearby Cu refinery contained from 141 to 550 ppm (DW) Se.[113]

In areas with low soil Se, applications of sodium selenite to the soil or as a foliage spray are proposed for correcting Se nutritional deficiencies. In view of the toxic properties of Se salts, however, these practices should be carefully controlled.

III. TELLURIUM

The crustal abundance of Te has not been widely studied and the scarce data show that its content in rocks ranges from 1 to 5 ppb, apparently being elevated in organic shales (Table 115). The geochemical behavior of Te resembles Se in some respects, therefore, its association with S and sulfide minerals, mainly in areas of Au, As, Sb, Ba, Hg, and Bi mineralization, has been observed.

Table 120
SELENIUM CONTENT OF CEREAL GRAINS
FROM DIFFERENT COUNTRIES (PPB DW)

Country	Cereal	Range	Mean	Ref.
Australia	Wheat	1—117	23	867
Canada	Barley	9—38	21	294
	Oats	4—43	28	294
Denmark	Barley	2—110	18	265
	Oats	3—54	16	265
	Rye	6—72	16	265
	Wheat	4—87	21	265
Egypt	Wheat	140—430	340	213
Finland	Barley	<10—50	—	411
	Wheat	100—170	—	411
	All grains[a]	2—85	7	371
France	Barley	27—42	33	37
	Oats	20—44	35	37
	Wheat	30—53	36	37
West Germany	Oats	70—140	110	578
	Rye	160—250	210	578
	Wheat	190—200	200	578
Japan	Oats[a]	8—17	—	765
	Rye[a]	—	22	765
Norway	Barley	—	8	247
	Oats	—	10	247
	Wheat	1—169	33	446
	All grains[a]	2—29	—	247
Sweden	All grains[a]	4—46	13	371
U.S.	Barley	200—1800	450	200
	Oats	150—1000	480	200
	Wheat	280—690	490	492
	Corn[a]	10—2030	87	371

[a]   Animal fodder.

Table 121
SELENIUM CONTENT OF FOOD PLANTS
FROM THE U.S. (PPB)

Plant	Tissue sample	FW basis (a)	DW basis (b)
Sweet corn	Grains	—	11
Cabbage	Leaves	8	150
Lettuce	Leaves	1.6	57
Snap bean	Pods	—	28
Carrot	Roots	5.3	64
Onion	Bulbs	—	42
Potato	Tubers	10	11
Tomato	Fruits	—	36
Apple	Fruits	1.1	2.6
Orange	Fruits	3	7.7

Note: Sources are as follows: a, 574; and b, 705.

Te shows variable valencies of +2 to +6 and, like Se, forms di- and trivalent oxides. During weathering Te is oxidized to slightly mobile tellurites, and most commonly is sorbed by Fe hydroxides. The accumulation of Te in coals (0.02 to 2 ppm) indicates Te sorption also by organic matter.

The biological cycling of Te is known to resemble that of Se, and the microbial metabolism of Te also seems to be similar to that of Se. Weinberg[856] stated that bacteria capable of methylating As and Se can also methylate Te, whereas the reduction of tellurite to Te can readily occur under the influence of a variety of microorganisms.

Apparently, Te occurs in plant tissues at concentrations lower than those of Se. Bowen[94] cited high Te accumulations, from 2 to 25 ppm (DW), in a few plants from Te-rich soils. Schroeder et al.[697] also reported a relatively high concentration of Te in soils (0.5 to 37 ppm) and in plants (0.7 to 6 ppm DW). Oakes et al.[574] gave the range in Te content of vegetables as <0.013 to 0.35 ppm (FW), being the lowest in apple fruits. Some high values given for the Te content of soils and plants may be errors and these values require further study. The garlicky odor of some plants is caused by vapors of dimethyl telluride. Concentrations of Te in onion and garlic are reported to be as high as about 300 ppm (DW).[697]

IV. POLONIUM

Po occurs in the geosphere and biosphere as natural radioisotopes, including ^{210}Po and several other short-lived radionuclides, all associated with the radioactive decay within uranium chains U-Ra-Pb.

The natural abundance of ^{210}Po in soil is reported to range from 8 to 220 Bq kg^{-1} and in land plants to range from 8 to 12 Bq kg^{-1} (DW).[94] Elevated contents of this radionuclide in some vegetables and tobacco plants can be derived from either the soil or the air and apparently reflect pollution with radionuclides.[549,710]

V. CHROMIUM

A. Soils

The terrestrial abundance of Cr indicates that it is associated mainly with ultramafic and mafic rocks in which Cr can be as concentrated as 0.X% (Table 115). The Cr content of acid igneous and sedimentary rocks is much lower and commonly ranges from 5 to 120 ppm, being the highest in argillaceous sediments.

Cr shows highly variable oxidation states (from +2 to +6) and it is also known to form complex anionic and cationic ions, e.g., $Cr(OH)^{2+}$, CrO_4^{2-}, CrO_3^{3-}. Naturally occurring Cr compounds have valences of +3 (chromic) and +6 (chromate). Highly oxidized forms of Cr are much less stable than Cr^{3+}.

Most of the Cr^{3+} is present in the mineral chromite ($FeCr_2O_4$) or in other spinel structures, substituting for Fe or Al. In general, Cr^{3+} closely resembles Fe^{3+} and Al^{3+} in ionic size and in geochemical properties.

Chromite, the common Cr mineral, is resistant to weathering and therefore accounts for most of the Cr in residual material. However, under progressive oxidation Cr forms the chromate ion (CrO_4^{2-}), which is readily mobile and also is easily sorbed by clays and hydrous oxides.

The behavior of soil Cr has been extensively studied by Bartlett and Kimble,[51] Bartlett and James,[50] Cary et al.,[122] Bloomfield and Pruden,[82] and Grove and Ellis.[289] It has been shown that most of the soil Cr occurs as Cr^{3+} and is within the mineral structures or forms of mixed Cr^{3+} and Fe^{3+} oxides. Since Cr^{3+} is slightly mobile only in very acid media, and at pH 5.5 is almost completely precipitated, its compounds are considered to be very stable in soils. On the other hand, Cr^{6+} is very unstable in soils and is easily mobilized in both acid and alkaline soils (Figure 55).

El-Bassam et al.[205] emphasized that Cr behavior is governed by both soil pH and redox-

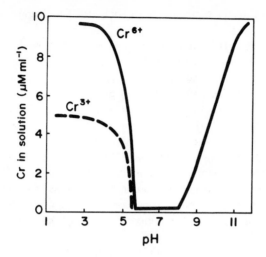

FIGURE 55.   Solubility of trivalent and hexavalent Cr
as a function of pH.[51]

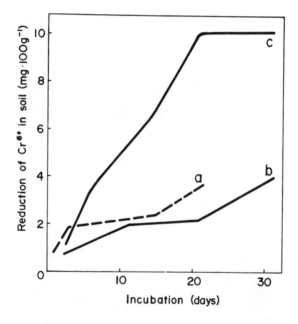

FIGURE 56.   Reduction of Cr^{6+} by soil (pH 6.7) during the
incubation. (a) Aerobic; (b) anaerobic; (c) anaerobic with added
dry alfalfa material.[82]

potential. Under the same redox-potential of 500 mV, Cr^{3+} predominated at pH 5, $Cr(OH)_3$
was formed between pH 5 to 7, and CrO_4^{2-} occurred at pH >7. Griffin et al.[284] found that
adsorption of Cr by clays is also highly pH dependent, and while Cr^{6+} adsorption decreased
as pH increased, the adsorption of Cr^{3+} increased as pH increased. The behavior of Cr in
soils may be modified by organic complexes of Cr; however, the dominant effect of organic
matter is the stimulation of the reduction of Cr^{6+} to Cr^{3+} (Figure 56).

The ready conversion of soluble Cr^{6+} to insoluble Cr^{3+} under normal soil conditions is
of great importance because it is responsible for the low Cr availability to plants. Although

the reduction of Cr^{6+} and Cr^{4+} is commonly reported, the oxidation of Cr^{3+} in soils has also been observed, apparently as an effect of the oxidizing ability of Mn compounds.[50]

The soil Cr is inherited from parent rocks and therefore its higher concentration is in soil derived from mafic and volcanic rocks (Tables 117 and 122). Soils on serpentines in particular are known to contain as much as 0.2 to 0.4% Cr. Sandy soils and histosols are usually poorest in Cr. The grand mean Cr content is calculated to be 54 ppm for surface soils of the U.S. and 65 ppm for world-wide soils.

The immobility of soil Cr may be responsible for an inadequate Cr supply to plants. Cr is of nutritional importance because it is a required element in human and animal nutrition. However, as Cary et al.[122] stated, the Cr added to soil appeared to be very inefficient in terms of its recovery by food crops, although it caused an increase of this element in various plants.

Readily soluble Cr^{6+} in soils is toxic to plants and animals. Therefore, the variability in the oxidation states of Cr in soils is of great environmental concern.

The Cr content of surface soil is known to increase due to pollution from various sources, of which the main ones are several industrial wastes (e.g., electroplating sludges, Cr pigment and tannery wastes, leather manufacturing wastes) and municipal sewage sludges. The Cr added to soils is usually accumulated at the thin top layer. El-Bassam et al.[205] found that after 80 years of irrigation with sewage sludge containing 112 ppm Cr, the metal concentration in soil increased from 43 to 113 ppm. Other authors also reported a high Cr accumulation for surface horizons of sludged farmland, where the highest Cr levels ranged from 214 to 399 ppm.[59, 176,314]

Chaney et al.[130] extensively discussed the Cr hazard in biological waste management and stated that the food chain is well protected from an excess of Cr by the "soil-plant barrier". This statement, however, is not fully supported by findings of Diez and Rosopulo,[176] who reported the ready availability of Cr from soils amended with sewage sludge.

Liming, P application, and organic matter are known to be effective in reducing chromate toxicity in Cr-polluted soils. If contamination of soil is by Cr^{6+}, acidification and then reducing agents (e.g., S and leaf litter) could be used to speed the Cr^{6+} reduction. After the reduction, liming to further precipitate Cr^{3+} compounds might be advisable.[289]

B. Plants

There is no evidence yet of an essential role of Cr in plant metabolism, although Mertz[534] has reviewed the positive effects on plant growth of Cr applications to soils having a low soluble Cr content. The Cr content in plants is controlled mainly by the soluble Cr content of the soils. Most soils contain significant amounts of Cr, but its availability to plants is highly limited. However, the addition of Cr to soil affects the Cr content of plants, and the rate of Cr uptake by plants is dependent on several soil and plant factors. Usually a higher Cr content is observed in roots than in leaves or shoots, whereas the lowest concentration is in grains (Figure 57). Several native plants, mainly those from areas of serpentine or chromite deposits, can accumulate as much as 0.3% (DW) or 3.4% (AW) Cr.[535,613] The elevated Cr content of soils in such areas is known to be responsible for the poor growth of forest trees.[855]

A low rate of Cr uptake by plants from the soluble fraction of this metal is related to the mechanism of uptake by roots. Apparently, root tissues are not capable of stimulating the reduction of Cr^{3+} to readily soluble Cr^{2+}, which is the key process in Fe absorption by plants.[122,788] The form most available to plants is Cr^{6+}, which is the most unstable form under normal soil conditions. Nevertheless, the mechanisms of absorption and translocation of Cr in plants seem to be similar to those of Fe, which is reflected in a fairly stable Cr/Fe ratio in plant tissues.[122] Tiffin[789] concluded that Cr is transported in plants as anionic complexes which have been identified in the extracts of plant tissues and in xylem fluid. The presence of trioxalatochromate in plant leaves has been reported.[798]

Table 122
CHROMIUM CONTENT OF SURFACE SOILS OF DIFFERENT COUNTRIES (PPM DW)

Soil	Country	Range	Mean	Ref.
Podzols and sandy soils	Austria	1.4—3.5	—	6
	Canada	2.6—34	—	243
	Madagascar	—	110	557a
	New Zealand	47—530[a]	—	861
	Poland	30—91	51	665
	U.S.S.R.	18—25	21	493
Loess and silty soils	Bulgaria	77—128	—	255
	Chad	180—300	—	39
	New Zealand	31—160[a]	—	861
	Poland	21—38	29	268
	U.S.S.R.	—	84	493
Loamy and clay soils	Austria	23—24	—	6
	Bulgaria	107—122	115	558
	Canada	4—46	19	243
	Chad	100—200	—	39
	New Zealand	70—1100[a]	—	861
	Poland	35—81	58	665
	U.S.S.R.	—	51	493, 631, 714
Soils on glacial till	Denmark	—	12	801
Fluvisols	Austria	13—30	16	6
	Bulgaria	—	91	558
	Madagascar	—	190	557a
	U.S.S.R.	—	55	4, 493
Gleysols	Austria	—	19	6
	Poland	27—100	57	665
	U.S.S.R.	—	85	631
Rendzinas	Austria	—	38	6
	Ireland	35—50	—	236
	Madagascar	—	95	557a
Kastanozems and brown soils	Austria	11—31	19	6
	U.S.S.R.	—	72	714
Ferralsols	Burma	69—331	—	575
	Chad	100—280	—	39
	Madagascar	130—540	—	557a
Solonchaks and solonetz	Burma	81—110	82	575
	Chad	25—80	—	39
	Madagascar	—	215	557a
	U.S.S.R.	78—99	88	351
Chernozems	Bulgaria	116—173	153	558
	U.S.S.R.	71—195	121	4, 351, 714
Meadow soils	U.S.S.R.	30—110	—	631, 714
Histosols and other organic soils	Canada	4—39	15	243
	Denmark	1.8—10	7	1, 801
	U.S.S.R.	—	8	493
Forest soils	Bulgaria	152—1384	—	558
	U.S.S.R.	—	54	631, 714
Various soils	Bulgaria	71—1085	221	558
	Great Britain	—	69	818
	Canada	11.6—189	50	409, 540
	Canada	10—100	43[b]	521
	Chad	4—80	—	39
	Denmark	—	15	801
	Japan	3.5—810	50	395

Table 122 (continued)
CHROMIUM CONTENT OF SURFACE SOILS OF
DIFFERENT COUNTRIES (PPM DW)

Soil	Country	Range	Mean	Ref.
	Poland	28—107	60	183
	West Germany	9—57	28	325, 390

a   Soils derived from basalts and andesites.
b   Data for whole soil profiles.

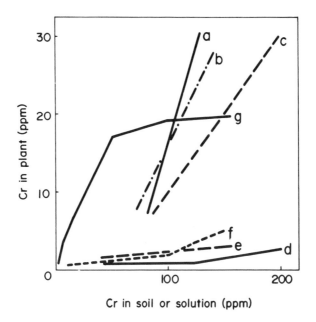

FIGURE 57.   Concentration of Cr in plant tissues as a function of Cr content of soil. (a) Potato stalks; (b) corn leaves; (c) wheat straw; (d) wheat grains; (e) barley grains; (f) tomato tops; (g) tomato roots.[122,176]

The Cr content of plants has received much attention since the relatively recent discovery that Cr participates in glucose and cholesterol metabolism, and therefore is essential to man and animals.[696] There is not much literature on Cr in plants. Common levels of Cr found in plant material are usually in the order of 0.02 to 0.2 ppm (DW); however, a relatively great variation is observed in the Cr content of food plants (Table 123). Concentrations of Cr in plants vary widely for kinds of tissues and stages of growth, and the trend in Cr variation appears to be irregular, as can be concluded from the data given by Mertz et al.[535] for various trees.

Although stimulating effects of Cr on plants have been observed by several authors, the phytotoxicity of Cr has been often reported, especially in plants on soils developed from ultrabasic rocks. Anderson et al.[23] reported toxicity in oats having a Cr content of 49 ppm (DW) where grown on soil containing 634 ppm Cr. Turner and Rust[810] observed initial

Table 123

CHROMIUM CONTENT OF PLANT FOODSTUFFS AND FORAGE PLANTS (PPM)

Plant	Tissue sample	FW basis	DW basis	AW basis
Buckwheat	Seeds	—	0.03(a)	—
Wheat	Grains	—	0.2(b), 0.014(a,c)	—
Sweet corn	Grains	0.037(d)	0.15(d)	5.7(d)
Beans	Pods	0.0087(e)	0.15(d)	2.3(d)
Cabbage	Leaves	0.0013(e)	—	<1.5(d)
Lettuce	Leaves	0.008(e)	—	<1.5(d)
Onion	Bulbs	0.0021(d)	0.021(d)	0.5(d)
Carrot	Roots	0.018(e)	—	<1.5(d)
Potato	Tubers	0.018(e)	0.021(d)	0.49(d)
Cucumber	Fruits	0.064(e)	—	—
Tomato	Fruits	0.0039(d)	0.074(d)	0.62(d)
Apple	Fruits	0.003—0.008 (e)	0.013(d)	0.70(d)
Orange	Fruits	0.004(d)	0.029(d)	0.80(d)
Vegetables	Edible parts	—	0.05—8.0(f)	—
	Edible parts	—	0.02—14.0(g)	—
Alfalfa	Tops	—	0.101(a), 0.46—0.91(h)	—
Grass	Tops	—	0.6—3.4(i)	—
	Tops	—	0.11— 0.35(j)	—

Note: Sources are as follows: a, 122; b, 267; c, 860; d, 705; e, 574; f, 197; g, 94; h, 200; i, 27; and j, 388.

symptoms of Cr toxicity with the addition of as little as 0.5 ppm Cr to the nutrient culture, and 60 ppm to the soil culture. These Cr additions resulted in decreased concentrations of almost all major nutrients in tops and of K, P, Fe, and Mg in roots. The antagonistic interaction between Cr and Mn, Cu, and B has also been reported by Turner and Rust,[810] and this can be related to interferences both within the soil medium and in the plant tissues.

The toxicity of Cr depends on its oxidation stage, but is also related to readily available forms of chromate. While a Cr_2O_7 addition at the 10^{-5} N concentration level decreased plant growth by about 25%, the same level of $Cr_2(SO_4)_3$ was without any effect on growth, but the Cr additions resulted in 2.2 and 1.3 ppm Cr, respectively, in bush bean leaves.[841] The phytotoxic concentrations of Cr in tops of plants were reported as follows (DW basis): 18 to 24 ppm in tobacco, 4 to 8 ppm in corn, 10 ppm in barley seedlings, and 10 to 100 ppm in rice.[171,279,395]

Symptoms of Cr toxicity appear as wilting of tops and root injury; also chlorosis in young leaves, chlorotic bands on cereals, and brownish-red leaves are typical features. Increased levels of Cr in the nutrient solution (up to 10^4 μM) are reported to disorganize the fine structure of chloroplasts and the chloroplast membranes of *Lemna minor* (duckweed).[53]

Anthropogenic sources of Cr are responsible for the elevated content of this metal in plants. Czarnowska[158] reported mean Cr concentrations in grass near city streets to be as high as 17 ppm (DW). Kitagishi and Yamane[395] gave the range in Cr in lichens collected within urban areas to be 5 to 10 ppm (DW). Folkeson[239] found 5 ppm Cr (DW) in mosses near peat-fired plants and 9 ppm Cr in mosses from the edge of a waste heap of the ash, as contrasted with background concentrations of 1 to 1.5 ppm. Gough and Severson[278] found 500 ppm Cr in ash of sagebrush from the vicinity of a P fertilizer factory.

Wastes containing increased Cr concentrations are reported by Chaney et al.,[130] not to be hazardous to human health; however, there is evidence of Cr toxicity to livestock grazing grass with an elevated Cr content due to rich Cr sewage sludge amendment.[381]

VI. MOLYBDENUM

A. Soils

The terrestrial abundance of Mo shows its association with granitic and other acid magmatic rocks. The common range of Mo in these rocks is 1 to 2 ppm, while in organic-rich argillaceous sediments the Mo content may be above 2 ppm (Table 115). Mo behaves both like a chalcophile and a lithophile element, and its geochemistry in the surface environments is related mainly to anionic species. The primary mineral of Mo^{4+}, molybdenide (MoS_2), is known to contain most of the terrestrial Mo and is associated with Fe and Ti minerals.

During weathering, Mo sulfides are slowly oxidized and yield mainly the MoO_4^{2-} anion which dominates in neutral and moderate alkaline pH ranges, and $HMoO_4^-$ which occurs at lower pH values. However, easily mobile anions are readily coprecipitated by organic matter, $CaCO_3$, and by several cations such as Pb^{2+}, Cu^{2+}, Zn^{2+}, Mn^{2+}, and Ca^{2+}. Also differential adsorption of Mo by Fe, Al, and Mn hydrous oxides contributes to the retention of Mo in surficial deposits. All these reactions are highly dependent on pH and Eh conditions and therefore the net results of Mo migration during weathering may be poorly predictable.

The Mo content of soils usually resembles that of their parent rocks, and ranges from 0.013 to 17.0 ppm in world soils (Table 124). Kubota[435] gave the range of Mo in U.S. soils as from 0.08 to over 30 ppm and the median concentration as slightly more than 1 ppm (Table 125). The world mean of 2.0 ppm calculated from the data presented in Table 124 appears to be reasonable and is supported by the value given by Wedepohl.[855] In general, soils derived from granitic rocks and from some organic-rich shales are likely to contain large amounts of Mo.

200 *Trace Elements in Soils and Plants*

Table 124
MOLYBDENUM CONTENT OF SURFACE SOILS OF DIFFERENT COUNTRIES
(PPM DW)

Soil	Country	Range	Mean	Ref.
Podzols and sandy soils	Australia	2.6—3.7	—	792
	Canada	0.40—2.46	1.5	244
	New Zealand	1—2[a]	—	861
	Poland	0.2—3.0	—	382
	Yugoslavia	0.17—0.51[b]	—	687
	U.S.S.R.	0.3—2.9	1.5	279, 914
Loess and silty soils	New Zealand	2.2—3.1[a]	—	861
	Poland	0.6—3.0	—	381
	U.S.S.R.	1.8—3.3	2.2	274
Loamy and clay soils	Great Britain	0.7—4.5	2.5	819
	Canada	0.93—4.74	1.7	736, 244
	Mali Republic	0.5—0.75	—	39
	New Zealand	2.1—4.2[a]	—	861
	Poland	0.1—6.0	—	382
	U.S.S.R.	0.6—4.0	2	274, 914
Fluvisols	India	0.4—3.1[b]	1.6	645, 730, 455
	Mali Republic	0.44—0.65	—	39
	Yugoslavia	0.35—0.53[b]	—	687
	U.S.S.R.	1.8—3.0	2.4	914
Gleysols	Australia	2.5—3.5	—	792
	Dahomey	—	3.0	617
	India[b]	1.1—1.8	—	544, 645
	Ivory Coast	0.18—0.60	—	650
	Yugoslavia	0.52—0.74[b]	—	687
	U.S.S.R.	0.6—2.0	1.3	914
Rendzinas	India	1.4—1.8	—	455
	Ireland	—	1	236
	Poland	1—3	—	382
	Yugoslavia	0.76—1.03[b]	—	687
	U.S.S.R.	0.6—1.9	1.4	274, 914
Kastanozems and brown soils	Australia	3.5—6.9	—	792
	Ireland	<1—1	—	236
	U.S.S.R.	0.4—2.8	1.3	343, 914
Ferralsols	India	1.3—11.6	—	455, 772
	Ivory Coast	0.4—10.0	—	650
	Madagascar	2.5—17.0	—	557a
Solonchaks	U.S.S.R.	0.9—5.7	2.4	12, 914
Chernozems	U.S.S.R.	1.6—4.6	2.6	274, 914
Meadow soils	U.S.S.R.	1.0—3.9	2.0	12, 274, 914
Histosols and other organic soils	Canada	0.69—3.2	1.9	244
	U.S.S.R.	0.3—1.9	1.2	914
Forest soils	Bulgaria	0.3—4.6	—	39, 188
	China	—	2.2	225
	U.S.S.R.	0.2—8.3	3.1	420
Various soils	Great Britain	1—5	1.2	785, 818
	India	0.013—2.5	—	772

Table 124 (continued)
MOLYBDENUM CONTENT OF SURFACE SOILS OF DIFFERENT COUNTRIES
(PPM DW)

Soil	Country	Range	Mean	Ref.
	Japan	0.2—11.3	2.6	395
	U.S.S.R.	0.8—3.6	2.2	914

[a]   Soils derived from basalts and andesites.
[b]   Data for whole soil profiles.

Table 125
MOLYBDENUM CONTENT OF
SURFACE SOILS OF THE U.S.
(PPM DW)[434,435]

Soil	Range	Mean
Soils over shales	0.3—3.3	2.13
Soils over volcanic ash	0.4—1.8	1.08
Soils on loess	0.75--6.40	2.53
Soils on lacustrine deposits	1.2—7.15	4.14
Alluvial soils	0.4—2.8	0.88
Alluvial soils[a]	1.5—17.8	5.8
Soils on glacial deposits	0.08—4.68	1.22
Soils on granitic alluvium	0.13—0.50	0.35
Soils on calcareous alluvium	0.3—2.0	1.25
Various soils	0.8—3.3	2.0

[a]   Soils from areas of the western states of Mo toxicity
to grazing animals.

The behavior of Mo in soils has been extensively studied because it has a rather unique position among other micronutrients in that it is least soluble in acid soils and readily mobilized in alkaline soils (Figure 58). Vlek and Lindsay[827] and Lindsay[477] studied the behavior of different Mo minerals and concluded that soil solubility of Mo is very close to that of $PbMoO_4$-soil-Pb systems. This mineral (wulfenite), however, cannot be expected to be the most common Mo compound in soil. It seems most likely that a great proportion of soil Mo is associated with organic matter and Fe hydrous oxides.

Krauskopf,[427] Norrish,[570] and Lindsay[477] have reviewed findings on Mo behavior in soils and have concluded that, in the inorganic forms, Mo is associated mainly with Fe oxides, probably as an adsorbed phase. The molybdate adsorbed on freshly precipitated $Fe(OH)_3$ is readily exchangeable, but as the precipitate ages the Mo becomes less soluble and ferrimolybdite ($Fe_2(MoO_4)_3 \cdot 8H_2O$) or other slightly soluble Fe-Mo semicrystalline forms may occur.

The solubility, and thus availability of Mo to plants, is highly governed by soil pH and drainage conditions. Mo from wet alkaline soils is most easily taken up, but the geochemical processes involved in this phenomenon are not completely understood. Apparently, this uptake is related to the high activity of MoO_4^{-2} in an alkaline medium (Figure 58), as well as to an ability to form soluble thiomolybdates under reducing conditions (e.g., MoS_4^{2-}, $MoO_2S_2^{2-}$). Lindsay[477] assumed that plants growing in a reducing environment are capable

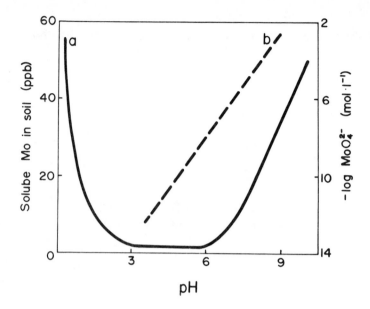

FIGURE 58.   Solubility* of Mo (a) and activity of Mo (b) as a function of soil pH.* Soil containing 1.6 ppm Mo leached with HCl and KOH at the presence of 1 μ*M* CaCl$_2$.[384,475]

of transporting O$_2$ through the stem to the roots, and therefore Mo compounds close to the root surface become more oxidized and thus more soluble.

On acid soils (pH <5.5) low in Mo, and especially on those with a high Fe oxide level, Mo is hardly available to plants. The low availability of Mo that occasionally appears in peat soils seems to be effected by strong fixation of Mo^{5+} by humic acid following the earlier reduction of the MoO$_4^{2-}$ as stated by Mengel and Kirby.[531] On the other hand, organic-rich soils can supply adequate amounts of Mo to plants due to a slow release of this element from organic bound forms.

Liming of acid soils is a common practice to increase Mo availability to plants. However, at higher rates of liming the solubility of Mo may decrease due to its adsorption by CaCO$_3$.[772] The application of Mo salts also increases the available Mo pool in soils, and it is preferable to liming when an increase in soil pH is not desired. Great caution must be taken in Mo fertilization, as this can result in high Mo levels in fodder. The method for assessing the need for Mo application has been extensively described in several textbooks. Ammonium oxalate solution is currently used in many laboratories to extract the available Mo from soil, as well as hot water leaching.[151]

Soils in arid and semiarid regions, especially ferralsols, generally have higher Mo contents. However, in humid and temperate regions, soils over Mo-rich material may also contain hazardous amounts of Mo. These soils, and soils polluted with Mo, need amelioration to depress the Mo availability to plants. The application of S is most effective, whereas increasing the soil P is known to stimulate Mo uptake by plants.

Industrial pollution (mining, smelting, processing of metals, and oil refining) may be responsible for elevated Mo concentrations in soils. Hornick et al.[324] found 35 ppm Mo in the soil surrounding a Mo processing plant. Some sewage sludges contain elevated amounts of Mo (up to 50 ppm) and may have an impact by both enriching the total Mo content and increasing the solubility of soil Mo.[314,452] Also fallout or the application of fly ash from some coal-fired power plants should be considered as potential sources of Mo, since even if only small amounts of Mo are added, its availability is increased by the alkaline reaction of the fly ashes.

B. Plants

1. Absorption and Biochemical Functions

Mo is an essential micronutrient, but the physiological requirement for this element is relatively low.

Plants take up Mo mainly as molybdate ions, and its absorption is proportional to its concentration in the soil solution. Although there is no direct evidence, there is a suggestion of the active uptake of Mo.[548] Mo is moderately mobile in plants, but the form of translocated Mo is unknown. Tiffin[788] discussed the possibility of organic complexing, mainly of the Mo-S amino acid complex that was found in the xylem fluid.

Mo, the essential component of nitrogenase and nitrate reductase, is also present in other enzymes (oxidases) that catalyze diverse and unrelated reactions. The basic enzymatic role of Mo is related to its function as redox carrier and is apparenly reflected in the valency change between Mo^{6+} and Mo^{5+}.[564]

Normally there is 1 ppm (DW) or less of Mo in leaf tissues, whereas nodulated roots contain several times these concentrations. Most of this Mo apparently is in the nitrate reductase of root and shoot and in the nitrogenase of the nodule bacterioids. Two Mo-containing enzymes in N metabolism are involved in either N^2 fixation or NO_3 reduction. Thus, the requirement of plants for Mo appears to be related to the N supply; plants supplied with NH_4-N have less need for Mo than those utilizing NO_3-N.

Mo is known to be essential to microorganisms, and some bacteria species are able to oxidize molybdenite in soils. Rhizobium bacteria and other N-fixing microorganisms have an especially large requirement for Mo.

Since the most important function of Mo in plants is NO_3 reduction, a deficiency of this micronutrient causes symptoms similar to those of N deficiency. Some plants, however, show also more specific Mo deficiency symptoms (e.g., "whiptail" in cauliflower) (Table 27).

A deficient Mo content of plants depends on various factors, in particular, on interactions with other elements. Requirements for Mo are generally met at concentrations within the range of 0.03 to 0.15 ppm (DW) in tissues of most plants; only some leguminous crops require more Mo (Table 25).

Although easily soluble soil Mo is also readily taken up by plants and some plant species are known to accumulate much Mo, its toxicity symptoms in plants under field conditions are very rare, whereas toxicities to animals feeding on forages high in this element are well known. The only relatively high concentration of Mo that was toxic, 135 ppm (DW), was reported by Davis et al.[171] for young spring barley.

Crop response, particularly of leguminous crops, to Mo application has been widely reported for soil conditions throughout the world, especially for acid and ferralitic soils. While correction of Mo deficiencies may be accomplished by Mo application as soil, foliar, or seed treatments, care must be taken in its use since Mo may be very toxic, even at quite low concentrations, in fodder plants. Therefore, the preferable control of Mo deficiency is liming the soil to a pH of around 6.5.

2. Interactions with Other Elements

Several complex interactions between Mo and other elements are observed within plant tissues and also in the external root media (Figures 16). The most important interactions are those between Cu, Mo, and S, which are differentially governed by diverse factors. The Mo-Cu antagonism in plants is strongly related to N metabolism (Chapter 6, Section V.B.3.). Mo-S relations may have antagonistic or stimulative effects on Mo uptake by plants, depending on the rate of application of S-containing fertilizers.[581] However, a wide range of combinations between Cu, Mo, and S in herbage and, further, in animal nutrition, may occur. Gartell[257] has reviewed this topic and made a few generalizations showing that soil

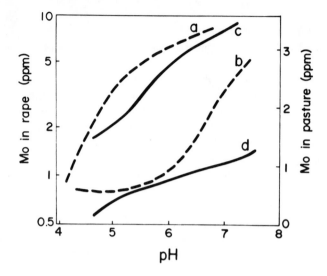

FIGURE 59.   The effect of soil pH on the Mo content of (a) pasture
on peat bog soil; (b) pasture on mineral soil; (c) rape, shoots; and (d)
rape, roots.[275,384]

factors that increase the availability of Mo to plants usually have inhibitory effects on the
Cu uptake by plants and that the physiological barrier to Mo uptake by plants is much less
effective than that to Cu uptake. Nutritional effects of Cu/Mo ratios in pastures on animal
health is controlled by the sulfate concentration in plants and an increased level of SO_4^{2-}
may reduce Cu absorption with even small amounts of Mo. It has been established that soils
with 5 ppm Mo can be associated with growth retardation and lower reproduction in cattle
due to the Mo-Cu nutritional relationship.

Mo-Mn antagonism resulting from soil acidity influences the availability of these elements,
therefore, liming can correct both Mo deficiency and Mn toxicity. Mo-Fe interactions are
demonstrated as low Mo availability in Fe-rich soils, whereas increased Mo levels may
induce Fe deficiency, or accentuated Mn-induced Fe chlorosis. Although mechanisms of the
various interactions are not well understood, the formation of Fe-Mo precipitates within root
tissues may be responsible for low Mo translocation.[581]

Mo-W and Mo-V metabolic interactions are not precluded, since a substitution is possible
between these elements in biochemical processes. A Mo-P interaction is often demonstrated
as an enhancing effect of P on Mo availability in acid soils, apparently due to both a higher
solubility of the phosphomolybdate complex, as well as to a higher Mo mobility within
plant tissues. However, reported effects of P fertilizers on Mo availability are contradictory,
and while the ordinary SO_4-containing superphosphate reduced Mo uptake, concentrated
superphosphate increased Mo uptake.[3] Thus, Mo-P interactions are variable and highly
governed by diverse soil factors and are also related to plant metabolic processes. Mo-Ca
interactions are complex and highly cross-linked with the range of the soil pH.

3. Concentrations in Plants

No simple relationship is apparent between the total Mo content of soils and plants,
although some authors have reported a linear relationship for Mo content of herbage and its
total concentration in soil. The Mo concentrations in plants closely reflect the soluble Mo
pool, for Mo seems to be very readily absorbed by plants when present in soluble forms.
The Mo uptake by plants is a function of soil pH (Figure 59). Usually a positive correlation
is observed between the relative Mo uptake and the soil pH.[784] This has been clearly
demonstrated by Doyle et al.,[193] who found abnormally high Mo values (up to 52 ppm,

Table 126
MOLYBDENUM CONTENT OF GRASSES AND
LEGUMES FROM AREAS WHERE Mo TOXICITY IN
GRAZING ANIMALS WAS NOT OBSERVED (PPM DW)

Country	Grasses Range	Grasses Mean	Legumes Range	Legumes Mean	Ref.
Bulgaria	—	—	0.04—0.32	0.18	220
Canada	0.4—8.0	1.3	—	—	192
Czechoslovakia	—	0.3	—	—	154
Finland	0.14—0.80	0.38[a]	—	—	221
	0.23—0.91	0.45	0.20—1.3	0.70	591, 727
East Germany	0.08—1.04	0.33	0.21—5.0	0.80	31, 65
West Germany	0.42—0.88	0.70	—	—	596
Great Britain	0.5—4.0[b]	—	—	—	784
	0.25—1.47	0.56	—	—	543
Ireland	0.18—0.77	—	0.28—0.52	—	235
Japan	0.04—3.05	0.72	0.01—3.64	0.92	770
Poland	<0.02—1.68	0.33	0.02—3.56	0.50	915
Sweden	0.2—4.8	1.4	0.3—20.5	2.5	384
U.S.[c]	0.5—2.0	—	<0.7—15.0	1.8	710
U.S.	—	—	0.01—3.46	0.73	435
U.S.S.R.	—	—	1.31—3.61	2.3	337

[a]   Grass heavily fertilized with N.
[b]   Pasture herbage.
[c]   Calculated from AW basis.

mean 11 ppm) in native plants grown on neutral or alkaline soils, whereas on Mo-rich acid soils, and Mo-low soils, Mo concentrations in the same varieties of plants averaged, respectively, 0.9 and <0.2 ppm.

Some native plants, particularly leguminous species, are known to accumulate as much as about 500 ppm (AW) Mo, or about 350 ppm (DW), without showing toxicity symptoms.[556,618] The Mo content of forages is of special concern, therefore, most available data are on the Mo status in grasses and legumes. Mean Mo levels range from 0.33 to 1.5 ppm (DW) in grasses and from 0.73 to 2.3 ppm in legumes from different countries. In areas where Mo toxicity was observed in grazing animals, the Mo content of grasses ranged from 1.5 to 5.0 ppm (DW) and in legumes from 5.2 to 26.6 ppm (DW) (Tables 126 and 127). In pasture, Mo is known to vary with different stages of plant development and with periods of the growing season. The reported Mo content was high in the spring and autumn seasons.[384]

Plant foodstuffs contain variable amounts of Mo within the range 0.0018 to 1.23 ppm (FW), 0.07 to 1.75 ppm (DW), and 0.53 to 30 ppm (AW), with legume vegetables being in the upper range and fruits being in the lower range (Table 128). The Mo content of cereal grains averages 0.49 ppm (DW) and does not show a great variation under widely ranging field conditions (Table 129).

The ready availability of Mo causes a great increase in uptake when plants are grown in contaminated sites. Karlsson[384] found up to 200 ppm DW of Mo in pasture plants from the vicinity of a metallurgical factory. Hornick et al.[324] reported that plants grown on the Mo-polluted soil near a Mo processing plant accumulated this element in concentrations ranging from 124 to 1061 ppm (DW) in lettuce and cabbage, respectively. Furr et al.[250] found that beans and cabbage grown on soil amended with municipal sludge ashes had an increased Mo content up to 18 and 19 ppm (DW) contrasted to the control values of 0.8 and 0.4 ppm, respectively. Thus, elevated Mo concentration in soils resulting from industrial pollution or

Table 127
MOLYBDENUM CONTENT OF FORAGE
PLANTS FROM AREAS WHERE Mo TOXICITY
IN GRAZING ANIMALS WAS OBSERVED
(PPM DW)

Country	Grasses Range	Grasses Mean	Legumes Range	Legumes Mean	Ref.
Canada	0.6—17.0	4.0	1.0—20.0	5.2	192
	2.4—12.0	5.0[a]	4.8—6.0	5.4[a]	237
Great Britain	0.1—7.2	1.5[b]	—	—	784
Sweden	1—234[b]	—	—	—	384
U.S.	0.7—6.8	3.7	18.9—39.6	26.6	434
U.S.S.R.	10—50[b]	—	—	—	420

[a]   Samples from the periphery of a Mo ore body.
[b]   Pasture herbage.

Table 128
MOLYBDENUM CONTENT OF PLANT FOODSTUFFS (PPM)

Plant	Tissue sample	FW basis	DW basis	AW basis
Sweet corn	Grains	0.045(a)	0.18(a)	6.9(a)
Kidney bean	Seeds	—	0.9—1.6(b)	—
Pea	Seeds	—	1.2—1.75(c)	—
Snap bean	Pods	0.23(a)	2.1(a)	30.0(a)
Carrot	Roots	<0.08(d)	0.04(e)	<7.0(a)
Sugar beet	Roots	—	0.45—0.75(e)	—
	Leaves	—	0.39(e)	—
Lettuce	Leaves	0.005(d)	0.074(a)	0.53(a)
Cabbage	Leaves	<0.099(d)	0.85(a)	9.1(a)
Potato	Tubers	0.047(a)	0.25(a), 0.15(e)	5.9(a)
Onion	Bulbs	0.024(a)	0.24(a), 0.16(e)	5.6(a)
Cucumber	Fruits	<0.087(d)	0.82(a)	8.3(a)
Tomato	Fruits	0.042(a)	0.82(a)	6.8(a)
Apple	Fruits	0.0018(d)	0.07(a)	3.9(a)
Orange	Fruits	0.014(d)	0.11(a)	3.1(a)
Tea	Leaves	—	0.2—0.3(f)	—

Note: Sources are as follows: a, 705; b, 589; c, 337; d, 574; e, 727; and f, 910.

agricultural practices may be locally responsible for the pronounced abnormal Mo content of plants.

VII. TUNGSTEN

The distribution of W in the earth's crust shows that its concentration seems to increase with increasing acidity of magmatic rocks and with increasing clay content of sedimentary rocks (Table 115). The common W contents of acid granitoids and of argillaceous sediments range from 1 to 2 ppm, whereas in mafic rocks and sandstones or limestones the W range is 0.5 to 1.1 ppm.

All W minerals are only slightly soluble and hence have low mobility in hypergenic environments. Shcherbina[716] reported that WO_4^{2-} as well as W-complexed compounds may

Table 129
MOLYBDENUM CONTENT OF CEREAL
GRAINS FROM DIFFERENT COUNTRIES
(PPM DW)

Country	Plant	Range	Mean	Ref.
Canada	Barley	—	0.29	293
	Oats	—	0.41	293
	Wheat	—	0.18	293
Czechoslovakia	Barley[a]	0.28—0.41	0.32	563
	Oats	0.59—0.84	0.69	563
	Wheat[b]	0.18—0.42	0.32	563
Finland	Barley	0.19—0.49	0.33	727
	Oats	0.21—0.59	0.40	727
	Rye	0.23—0.57	0.40	727
	Wheat[b]	0.14—0.38	0.26	727
Norway	Barley	0.02—0.59	0.16	446
	Wheat	0.07—1.09	0.29	446
Poland	Wheat[b]	0.20—0.60	0.35	267, 335
Sweden	Barley	0.54—1.00	0.79	384
U.S.	Barley	0.58—2.4	0.92	200
	Oats	0.28—1.9	0.88	200
	Wheat[a]	0.08—1.1	0.49	200
	Wheat[b]	0.40—1.1	0.64	200
U.S.S.R.	Barley	0.69—0.75	0.72	337
	Oats	0.42—0.62	0.50	337
	Rye	0.12—1.31	0.50	337, 501
	Wheat	0.26—1.31	0.64	337

[a] Spring wheat.
[b] Winter wheat.

be readily transported in a particular geochemical environment. Presumably, the geochemical behavior of W resembles that of Mo.

There is a scarcity of data concerning the occurrence of W in soils. Ure et al.[819] found a range of W from 0.68 to 2.7 ppm in Scottish soils derived from different parent material, with the highest value for soil over quartz-mica schist and the lowest in the soil over serpentine. Furr et al.[250] reported 1.2 and 2.5 ppm W in two soils of the U.S. These authors gave the W content of two sludge ashes as 43 and 186 ppm.

Apparently, like Mo, W is easily available to plants under certain conditions, therefore, its concentrations are likely to be elevated in plants growing on soil overlying W ore bodies. Wilson and Cline[882] found that barley grown in soils with added ^{185}W removed a large proportion of that radionuclide and that the lower uptake from acid soil suggested that plants probably take up an anionic form, WO_4^{2-}. Also Bell and Sneed[62] reported a high accumulation of ^{185}W, released from nuclear reactions, in plant roots.

Shacklette et al.[710] reported that Curtin and King[916] found the range in W concentrations in trees and shrubs of the Rocky Mountains in Colorado to be 5 to 100 ppm (DW), with the highest value in juniper stems. Connor and Shacklette[145] found a W range from 30 to 70 ppm in ash of single tree samples. Duke[197] reported the range of W in various food plants to be <0.001 to 0.35 (DW). Bowen[94] gave the overall range of W as 0.0 to 0.15 ppm (DW) in land plants, 0.01 to 0.15 in edible parts of vegetables, and 0.02 to 0.13 ppm in lichens and bryophytes. Furr et al.[249] found W to range from 0.7 to 3.5 ppm (DW) in vegetables growing on soil amended with fly ash.

Gough et al.[279] reviewed the possible toxicity of W to plants, which appears to be moderate. Plants growing in a mineralized zone contained up to 18 times the background W value of 2.7 ppm without showing toxicity symptoms.

There are some suggestions of a possible biological significance of W in plants. Nicholas[564] mentioned that when W is subtituted for Mo in the nitrate reductase enzyme, it has no catalytic activity. Apparently, W^{6+} is more stable than Mo^{6+} and does not readily change its valence. Zajic[898] described the antagonistic influence of W on Mo activity in N-fixation, but this effect is differentially governed by the pH of the media.

Chapter 12

ELEMENTS OF GROUP VII

I. INTRODUCTION

Geochemical and biochemical characteristics of the elements in Group VII are widely diverging. Trace elements of the Subgroup VIIa, the halogens (F, Br, I), are much less common in the biosphere than Cl which some authors tend to include also as a trace element. The last element of this subgroup, astatine (At), occurs as radioactive isotopes of very short lives. In nature halogens form simple anions or, except F, combine with oxygen anions and play a significant role in biochemical processes. These anions are highly mobile in the earth's crust although they are considered to reveal lithophile properties.

Of the Subgroup VIIb, Mn is the only element essential to living organisms and is common in natural environments. Technetium (Tc) does not occur in nature because all of its radioactive isotopes are short-lived. Metallic Tc is a product of fission and reveals chemical features similar to Re. Re is a dispersed element in nature and, like Mn, has variable valency and shows lithophile or chalcophile characters.

II. FLUORINE

A. Soils

F is a typical lithophile element under terrestrial conditions, and its larger concentration (850 to 1200 ppm) is found in intermediate and acid siliceous igneous rocks (Table 130). In sediments F is known to be associated with clay fractions and therefore is likely to occur in larger amounts in argillaceous deposits.

There are not many stable F minerals; the most common are topaz ($Al_2(F,OH)_2SiO_4$) and fluorite (CaF_2). Free F may sometimes occur in rocks in gaseous nebulae. Compounds of F are also important constituents of magmatic gases and volcanic exhalations.

F reveals an affinity to replace hydroxyl groups in minerals, and these reactions have resulted in fluoroapatite ($Ca_{10}(PO_4)_6F_2$), the most common F mineral, and have also been responsible for increased amounts of F in amphiboles, micaceous minerals, etc. A strong association of F with phosphates is also observed in both primary and secondary minerals.

During weathering F combined with siliceous minerals remains in residual materials. Fluoroapatite and fluorite dissolve slowly, while cryolite (Na_3AlF_6) and similar minerals are readily soluble. Mobile F is easily sorbed by clays and phosphorites.

The behavior of F in soils has been studied by Larsen and Widdowson,[461] Perrott et al.[607] Chhabra et al.,[137] and Omueti and Jones.[584] The results obtained show that the mobility of F in soils is complex and that the predominant factors controlling the level of this ion in the soil solution are the amount of clay minerals, the soil pH, and the concentrations of Ca and P in soils. In general, the greatest adsorption of F by soil mineral components is either at the distinct acid range of pH, or at about pH 6 to 7 (Figure 60).

Fluoroapatite is considered to be the most common form of F in soils. However, several fluorides (e.g., CaF_2, AlF_3) and alumino-silicates (e.g., $Al_2(SiF_6)_2$) are also reported to occur in soils. F is also known to form complex ions with Al (AlF^{2+}, AlF_2^+, AlF_4^-), therefore, the mobile F^- in the soil solution might control the activity of Al^{3+} in soils.[477]

As F is known to readily replace hydroxyls of clays, the clay minerals, illites in particular, are believed to be the seat of most of this element in soils. The range of F content of illite and chlorite mineral groups is reported as 0.1 to 2.3% by Thomas et al.[781]

Under natural soil conditions F is slightly mobile, but it is not accumulated in the surface horizon, especially of acid soils. The high solubility of F in acid soils is attributed to the

Table 130
FLUORINE, BROMINE, IODINE, AND MANGANESE IN MAJOR ROCK TYPES (PPM) (VALUES COMMONLY FOUND, BASED ON VARIOUS SOURCES)

Rock type	F	Br	I	Mn
Magmatic rocks				
Ultramafic rocks	50—100	0.2—1.0	0.01—0.50	850—1500
Dunites, peridotites, pyroxenites				
Mafic rocks	300—500	0.5—3.0	0.08—0.50	1200—2000
Basalts, gabbros				
Intermediate rocks	500—1200	1—4	0.3—0.5	500—1200
Diorites, syenites				
Acid rocks	520—850	0.3—4.5	0.2—0.5	350—600
Granites, gneisses				
Acid rocks (volcanic)	300—700	0.2—1.0	0.1—0.5	600—1200
Rhyolites, trachytes, dacites				
Sedimentary Rocks				
Argillaceous sediments	500—800	5—10	1.0—2.2	400—800
Shales	500—800	6—10	2—6	500—850
Sandstones	50—270	1—5	0.5—1.5	100—500
Limestones, dolomites	50—350	6 ·	0.5—3.0	200—1000

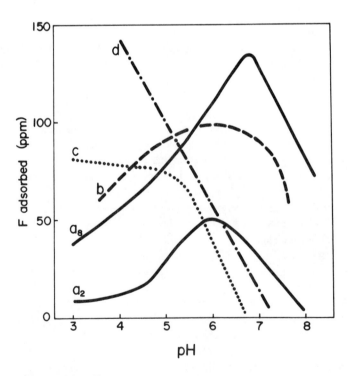

FIGURE 60.   Adsorption of F by soil and clay as a function of the pH. (a_2 and a_8) Soil equilibrated against 2 and 8 mg F^- per 1-ℓ solution, respectively; (b) bentonite; (c) bauxite; (d) kaolinite. All clays equilibrated against 2 mg F^- per 1-ℓ solution.[583]

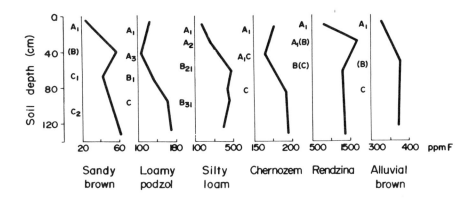

FIGURE 61. Distribution of F in the profiles of different soils developed under humid climate. (Letters indicate genetic soil horizons.)

occurrence of readily soluble fluorides such as NaF, KF, NH_4F, whereas AlF_3 is known to be of low solubility. Thus, the increasing F content with depth reflects the response to the soil pH; however, the enriched clay content usually contributes more to the F distribution in soil profiles. This relationship with clay content has been reported by Piotrowska and Wiacek[619] and Omueti and Jones.[584]

In calcareous soils the formation of slightly soluble CaF_2 and F complexes with Fe, Al, and Si are responsible for the low migration of this element. In sodic soils, on the other hand, a high exchangeable Na affects increased solubility of F.

The F concentration in soils is inherited from parent material, whereas its distribution in soil profiles is a function of soil-forming processes, of which the degree of weathering and clay content are the most pronounced. Apparently, F has been lost from the surface horizons of most soils (Figure 61) and this reflects partly a low F affinity for organic matter. Omueti and Jones[584] gave the range in F concentrations in organic matter of surface horizons to be as low as 0.03 to 0.12 ppm.

The average F content of world-wide soils has been calculated to be 320 ppm and to be 360 ppm for U.S. soils (Tables 131 and 132). In general, the lowest F amounts are in sandy soils in a humid climate, whereas higher F concentrations occur in heavy clay soils and in soils derived from mafic rocks. The range for most normal soils seems to be from 150 to 400 ppm, but the overall variation is much broader, and in some heavy soils F levels above 1000 ppm have been found. Much higher levels of F in uncontaminated soils are reported for provinces of endemic fluorosis.[73]

Naturally occurring F in soils is slightly available to plants. Hall[300] found that in particular tropical soils, organically bound F (monofluorinated compounds) may occur. These compounds are shown to be highly available to plants, and highly toxic to animals. Their origin, however, is not yet clear, but presumably they are synthesized by certain microorganisms.

The F-emitting industrial sources are mostly Al smelters and phosphate fertilizer factories, but also coal combustion, steel works, brick yards, and glass works may significantly contribute to the total F pollution. Moreover, artificially high soil F levels can occur through contamination by the application of phosphate fertilizers or sewage sludges, or from pesticides (Table 133). Assuming a high F content of phosphate fertilizers (Table 5), an input of F to arable soils may be of ecological importance.

The F compounds added to soils by pollution are usually readily soluble and thus available to plants. Most of the added F to soil is either effectively fixed by soil components (clays, Ca, and P) or readily removed from light soils by water. Nevertheless, easily soluble F-bearing fertilizers (e.g., potassium fluoroborate) or sewage sludges may effect a remarkable increase in the bioaccumulation of F from soils.[166,857]

Table 131

**FLUORINE CONTENT OF SURFACE SOILS OF DIFFERENT
COUNTRIES (PPM DW)**

Soil	Country	Range	Mean	Ref.
Podzols and sandy soils	Poland	20—150	83	619
	Sweden	41—198	90	568
Loess and silty soils	Poland	122—228	175	619
Loamy and clay soils	Poland	250—750	418	619
	Sweden	248—657	450	568
	Great Britain	110—700	462	819
Fluvisols	U.S.S.R.	1175—1360	—	73
Soils on mafic rocks	Great Britain	470—680	566	819
Solonchaks	U.S.S.R.	444—1024	—	73
Chernozems	U.S.S.R.	454—1194	—	73
Histosols and other organic soils	Sweden	42—123	73	568
	U.S.S.R.	333—335	—	73
Forest soils	U.S.S.R.	463—652	560	568
Various soils	Great Britain	42—490	266	818
	West Germany	80—1100	460	855
	Japan	260—520	370	855
	U.S.S.R.	30—320	200	855

Table 132

**FLUORINE CONTENT OF SURFACE SOILS OF THE
U.S. (PPM DW)**

Soil	Range	Mean
Sandy soils and lithosols on sandstones	<10—1100	205
Light loamy soils	30—810	355(a)
Loess and soils on silt deposits	30—600	395
Clay and clay loamy soils	<10—800	410(a)
Alluvial soils	<10—1200	465
Soils over granites and gneisses	20—540	285
Soils over volcanic rocks	130—800	405
Soils over limestones and calcareous rocks	<10—840	360
Soils on glacial till and drift	120—600	440
Light desert soils	110—810	415
Silty prairie soils	120—600	425
Chernozems and dark prairie soils	<10—940	350
Organic light soils	<10—300	250
Forest soils	20—900	305
Various soils	<10—1050	300(a)

Note: Sources are as follows: 200, 706; a, 262.

The most important hazard of F contamination in soils concerns changes in soil properties due to the great chemical activity of hydrofluoric acid which is temporarily formed from both solid and gaseous F pollutants. Bolewski et al.[85] reported the decomposition of clays and other silica minerals in soil having heavy F pollution. They also reported the destruction of humic mineral complexes resulting in a significant loss of organic matter in soils. The reduction of enzymatic activity of some soil microorganisms with the addition of NaF has been reported by Russel and Świecicki.[670] Thus, the availability of F in polluted soils is a function of a number of soil characteristics and is not the only cause of limited plant growth on those soils.

Presently great attention is given to the possible formation of highly toxic organic F

Table 133
FLUORINE CONTAMINATION OF SURFACE SOILS
(PPM DW)

Site and pollution source	Maximum or range of content	Country	Ref.
Old mining area	2000	Great Britain	166
Al-processing industry	1350	Czechoslovakia	500
	1500—3200	Poland	85
Other metal-processing industry	305—345	Poland	512
P fertilizer manufacturer or application	308—2080	Canada	478, 782
	385[a]	Poland	620

[a] After 10 years of P fertilizing, against the background value of 296 ppm F.

compounds that can be synthesized by both higher plants and microorganisms under various soil conditions.[288] Athough a high level of F in soils is not in itself harmful to plants, F-polluted soils should be ameliorated for a proper growth of vegetation. The application of materials which increase the soil pH and the sorption capacity or fixation ability for F ions is known to improve soil properties and to limit F uptake by plants.

B. Plants

1. Absorption and Transport

The availability of F to plants usually is not related closely to the total or soluble F content of a soil. However, under certain soil and plant conditions the F content of plants seems to reflect its occurrence in soils.[723] Bieliyakova[73] gave the ratio of F in plant ash to F in topsoils as 0.2 and 0.6 for cultivated and natural vegetation, respectively. These values indicate a relatively low F bioavailability.

The soluble F fraction in soil is taken up passively by roots and apparently is easily transported in plants. This statement is supported by the observation of a ready F uptake from fields irrigated with F-containing waters.

Vandeputte[917] found that F absorption from the solution media by alfalfa and wheat is positively correlated with F concentration. He obtained results that indicated a very high accumulation in the grains (up to 3250 ppm DW at 50 ppm F^- in solution). This phenomenon reflected a possible influence of the formation of an F complex that may be more easily taken up by plants than are F ions.

Although it has been shown that plants can take up F quite easily from polluted soils, the bioavailability of soil F is of much less significance than that from airborne compounds. The effects of atmospheric F depositions, both in soils and on plant surfaces, have been extensively studied. The results, as summarized by Groth[288] and Weinstein,[857] have indicated that the deposition of airborne F on soil has little or no effect on the F content of plants, but that this conclusion is not a unanimous one. However, when F is present as both an air pollutant and a soil pollutant, the F uptake by plants from air is much more significant than their uptake from soil. Several factors affect plant accumulation of airborne F, but the most pronounced are atmospheric F concentration and the duration of exposure (Figure 62).

Foliar uptake of gaseous F apparently follows the path of the penetration of other gases into the leaf. Chamel and Garbec[125] reported that F penetration through the cuticle is slight; however, under natural conditions most leaf surfaces have open stomata and also show breaks and punctures that facilitate the F uptake by foliage. Soluble F compounds are also known to be absorbed through the cuticle when deposited on the leaf surface. When F is

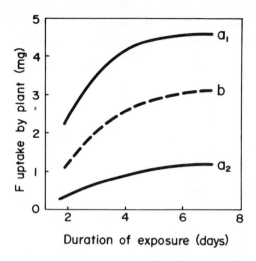

FIGURE 62.   F uptake by plants as a function of the
F contamination in air and of exposure time. (a_1 and a_2)
Beans (tops) at the F concentration of 11 and 5 $\mu g\ m^{-3}$,
respectively; (b) tomato plants at F concentration of 5
$\mu g\ m^{-3}$.[381]

accumulated in vegetative tissues during exposures to gaseous HF, it is not translocated to the developing grains.[499] Biological factors such as plant species, stage of development, and others are very important determinants of F accumulation.

2. Biochemical Roles

Centuries ago symptoms of F toxicity to animals were observed after volcanic eruptions. During the last century the role of plants as a sink for emitted F and in its transfer to animals was recognized. Contemporary investigation on grasses collected immediately after the eruption of Hekla volcano in Iceland in 1970 showed that their F content was 4300 ppm (DW), and that it rapidly decreased during 40 days to a concentration of less than 30 ppm.[598]

At present, F is considered to be the most hazardous and the most phytotoxic trace pollutant among the common air pollutants such as O_2, SO_2, and NO_x. The phytotoxicity of airborne F is influenced by ecological and biological factors and also by the physical and chemical characteristics of the pollutant.

The literature on the biological aspects of plant pollution by F has recently grown at an astonishing rate. Among other excellent monographs, the effects of atmospheric F on plants have been extensively reviewed by Groth,[288] Kluczyński,[401] and Weinstein.[857]

There is no evidence of F phytotoxicity when this element is absorbed by roots, whereas airborne F, especially when it occurs as hydrofluoric acid, is highly toxic. The most emphasized effects of F on plant metabolism are related to:

1.   Oxygen uptake decrease
2.   Respiratory disorder
3.   Assimilation decrease
4.   Reduction in chlorophyll content
5.   Inhibition of starch synthesis
6.   Inhibition of pyrophosphatase function
7.   Altered metabolism of cell organelles
8.   Injured cell membranes
9.   Disturbance of DNA and RNA
10.   Synthesis of fluoroacetate, a most hazardous F compound

Although Shkolnik[718] and Weinstein[857] have reviewed studies on plant growth stimulated by F, neither a plant requirement for this element nor its essential role in plant metabolism has been established. As Weinstein[857] described, the apparent stimulation of some isoenzymes (e.g., acid phosphatase, dehydrogenase) by HF fumigation may, in fact, be the result of the inhibition of other enzymes.

Interactions between F and other elements are not unanimously agreed upon. There is some evidence that an increased F content is likely to inhibit the absorption of several nutrients, whereas also a synergistic effect, mainly to P uptake, has been observed (Table 31). In general, a sufficient supply of major nutrients increased plant tolerance to F.[47]

Mutual pollution by SO^2 and F is reported to cause more significant damage to pine trees than changes induced separately by these pollutants.[663] The reactions of plants exposed to F pollution, even before any visible symptoms of F toxicity occur, are retarded growth, inhibited reproduction, and yield reduction. However, the most significant effect of increased F concentrations in plants is toxicity to animals. Thus, the F content of forage crops seems to be of greater ecological concern than that of other plants. However, recent findings of increased F levels in food plants, especially of fluoro-organic residues in several vegetables, indicate a possible toxicity to man.[539]

The best-documented plant responses to F are foliar symptoms such as chlorotic and necrotic lesions and deformation of fruits. These symptoms are not specific, and mimicking symptoms induced by other agents as well as the increased susceptibility of F-polluted plants to microbial diseases have been observed by Treshow[805] and Weinstein.[857]

Plants exhibit a broad range of tolerances to foliar injury by F pollution. Several lists of the tolerant and susceptible plant species are given in the literature. Commonly listed as tolerant are asparagus, bean, cabbage, carrot, and willow; while barley, corn, gladiolus, apricot, pine, and larch are classified as susceptible. However, a great variability in plant response to F accumulation in tissues is observed even between cultivars or genotypes of the same species. The response of plants is also highly dependent on several environmental and biologic factors. Based on Weinstein's report,[857] it can be generalized that susceptible plants would be injured by foliar F concentrations that ranged between 20 and 150 ppm (DW), intermediate plants can probably tolerate an F content in excess of 200 ppm, and highly tolerant plants do not exhibit injury below 500 ppm.

The effects of plant F on humans and, particularly, on animals is of the greatest concern. Opinions vary as to the toxic threshold values of F in forage, but 30 to 40 ppm (DW) is commonly reported.[166,288,766] More tolerant animals (e.g., lambs, turkeys) may tolerate a higher F concentration in their rations.

3. Concentrations in Plants

The F content of plants has been investigated for a number of reasons such as assessing its hazard to grazing animals, the diagnosis of plant injury, and for monitoring airborne F pollutants. Most of the information on F accumulation in plants has been determined for forage crops and native vegetation. Data on F concentrations in food plants are rather limited (Table 134). Higher concentrations of F have usually been reported for the aerial parts of plants. The F content of plants grown in uncontaminated areas is very unlikely to exceed 30 ppm (DW).

According to the general opinion, plant F seems to be positively correlated with the concentration of F in rainwater. Davison et al.[173] discussed a method of the prediction of F concentrations in pasture using multiple regression techniques and including data on aerial deposition of F and rainfall. Several plants, especially forage vegetation, when growing in polluted areas are reported to contain large amounts of F (Table 135).

Table 134
FLUORINE CONTENT OF FOOD AND FORAGE PLANTS (PPM DW)

		F range or mean content as given for different countries			
Plant	Tissue sample	West Germany[710]	Sweden[568]	U.S.[288]	Other countries[73,146,173,381]
Barley	Grains	1.7	0.5—5.5	<1—2	—
Oats	Grains	0.5	0.2—0.9	<1—2	—
Wheat	Grains	1	0.4—1.4	<1—2	—
Corn	Grains	0.2—0.4	—	—	—
Cabbage	Leaves	1.5	—	—	311[a]
Lettuce	Leaves	4.4—11.3	—	—	—
Spinach	Leaves	1.3—28.3	—	—	—
Carrot	Roots	2	—	—	—
Beet, red	Roots	4—7	—	—	<300[a]
Onion	Bulbs	3	—	—	<300[a]
Potato	Tubers	1.5—3	0.1—1.1	—	69[a]
Apple	Fruits	1.3—5.7	—	—	7
Pear	Fruits	2.1—4.4	—	—	—
Peach	Fruits	0.21	—	—	—
Alfalfa	Aerial parts	—	1.5—3.2	1—9	—
Clover	Aerial parts	6.7	2.8—7.8	—	—
Grass	Aerial parts	6.8	—	3—6	5—18, 310[a]
Forages	Aerial parts	—	0.3—1.3	4—17	36—98

[a] AW basis.

Table 135
EXCESSIVE LEVELS OF FLUORINE IN PLANTS GROWN IN CONTAMINATED SITES (PPM DW)

Site and pollution source	Plant and part	Maximum or range of content	Country	Ref.
Mine waste	Grass, tops	130—5450	Great Britain	146
Al-processing industry	Vegetation, foliage	396	Canada	478
	Shrub[a], leaves	150—500	Australia	557
	Cereals, grains	14—36	Poland	63
	Clover, tops	14—173	Poland	768
	Grass, tops	1330	Czechoslovakia	500
	Grass, tops	75—340	Poland	381
	Birch, leaves	230	Norway	261
	Pine, needles	48	Norway	261
	Lichens	27—241	Great Britain	606
Brick kilns	Pasture plants	160	Great Britain	166
Fiberglass plant	Vegetation, foliage	945	Canada	478
Phosphate rock processing	Tree,[b] foliage	71—900	Canada	782
	Vegetation, foliage	70	Canada	478
	Sagebrush, tops	100—360[c]	U.S.	278
Fumigation with F	Alfalfa	1327	—	857
	Grass	496	—	857

[a] *Melaleuca nodosa.*
[b] Balsam fir.
[c] AW basis.

Table 136
BROMINE CONTENT OF SURFACE SOILS OF
DIFFERENT COUNTRIES (PPM DW)

Soil	Country	Range	Mean	Ref.
Podzols	Poland	7.9—8.2	—	181
Volcanic ash soils	Japan	50—104	—	897
Gleysols	Great Britain	—	34	875
Brown earths	Great Britain	—	52	875
Forest soils	Japan	68—130	—	897
Various soils	Great Britian	10—515	50	818, 875
	Norway, Northern	16—100	45	445
	Norway, Eastern	5—14	7	445
	U.S.	<0.5—6	—	706
	U.S.S.R.	1—38	11	855

III. BROMINE

A. Soils

The common abundance of Br in the earth's crust varies within the range of 0.2 to 10 ppm, being the highest in argillaceous sediments (Table 130). Most Br compounds are similar to their corresponding Cl compounds, and therefore the close correlations of Cl and Br in various rock types have often been reported. The Br/Cl ratio is an important factor for defining various geological units.[855]

There are very few naturally occurring Br minerals (e.g., AgBr, Ag(Br,Cl,I)), and several polyhalides which are known to be unstable in hypergenic environments. Br is a very easily volatile element and its salts are readily soluble. Therefore, its geochemistry is closely related to water chemistry and to evaporite deposits.

Coal and organic matter are known to accumulate Br, and a strong correlation of Br with organic carbon has been reported for both superficial sediments and soils.[855,875] Låg and Steinnes[445] assumed, however, that Br enrichment in top soil horizons is principally an effect of its precipitation with rain. This conclusion is supported by the findings of Yuita et al.[897] that a considerable part of the Br accumulated in soils of the temperate monsoon climate is directly due to the long retention period of Br from rainwater by soils. In spite of the observed sorption capacity of organic matter and clays for Br, it is known as the element most easily leached from soil profiles, and its transportation to ocean basins in large amounts is often reported.

The few studies of the Br contents of soils show that its common range varies from 5 to 40 ppm (Table 136). The highest Br amounts are reported for Japanese soil derived from volcanic ash or andesite and also different Norwegian soils from the north.[445,890] In both cases the source of the large amounts of Br in topsoils is apparently related to atmospheric inputs either from volcanic exhalation and sea evaporation or from human activities.

The main anthropogenic source of Br is its release in automobile exhaust, but some Br used agriculturally as a soil fumigant (methyl bromide) and as a component of K fertilizers may also add to the total soil Br. Although the amount of methyl bromide that is transformed to Br^- in soil is reported to be relatively small, it may significantly increase the Br concentration in the solution of treated soil.[106]

B. Plants

Marine plants, in general, contain more Br than do land plants, and may concentrate this element up to the range of X000 ppm (DW). Although Br is reported to occur in all plant tissues, it is not yet known whether it is essential for plant growth.

The natural Br content of plants seems not to exceed about 40 ppm, and some higher

Table 137
BROMINE CONTENT OF PLANT FOODSTUFFS AND FORAGE PLANTS (PPM DW)

Plant	Tissue sample	Mean or range of content	Ref.
Barley	Grain	5.5	710
	Grain	2.1—6.4[a]	441
Oats	Grain	3.1	710
Triticale	Grain	33.0	492
Corn	Grain	0.9—1.7	462, 710
Beans	Seed	15	197
Peas	Seed	3.3	462
Lettuce	Leaves	20—22	751, 753
Cabbage	Leaves	0.37[b]	574
Radish	Roots	24—26	751, 753
Celery	Stalk	17	753
Carrot	Roots	0.85[b]	574
Potato	Tubers	4.2—14.3	462, 710
Cucumber	Fruits	10—20	751, 753
Tomato	Fruits	10	751, 753
Apple	Fruits	0.002[b]	574
Orange	Fruits	0.04[b]	574
Fruits of Central America	Fruits	0.2—1.0	197
Legumes	Hay	2.1—6.4	710
Clover	Tops	19—52	875
Cheatgrass	Tops	2.8	462
Grass	Tops	17—119	875
Mushroom[c]	Edible parts	2—36	94

[a]   For inland and coastal districts, respectively.
[b]   FW basis.
[c]   Kind not specified.

values should apparently be related to pollution (Table 137). Stärk et al.[753] found that Br was usually higher in leaves than in roots and that it was easily soluble from tissues. Wilkins[875] concluded that the Br concentrations in herbage do not correlate with the concentrations in the soil or with soil properties such as type, drainage status, or pH. However, plants are known to take up Br readily when grown in soils enriched in Br. The method of Br⁻ transportation from soil to plants has not yet been described.

Yamada[890] found as much as 2000 ppm (DW) Br in plants grown in volcanic ash soil, and Staerk and Suess[751] reported Br concentrations from 267 ppm (DW) in celery roots to 9515 ppm (DW) in lettuce leaves that were cultivated in a greenhouse after fumigation with CH_3Br at the rate of 75 g m^{-3}. Methyl bromide and other Br organic compounds used as fumigants for soils, grain, and fruits may be serious sources of Br in human diets.

Br can substitute for part of the Cl⁻ requirement of plants, therefore, its excess is toxic to plants. Plant species differ in their tolerance to soil Br, but several vegetables and flowers are known as sensitive plants (potato, spinach, sugar beet, onion, carnation, and chrysanthemum). Symptoms of Br toxicity resemble excess salt effects, so chlorosis followed by leaf tip necrosis is the common feature. As Gough et al.[279] reviewed, citrus seedlings have been satisfactorally used as an indicator of Br toxicity since the reduction in their growth correlated positively with water-soluble soil Br. Plants resistant to Br toxicity (carrot, tobacco, tomato, celery, and melon) can accumulate more than 2000 ppm (DW) Br without showing any effects.

IV. IODINE

A. Soils

Geochemical characteristics of I resemble those of Br, however, its abundance in the earth's crust is less than that of Br. The I content of most rocks varies from 0.01 to 6 ppm, being the highest in shales rich in organic matter (Table 130).

I occurs as a minor constituent of various minerals, but does not form many separate minerals. The known I minerals include iodides of some metals such as AgI, CuI, $Cu(OH)(IO_3)$, and polyhalides, iodates, and periodates.

The high I content of some nitrate deposits, especially of Chile saltpeter (up to about 400 ppm, mean 200 ppm) has been the subject of much discussion. Apparently, the suggestion of an atmospheric origin seems to be most reasonable.[855]

All I compounds are readily soluble, therefore, weathering of rocks results in the release of much of their I content. Although I is known to be easily transported by waters to ocean basins, its great sorption by carbon, organic matter, and clays greatly influences I cycling.

The geochemistry of I, a biophile element, is strongly connected to its involvement in biological processes. The high I content of sediments and soils is mostly due to uptake of I by plankton or is due to fixation of I by organic matter. Like Br, the I content is reported to be closely correlated with organic carbon content of sediments. As Prince and Calvert[632] stated, sediments of reducing environments contain greater amounts of I than do oxidized sediments.

It is most unlikely that I occurs in soils in the form of I minerals. The association between I and organic matter, hydrous oxides of Fe and Al, and clay of the chlorite-illite group has been noted by several workers. However, as Selezniev and Tiuriukanov[700] and Whitehead[868] reported, organic matter is most responsible for I sorption in soil and therefore I is accumulated mainly in topsoil horizons.

The influence of soil reaction on the I status in soil is diverse. Soil acidity favors I sorption by soil components such as organic matter, hydrous oxides of Fe and Al, and illitic clays.[402,868] On the other hand, in alkali soils of arid and semiarid regions, I is known to be greatly accumulated (e.g., solanchak soils of Baraba Steppe, U.S.S.R., is reported by Anikina[30] to contain as much as 340 ppm I). This is, however, due both to salinity processes and to the low degree of I mobilization occurring under basic pH conditions.

The oxidation of iodide to iodate and further alteration to elemental I may occur in soils and also the exchange of volatile I compounds between soil and atmosphere is reported to be possible.[166] Several ionic forms (I^-, IO_3^-, I_3^-, IO^-, IO_6^{3-}, $H_4IO_6^-$) may occur in the aquatic phase of soil of which the first two are most common.[256]

The behavior of I in soils has been studied in relation to its availability to plants. Hartmans[308] and Whitehead[869] reported relatively few effects of chalk, N, and P on the uptake of I by plants. However, liming is known to reduce the solubility of iodides, iodates, and iodine in soils and thus to also reduce I bioavailability. When I was applied to peat soils, only about 4% of the added amount was taken up by plants.

The fact that soils contain several times as much I as do the parent rocks has been confirmed by numerous analyses. Also an accumulation of I in surface and/or subsurface soil horizons is usually reported. In gleyed water-rich soils, however, a higher I concentration in the lower soil horizons can be expected.

Studies of I distribution in soils have been closely related to the occurrence of endemic goiter. Data collected in Table 138 show that, on the whole, the I concentrations in soils range from 0.1 to around 40 ppm and that the grand mean is 2.8 ppm. However, in certain soils of islands (e.g., Ireland, Japan, New Zealand), a higher I accumulation, up to around 80 ppm, is reported.[166] Usually light soils of humid climatic regions are I poor, whereas high humus and clayed soils are I rich. This may vary greatly because the I level in soils

Table 138
IODINE CONTENT OF SURFACE SOILS OF DIFFERENT COUNTRIES
(PPM DW)

Soil	Country	Range	Mean	Ref.
Podzols and sandy soils	Israel	0.3—0.4	—	644
	Poland	0.8—10.0	—	139
	U.S.S.R.	—	1.1	913
Loess and silty soils	Israel	4.4—5.8	—	644
	U.S.S.R.	0.3—7.6	1.5	913
Loamy and clay soils	U.S.	<0.5—8.3	1.8	706
	U.S.S.R.	0.7—4.9	1.7	503, 913
Soils on glacial till	U.S.S.R.	<0.1—6.3	1.7	913
Fluvisols	Israel	10.6—11.6	—	644
	U.S.	<0.5—3.5	1.4	706
	U.S.S.R.	0.6—6.7	2.0	386, 503, 913
Volcanic ash soils	Japan	32—41	—	897
Rendzinas	Israel	4.1—4.9	—	644
	U.S.	<0.5—4.5	1.7	706
	U.S.S.R.	0.3—2.8	1.2	913
Kastanozems and brown soils	Israel	6.4—7.3	—	644
	U.S.S.R.	0.3—5.3	2.8	503, 621, 760, 913
Ferralsols	Israel	—	7.8	644
Solonchaks	U.S.S.R.	<0.1—6.0	2.1	503, 760
Chernozems	U.S.	<0.5—4.3	1.1	706
	U.S.S.R.	0.4—10.8	3.8	419a, 503, 621, 760, 913
Meadow soils	U.S.S.R.	0.4—3.4	1.7	503, 621, 760, 913
Histosols	Israel	3.3—3.7	—	644
	U.S.S.R.	1—10	4.6	386, 503, 760, 913
Forest soils	Japan	20—23	—	897
	Norway	9.1—23.5	13.1	444, 445
	U.S.S.R.	0.06—25.4	3.4	419a, 503, 621, 913
Various soils	Bulgaria	1.9—4.1	3.0	177
	Great Britain	0.06—2.8	—	818, 868
	Norway, northern	5.4—16.6	9.0	444, 445
	Norway, eastern	2.8—7.6	4.4	444, 445
	U.S.	<0.5—5.4	1.2	706
	U.S.S.R.	0.1—16.0	1.7	503, 760, 913

is known to be dependent on atmospheric precipitation. Also the distance from the sea and recent glaciations influence the soil I status. Since I released into the atmosphere from sea water seems to be a significant source of this element, soils from coastal districts are known to be enriched in this element.[445] Soils derived from recent glacial (Pleistocene) deposits are usually I poor because this element has not yet been highly accumulated from atmospheric precipitation.

The I levels in soils are likely to be elevated in areas of coal and/or kelp burning and around busy roads.[830] Also some sewage sludges applied on fields can add I to surface soil. Because I is readily leached from soils under humid temperate climates, its concentrations in surface horizons do not have a significant environmental impact. The release of ^{131}I from nuclear power facilities has recently become of great environmental concern.

B. Plants

I has not been shown to be essential to plants, and reports on stimulating effects on plant growth at low concentrations have not been explained. Mengel and Kirkby[531] wrote that the

Table 139
IODINE CONTENT OF FOOD AND FORAGE PLANTS (PPM DW)

Plant	Tissue sample	Country	Range	Mean	Ref.
Barley	Grain	Norway	0.005—0.038[a]	—	441
	Grain	U.S.	3.4—7.1	—	709
Snap beans	Pods	U.S.	5.7—9.5	—	709
Cabbage	Leaves	U.S.	9—10	—	709
Lettuce	Leaves	U.S.	—	<0.01[b]	709
Asparagus	Stems	U.S.	5.6—5.9	—	574
Carrot	Roots	U.S.	—	0.025[b]	709
Onion	Bulbs	U.S.	7.8—10.4	—	709
Potato	Tubers	U.S.	2.8—4.9	—	709
Apple	Fruits	U.S.	—	<0.003[b]	574
Orange	Fruits	U.S.	—	<0.01[b]	574
Garden vegetables	Edible parts	U.S.	2.8—10.4	6.6	709
Grass	Tops	Great Britain	0.10—0.28	0.20	166
	Tops	Finland	<1—4[c]	—	221
	Tops	East Germany	0.3—1.6	0.80	316
	Tops	U.S.	4.3—7.1	5.5	709
	Tops	U.S.S.R., Armenia	0.03—0.08	0.06	386
Legume	Tops, clover	Great Britain	0.14—0.44	0.31	166
	Tops, clover	East Germany	0.3—0.5	—	316
	Tops, vetch	U.S.S.R., Armenia	0.06—0.12	0.1	386
Mushrooms	Stalks and caps	U.S.	5.2—9.5	6.7	709

[a] For inland and coastal districts, respectively.
[b] FW basis.
[c] For low I and high I areas, respectively.

stimulating effect of I was observed at 0.1 ppm I in nutrient solutions, whereas toxic effects occur at an I concentration of 0.5 to 1.0 ppm. The toxic concentration is higher than the normal soluble I content of soils, therefore, I toxicity is seldom present in plants under natural field conditions.

Opinions seem to differ regarding the relation between the I content of plants and the I status of soils, but apparently the variation in the I content of plants appears to be generally unaffected by soil kind and type. In general, soluble forms of I seem to be easily available to plants, therefore, terrestrial plants contain much less I than do marine plants which are known to concentrate I from 53 to 8800 ppm (DW).[709] The mechanism of I uptake by plants is not understood. It has been shown, however, by Selezniev and Tiuriukanov[700] that organically bound I is scarcely available to cultivated plants, but that after the decomposition of organic matter by bacteria, soil I becomes available.

Shacklette and Cuthbert[709] studied I distribution among a variety of plant groups from various soils and stated that although I contents of individual plant species may vary considerably, the range in amount seems to be a species characteristic. Generally, vegetables and fleshy mushrooms contain more I than do other land plants (Table 139). Some workers have found a higher I content in tops than in roots and also found a seasonal variation in I levels.[166,230,308] I levels are reported to be the lowest during the summer season.

Plants are capable of absorbing I directly from the atmosphere both through the cuticle and as adhesive particles on the surface of hairy leaves. The atmospheric I can contribute largely to the I content of plants. Gurievich[295] reported that in the Baltic regions higher plants accumulated I to as much as 40 to 50 ppm (DW) and mosses contained 360 to 410 ppm (DW).

Adequate I levels in food and feed plants are required in animal nutrition and in human diets, therefore, some workers have investigated I as a fertilizer for soil or foliar application.[166] These techniques, however, do not appear to be of any practical importance.

The I toxicity to plants due to pollution is not often reported. As Gough et al.[279] reviewed, large applications of kelp to soil as fertilizers (practiced on some coastal fields) may cause symptoms of I toxicity which are similar to those caused by Br excess — margin chlorosis in the older leaves, while the younger leaves become colored dark green.

V. MANGANESE

A. Soils

Mn is one of the most abundant trace elements in the lithosphere, and its common range in rocks is 350 to 2000 ppm (Table 130). Its highest concentrations are usually associated with mafic rocks.

Mn forms a number of minerals in which it commonly occurs as the ions Mn^{2+}, Mn^{3+}, or Mn^{4+}, but its oxidation stage $+2$ is most frequent in the rock-forming silicate minerals. The cation Mn^{2+} is known to replace the sites of some divalent cations (Fe^{2+}, Mg^{2+}) in silicates and oxides.

During weathering Mn compounds are oxidized under atmospheric conditions and the released Mn oxides are reprecipitated and readily concentrated in the form of secondary Mn minerals. The behavior of Mn in surficial deposits is very complex and is governed by different environmental factors, of which Eh-pH conditions are most important.

The complex mineralogical and chemical behavior of Mn results in the formation of the large number of oxides and hydroxides which give a continuous series of composition of stable and metastable arrangements of atoms. The physical features of Mn oxides and hydroxides, such as small size of crystals and consequently a large surface area, have important geochemical implications. Apparently, this is responsible for the high degree of association of Mn concretions with some heavy metals, in particular with Co, Ni, Cu, Zn, and Mo.

There is great progress in research on the chemical controls of the solution and deposition of Mn in both soils and sediments. The intensive studies on Mn behavior in soils have been carried out by a number of workers, and their findings are reviewed in Chapter 4, Section II.B.

McKenzie[524,526] has summarized the present knowledge relating to soil Mn. He stated that Mn is likely to occur in soils as oxides and hydroxides in the form of coatings on other soil particles and as nodules of different diameters. The nodules often exhibit a concentric layering that is suggestive of seasonal growth. The Mn concretions in soils are reported to accumulate Fe and several trace elements (Table 16).

Mn may form a number of simple and complex ions in the soil solution and also several oxides of variable composition (Figure 63). The Mn oxides in soils are mostly amorphous, but crystalline varieties have also been identified in several soils. As Norrish[570] stated, lithiophorite, $(Al, Li)MnO_2(OH)_2$, is most likely to occur in acid and neutral soils, while birnessite (unconfirmed composition, $Na_{0.7}Ca_{0.3}Mn_7O_{14}2.8H_2O$) was reported to be in alkaline soils. However, a number of other crystalline forms of Mn oxides was observed in soil horizons. Of all the Mn oxides, the most stable under oxidizing conditions are pyrolusite (βMnO_2), manganite ($\gamma MnOOH$), and hausmannite (Mn_3O_4). McKenzie[524] concluded that birnessite, lithiophorite, and hollandite are the most common crystalline forms, whereas todorokite and pyrolusite are less common.

All Mn compounds are very important soil constituents because this element is essential in plant nutrition and controls the behavior of several other micronutrients. It also has a considerable effect on some soil properties. The Mn compounds are known for their rapid oxidation and reduction under variable soil environments and thus oxidizing conditions may greatly reduce the availability of Mn and associated micronutrients, whereas reducing conditions may lead to the ready availability of these elements even up to the toxic range.

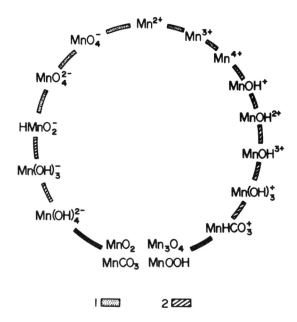

FIGURE 63. Ionic species and transformation of manganese compounds in soils. (1) Redox reactions; (2) redox and hydration reactions.

The solubility of Mn in soils is highly dependent on the pH and redox potential, therefore, the most common reactions occurring in soils are oxidation-reduction and hydrolysis. Although the solution of Mn species as a function of Eh-pH conditions is demonstrated on the diagrams by Lindsay,[477] the comparison of actual Mn levels in soil solutions with those predicted by the chemical equilibrium reactions has not met with success. The mixed and metastable composition of Mn oxides and hydroxides, organic complexes of Mn, and variable Eh-pH soil conditions are the main factors responsible for the lack of correspondence of actual and predicted Mn levels.

Because of the low solubility of Mn compounds in oxidizing systems at pH levels near neutrality, small shifts in the Eh-pH conditions can be very important in the Mn content of the soil solution. The abundance of soluble species of Mn in the soil solution is reported to range from 25 to 2200 $\mu g \ \ell^{-1}$ (Table 12). For solutions of neutral and acid soils, the Mn range is reported to vary from 1 to 100 $\mu M \ \ell^{-1}$.[320]

Hodgson et al.[320] reported that the soluble Mn in soil solutions is mainly involved in organic complexing. In the surrounding soil of plant roots the reduction of MnO_2 forms, and complexing by root exudates is apparently a significant factor controlling Mn mobility.[270] Cheshire et al.[136] found that Mn in topsoil was largely associated with fulvic acid, but the Mn^{2+} bound to these compounds was highly ionized.

Microbiological soil activity is also known to be largely responsible for the oxidation and reduction of Mn compounds, as well as for the formation of Mn concretions as described by Letunova et al.,[466] Bromfield,[103] Wada et al.,[836] and Aristovskaya and Zykina.[36] Zajic[898] and Weinberg[856] reviewed several microbiological processes which directly or indirectly affect the transformation of Mn compounds in soils.

The geochemistry of Mn hydroxides is closely related to the behavior of Fe hydroxides, and cross-interaction during redox reactions should be expected. As McKenzie[526] described, the formation of ferromanganese nodules in acid soils reflects this relationship.

The solubility of soil Mn is of significance since the plant supply of Mn depends mainly on the soluble Mn pool in the soil. In well-drained soils the solubility of Mn always increases

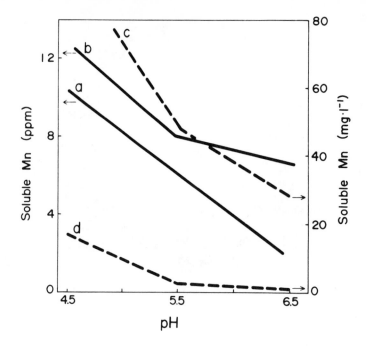

FIGURE 64.   Solubility of soil Mn at different pH values. (a) In 0.01 *M* CaCl₂ from bulk soil; (b) in 0.01 *M* CaCl₂ from rhizosphere soil; (c) in 0.01 *M* citrate solution; (d) in root exudates, after 72 hr.[270]

with the increase of soil acidity. However, the ability of Mn to form anionic complexes (Figure 63) and to complex with organic ligands may contribute to increased Mn solubility in the alkaline pH range (Figure 64).

Several extractants have been widely investigated for soil testing analysis. The best correlation with Mn uptake by plants was usually obtained for the water-soluble, the exchangeable, and the reducible fractions of soil Mn. It appears that most work has been done with the easily reducible Mn (the fraction extractable with hydroquinone), but the effects vary widely.

The soil Mn has been the object of much research, and a great proportion of the studies has been related to the Mn available to plants. Many data are also reported for total Mn in soils.

Mn is not distributed uniformly in soil substrata and, in addition to various nodules, is known to be also concentrated at certain spots which are usually enriched in several other trace elements. The variation of Mn content of surface soils rarely seems to be correlated with soil typology (Tables 140 and 141). However, higher Mn levels are often reported for soils over mafic rocks, for soils rich in Fe and/or organic matter, and for soils from arid or semiarid regions. Although Mn can be concentrated in various soil horizons, particularly in those enriched in Fe oxides or hydroxides, usually this element is also accumulated in top soils as the result of its fixation by organic matter.

On a world scale, the range of Mn varies from 10 to around 9000 ppm, and a maximum in the frequency distribution of values occurs approximately from 200 to 800 ppm. The grand mean calculated for world soils is 545 ppm, while for the U.S. soils the calculation is 495 ppm (Tables 140 and 141).

Mn has not been considered to be a polluting metal in soils, yet Hemkes et al.[314] reported the increase of Mn from 242 to 555 ppm (DW) in sludge-amended soil in 5 years. Grove and Ellis[289] found more water-soluble Mn in soil after fertilization with sludge, whereas

Table 140
MANGANESE CONTENT OF SURFACE SOILS OF DIFFERENT COUNTRIES
(PPM DW)

Soil	Country	Range	Mean	Ref.
Podzols and sandy soils	Australia	900—1000	—	792
	Austria	9—59	—	6
	Bulgaria	451—883	—	545
	Norway	—	165	443
	West Germany	25—200	—	689
	New Zealand	1200—1900[a]	—	861
	Poland	15—1535	—	91,378,382
	U.S.S.R.	135—310	217	74,432,493
Loess and silty soils	West Germany	775—1550	—	689
	New Zealand	1600—1800[a]	—	861
	Poland	110—1060	470	378,382
	U.S.S.R.	—	370	493
Loamy and clay soils	Austria	107—133	—	6
	West Germany	500—1500	—	689
	Mali Republic	75—600	—	39
	New Zealand	670—9200[a]	—	861
	Poland	45—1065	420	91,378,382
	U.S.S.R.	270—1300	475	74,345,432
Soils on glacial till	Denmark	—	268	801
Fluvisols	Austria	152—1030	554	6
	India	350—780	—	641
	Mali Republic	165—250	—	39
	Poland	150—1965	1085	378
	U.S.S.R.	—	240	493
Gleysols	Australia	1200—1900	—	792
	Austria	—	765	6
	Great Britain	—	530	876
	Madagascar	900—2650	—	557a
	Norway	—	80	443
	Poland	85—890	495	91,378
	U.S.S.R.	—	190	432
Rendzinas	Austria	—	1315	6
	Poland	50—7750	440	378,685
	West Germany	—	425	689
Kastanozems and brown soils	Australia	850—3150	—	792
	Austria	253—1675	511	6
	Bulgaria	2190—3907	—	545
	India	924—2615	—	78
	U.S.S.R.	390—580	460	74,343
Ferralsols	Australia	1350—4250	—	792
	India	925—2065	—	78
	Madagascar	850—3400	—	557a
	Spain	10—315[b]	—	253a
Solonchaks and solonetz	Madagascar	—	850	557a
	U.S.S.R.	265—1100	645	74,351,432
Chernozems	Poland	380—700	560	91,378
	U.S.S.R.	340—1100	745	4,346,351,432
Meadow soils	U.S.S.R.	690—1250	950	74,432
Histosols and other organic soils	Canada	240—540	338	243
	Denmark	43—200	100	1,801
	Poland	20—2200	150	382
	U.S.S.R.	510—1465	1005	74,432
Forest soils	China	—	840	225

Table 140 (continued)
MANGANESE CONTENT OF SURFACE SOILS OF DIFFERENT COUNTRIES
(PPM DW)

Soil	Country	Range	Mean	Ref.
	Hungary	120—600[b]	—	207
	U.S.S.R.	—	710	4,351
Various soils	Austria	190—600	388	6
	Canada	80—850	325	629
	Canada	100—1200[b]	520	521
	Denmark	—	279	801
	Great Britain	70—8423	1055	818,876
	West Germany	520—1800[b]	—	61
	Madagascar	680—3500	—	557[a]
	Romania	194—1870	755	43

[a] Soils derived from basalts and andesites.
[b] Data for whole soil profiles.

Table 141
MANGANESE CONTENT OF SURFACE SOILS OF THE
U.S. (PPM DW)[706]

Soil	Range	Mean
Sandy soils and lithosols on sandstones	7—2000	345
Light loamy soils	50—1000	480
Loess and soils on silt deposits	50—1500	525
Clay and clay loamy soils	50—2000	580
Alluvial soils	150—1500	405
Soils over granites and gneisses	150—1000	540
Soils over volcanic rocks	300—3000	840
Soils over limestones and calcareous rocks	70—2000	470
Soils on glacial till and drift	200—700	475
Light desert soils	150—1000	360
Silty prairie soils	200—1000	430
Chernozems and dark prairie soils	100—2000	600
Organic light soils	7—1500	260
Forest soils	150—1500	645
Various soils	20—3000	490

Diez and Rosopulo[176] observed a lower Mn uptake by plants from soil after sludge application. When Mn has accumulated in top soil due to Mn application over a long period of time toxic effects in some plants might be observed.

B. Plants

1. Absorption and Transport

Numerous studies have been carried out on Mn uptake by plants and on Mn distribution among plant tissues. All findings give ample evidence that Mn uptake is metabolically controlled, apparently in a way similar to that of other divalent cation species such as Mg^{2+} and Ca^{2+}. However, passive absorption of Mn is also likely to occur, especially in the high and toxic range of this metal in solution. Generally, Mn is known to be taken up and translocated within plants rapidly, therefore, it is likely that Mn is not binding to insoluble organic ligands either in root tissue or in xylem fluid.

Mn is reported to occur in plant fluids and extracts mainly as free cationic forms.[789,798] It appears, therefore, that Mn is likely to be transported as Mn^{2+}, but its complexing

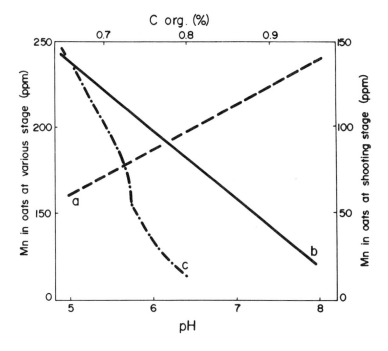

FIGURE 65. Influence of soil factors on the Mn content of oats. (a) Organic matter added as slurry, oats at various stages of growth; (b) pH adjusted with H_2SO_4 or CaO, oats at various stages of growth; (c) pH resulting from fertilization with different N-fertilizers, oats at shoot stage.[437,693]

compounds with organic molecules of 1000 to 5000 mol wt have also been found in phloem exudates.[822] Van Goor[820a] reported a much lower Mn concentration in phloem exudate than in leaf tissue and concluded that a slight transport of Mn through the phloem vessels is responsible for the low concentration of Mn in fruits, seeds, and storage roots.

Mn is preferentially transported to meristematic tissues, thus its concentration is mostly observed in young expanding tissue. Heenan and Campbell[309] reported that at a high Mn supply the leaves accumulated higher concentrations with age, but small amounts of Mn were translocated from old leaves when young expanding leaves were Mn deficient. Thus, Mn appeared to have a low mobility when the supply of the plant was limited. The Mn concentration fluctuates greatly within the plant parts and within the vegetative period. Scheffer et al.[688a] reported a relatively low Mn level in barley during intensive growth and further Mn accumulation in old leaves and sheaths.

It should be emphasized, however, that the Mn content of plants is not only an effect of plant characteristics, but also of the pool of available Mn which is highly controlled by soil properties. Generally, the most readily available Mn is in acid and flooded soil. More than a tenfold increase in the Mn content of lucerne (alfalfa) was observed in plants grown on flooded soil, compared to background values.[531] Therefore, the reducing ability of root exudates and of bacteria in the rhizosphere apparently is of direct importance in Mn nutrition of plants.[270]

Because Mn seems to be easily taken up by plants when it occurs in soluble forms in soils, the Mn content of plants should be a direct function of the soluble Mn pool in soils. And indeed, Mn concentration in plants shows a negative relationship with increasing soil pH and a positive relationship with soil organic matter (Figure 65).

2. Biochemical Functions

All plants have a specific requirement for Mn and apparently the most important Mn function is related to the oxidation-reduction process. The functions of Mn in plants have been widely reviewed by Shkolnik,[718] Boardman,[83] and Mengel and Kirkby.[531] Mn^{2+} is known to be a specific component of two enzymes, arginase and phosphotransferase, but this metal can also substitute for Mg in other enzymes. The mechanism by which Mn^{2+} activates several oxidases is not yet known precisely, but it appears to be related to the valency change between Mn^{3+} and Mn^{2+}.

Mn appears to participate in the O_2-evolving system of photosynthesis and also plays a basic role in the photosynthetic electron transport system. Apparently, the Mn fraction that is loosely bound in chloroplasts is associated with O_2 evolution, whereas the firmly bound Mn fraction is involved in the electron pathway in photosynthesis.

The role of Mn in the NO_2^- reduction step is not yet clear, but it appears to be a kind of indirect relationship between the Mn activity and N assimilation by plants. Adequate levels of available Mn are necessary in plant nutrition. Chloroplasts are the most sensitive of all cell components to Mn deficiency and react by showing structural impairment.[83] The deficiency symptoms occur first in younger leaves as interveinal chlorosis. At further stages necrotic spots on leaves and browning of roots appear. Plants deficient in Mn apparently are less frost hardy.[383,511] The growth of Mn-deficient plants is retarded, the turgor is reduced, and the affected leaves break. Crops and plant species differ in their susceptibility to Mn deficiency — the most sensitive are oats (gray speck symptom), peas (marsh spot symptom), sugar beet, and some fruit trees and bushes.

Although Mn deficiency is relatively common in certain crops grown on neutral and calcareous soils, diagnosis and correction of the deficiency is not well defined. Since soil analysis is not very reliable in diagnosing the Mn supply to plants, tissue tests should be considered together with soil and field observations. The correction of Mn deficiency in crops may be done by both soil and foliar application. The optimum rates and the method of Mn application is extensively reviewed in several textbooks.[556,649,847] A high rate of Mn application or its inappropriate form (e.g., Mn-EDTA for certain fields), as well as variable soil conditions, can easily result in toxic effects on plants.

Toxicity of Mn to some field crops might be expected on acid soils of pH around 5.5 or lower and with a high Mn level. However, the critical Mn content and unfavorable soil pH range depend upon several other environmental factors. Mn toxicity is also known to occur at higher pH levels in poorly drained (poorly aerated) soils. However, if acid soils are very low in total Mn, plants are not subjected to Mn toxicity. As Beckwith et al.[60] reported, flooding did not always increase Mn uptake by rice shoots, since flooding may also increase soil pH and therefore decrease Mn uptake.

Steam sterilization of greenhouse soils is known to increase the available Mn to levels toxic to certain plants. This phenomenon is closely linked with soil biological activity (Chapter 4, Section III.).

Foy et al.[241] described the physiology of Mn toxicity in plants and emphasized its complex character and interrelation with other elements. The response of plants to excessive Mn levels is highly controlled by differences between genotypes. Brown and Devine[105] stated that the control of tolerance to excess Mn appeared to be multigenic and was apparently related to Fe metabolism in plants. Legumes appear to be more sensitive, because Mn excess affects rhizobia nodule numbers and thus the efficiency of N fixation.

The most common symptom of Mn toxicity is Fe chlorosis. Leaf puckering, necrotic brown spots, and an uneven distribution of chlorophyll in older leaves are also symptoms of Mn toxicity. In severely injured plants, browning of roots occur. Plants resistant to Mn excess have an ability to accumulate Mn in root tissues or to precipitate MnO_2, which is deposited mainly within the epidermis. Also, an increased Fe uptake by these plants has been observed.[114,340]

3. Interactions with Other Elements

Mn is known to be involved in both biological and geochemical interactions (Figure 16). The most prominent geochemical interference is observed in the strong affinity of Mn oxides for Co. This reaction is so marked that most of the native Co in soils may be unavailable to plants in the presence of moderate amounts of Mn.[524,526] The strong absorption capacity of Mn oxides for other heavy metals may also highly govern availability of these metals to plants.

Mn-Fe antagonism is widely known and is observed mainly in acidic soils that contain large amounts of available Mn. In general, Fe and Mn are interrelated in their metabolic functions, and their appropriate proportion (the Fe/Mn ratio should range from 1.5 to 2.5) is necessary for the healthy plant. Alvarez-Tinaut et al.[16] reported that both deficient and normal Mn levels antagonize Fe absorption, but the reverse influence was true when Mn reached toxic concentration in plants. In certain field and crop conditions, both Mn or Fe toxicity can be remedied by Fe or Mn application.[241]

Interactions between Mn and other heavy metals are not confirmed, although there are reports of either antagonistic or synergistic effects of Mn on the uptake of Cd and Pb.[381] The interactions of Mn and P may be cross-linked with Fe-P antagonism or related to both the variation in the Mn-phosphate solubility in soils and the Mn influence on P metabolic reactions.[3,241] Depending on soil conditions, P fertilizers are known to either aggravate Mn deficiency in oats or to increase Mn uptake by other plants. These phenomena are closely related to the soil pH and soil sorption capacity.

Interactions of Mn and Si have been reported by several authors. An adequate Si supply to plants is reflected in the easy transport of Mn and a more homogeneous Mn distribution in the plant.[326] Plants deficient in Si are known to accumulate more Mn that Si-sufficient plants. There are several indications that an available Si supply reduces the Mn toxicity to plants.[241] Antagonistic effects of Ca and Mg on Mn uptake seem to have a complex character.

4. Concentrations in Plants

Loneragan[489] stated that Mn showed a particularly wide variation among plant species grown on the same soil, ranging from an average of 30 ppm (DW) in *Medicago trunculata* to around 500 ppm (DW) in *Lupinus albus*. Similarly, a wide range of Mn has been observed in forage plants as reported for different countries (Table 142). World-wide background contents of Mn range from 17 to 334 ppm in grass and from 25 to 119 ppm in clover.

Plant foodstuffs are also reported to contain variable amounts of Mn, being the highest in beet roots (36 to 113 ppm DW) and the lowest in tree fruits (1.3 to 1.5 ppm DW) (Table 143). The Mn content shows a remarkable variation for plant species, stage of growth, and different organs as well as for different ecosystems. A relatively small variation has been observed in the Mn content of cereal grains, which average from 15 to 80 ppm throughout the world (Table 144).

The critical Mn deficiency level for most plants ranges from 15 to 25 ppm (DW), whereas the toxic concentration of Mn to plants is more variable, depending on both plant and soil factors. Generally, most plants are affected by a Mn content around 500 ppm (DW) (Table 25). However, the accumulation above 1000 ppm (DW) also has been often reported for several more resistant species or genotypes.[279,345]

VI. RHENIUM

Re is a highly dispersed element in the earth's crust, but it is known to be associated with some granitoids and pegmatites. Its average abundance is estimated at about 5 ppb in magmatic rock and 0.5 ppb in sedimentary rocks.[552] Two ionic forms, Re^{4+} and Re^{6+}, reveal its similarity to Mo cations and they are likely to substitute for this element in

Table 142
MEAN LEVELS AND RANGES OF MANGANESE IN GRASS AND CLOVER AT THE IMMATURE GROWTH STAGE FROM DIFFERENT COUNTRIES (PPM DW)

Country	Grasses Range	Grasses Mean	Clovers Range	Clovers Mean	Ref.
Australia	67—187	120	33—43	38	266
Great Britain	79—160[a]	—	31—65[a]	—	67
Czechoslovakia	24—130	71	17—42	25	154, 562
Finland	41—144	77[b]	34—140	87[c]	388, 727
	—	—	33—205	119[d]	727
East Germany	51—128	82	29—200	53	31,65
West Germany	35—106	70	24—420	71	576, 596
Hungary	67—309	161	55—126	82	803
Ireland	77—116	86	18—39	26[e]	235
Japan	20—330	127	15—436	89	770
New Zealand	49—139	114	29—165	77	536
Poland	85—215	154	66—96	73	838
U.S.	80—1840	334[f]	—	—	15
U.S.S.R.	26—493	44	19—165	70	338
Yugoslavia	16—18	17	—	—	755

[a]   From freely and poorly drained soils, respectively.
[b]   Timothy.
[c]   Dry hay of red clover.
[d]   Fresh grown red clover.
[e]   Alfalfa.
[f]   Sample from grass tetany pasture.

Table 143
MEAN MANGANESE CONTENT OF PLANT FOODSTUFFS (PPM)

Plant	Tissue sample	FW basis Data source 574	FW basis Data source 705	FW basis Data source 395	DW basis Data source 705	DW basis Data source 381	DW basis Data source 727	AW basis Data source 705
Sweet corn	Grains	—	0.9	—	3.6	—	—	140
Bean	Pods	0.28	2.3	5.0[a]	21	—	21[a]	300
Cabbage	Leaves	1.2	1.1	2.6	14	28	—	150
Lettuce	Leaves	0.1	1.2	4.0[b]	29	—	—	210
Beet	Roots	—	—	—	—	36[c]	92, 113[d]	—
Carrot	Roots	0.15	1.0	—	8.5	14	28	120
Onion	Bulbs	0.6	1.6	—	16	17	24	390
Potato	Tubers	0.3	0.7	2.9	3.6	15	8	86
Tomato	Fruits	0.94	0.6	—	12	—	—	100
Apple	Fruits	0.01	0.2	—	1.3	—	—	74
Orange	Fruits	0.05	0.2	0.5[e]	1.5	—	—	43

[a]   Pea seeds or other pulses.
[b]   Spinach.
[c]   Red beet.
[d]   Sugar and red beet, respectively.
[e]   *Citrus unshiu* (Satsuma orange).

Table 144
MANGANESE CONTENT OF CEREAL GRAINS FROM
DIFFERENT COUNTRIES (PPM DW)

Country	Cereal[a]	Range	Mean	Ref.
Australia	Wheat	17—84	43	867
Great Britain	Barley	—	49	112
	Oats	—	94	112
Canada	Oats	—	76	867
Czechoslovakia	Barley (w)	12—16	15	562
	Oats	40—60	48	562
	Wheat (w)	23—52	34	562
Egypt	Wheat	7.5—24.2	13.7	213
Finland	Barley	18—30	24	727
	Oats	47—93	70	727
	Rye	31—47	39	727
	Wheat (w)	29—103	80	508, 727
	Wheat (s)	28—84	55	508
East Germany	Oats	22—45	—	65
	Wheat	30—44	—	65
Japan	Rice, unpolished	—	26	395
	Wheat, flour	—	2.3	395
Poland	Rye	14—23	17	424, 667
	Triticale	—	26	424
	Wheat (w)	10—50	28	267, 335
Sweden	Wheat (w)	22—38	31	21
U.S.	Rye	11—75	—	484, 492
	Triticale	—	55	492
	Wheat	32—38	35	492
U.S.S.R.	Barley	13—22	17	337
	Oats	23—76	36	337
	Rye	25—87	30	337, 501
	Wheat	16—46	36	337

[a] Spring, s; winter, w.

geochemical processes. Therefore, Re may be expected to be concentrated in Mo minerals. Higher amounts of Re are also observed in some minerals of the rare earth elements.

During weathering Re seems to be readily soluble as the anionic form (ReO_4^{-1}) and is known to be concentrated in certain sediments such as Cu-rich shales or black pyritic shales. Shacklette et al.[710] reviewed data on Re occurrence in plants and gave the range in Re concentrations in native vegetation of the U.S. as 70 to 300 ppm (AW).

Chapter 13

ELEMENTS OF GROUP VIII

I. INTRODUCTION

Group VIII metals form chemical subgroups (triads) with somewhat similar behavior. Geochemical properties of the first triad, Fe, Co, and Ni, are very similar, and due to the small differences in atomic radii, they are likely to form a wide range of mixed crystals.

The terrestrial abundance of Fe is so great that it is not considered as a trace element in rocks and soils. However, Fe plays a special role in the behavior of several trace elements. This metal is also in an intermediate position between macro- and micronutrients insofar as its content in plants is concerned.

Both Fe and Co play significant roles in the biochemistry of plants, but the function of Ni is still not clear. However, all these metals are essential to animals.

Further elements of Group VIII belong to the so-called "platinum metal" group. They are conventionally separated into the middle weight Pd subgroup (Ru, Rh, and Pd) and the heaviest Pt metals (Os, Ir, and Pt). All these metals are highly dispersed in the earth's crust and are known to be mostly inactive in geochemical and biochemical processes.

II. IRON

A. Soils

Fe is one of the major constituents of the lithosphere and comprises approximately 5%, being concentrated mainly in the mafic series of magmatic rocks (Table 145). However, the global abundance of Fe is calculated to be around 45%.

The geochemistry of Fe is very complex in the terrestrial environment and is largely determined by the easy change of its valence states in response to the physicochemical conditions. The behavior of Fe is closely linked to the cycling of O, S, and C.

The reactions of Fe in processes of weathering are dependent largely on the Eh-pH system of the environment and on the stage of oxidation of the Fe compounds involved. The general rules governing the mobilization and fixation of Fe are that oxidizing and alkaline conditions promote the precipitation of Fe, whereas acid and reducing conditions promote the solution of Fe compounds. The released Fe readily precipitates as oxides and hydroxides, but it substitutes for Mg and Al in other minerals and often complexes with organic ligands.

In soils Fe is believed to occur mainly in the forms of oxides and hydroxides as small particles or associated with the surfaces of other minerals. However, in soil horizons rich in organic matter Fe appears to be mainly in a chelated form.

Free Fe minerals that occur in soil are used as a key characterization for soils and for soil horizons. Fe minerals that are known to also form pedogenically are

1. Hematite, α-Fe_2O_3, occurs mainly in soils of arid, semiarid, and tropical regions and is most often inherited from parent materials.
2. Maghemite, γ-Fe_2O_3, is formed in highly weathered soils of tropical zones and most often occurs in concentrations accompanied by hematite, magnetite, or goethite.
3. Magnetite, Fe_3O_4, is mostly inherited from parent materials; in soils it is strongly associated with maghemite.
4. Ferrihydrite, $Fe_2O_3 \cdot nH_2O$, is apparently a common, but unstable soil mineral and is easily transformed to hematite in warm regions and to goethite in humid temperature zones.

Table 145

IRON, COBALT, AND NICKEL IN MAJOR ROCK TYPES (VALUES COMMONLY FOUND, BASED ON VARIOUS SOURCES)

Rock type	Fe (%)	Co (ppm)	Ni (ppm)
Magmatic Rocks			
Ultramafic rocks	9.4—10.0	100—200	1400—2000
Dunites, peridotites, pyroxenites			
Mafic rocks	5.6—8.7	35—50	130—160
Basalts, gabbros			
Intermediate rocks	3.7—5.9	1—10	5—55
Diorites, syenites			
Acid rocks	1.4—2.7	1—7	5—15
Granites, gneisses			
Acid rocks (volcanic)	2.6	15	20
Rhyolites, trachytes, dacites			
Sedimentary Rocks			
Argillaceous sediments	3.3—4.7	14—20	40—90
Shales	4.3—4.8	11—20	50—70
Sandstones	1.0—3.0	0.3—10	5—20
Limestones, dolomites	0.4—1.0	0.1—3.0	7—20

5. Goethite, α-FeOOH, is the commonest Fe mineral in soils over broad climatic regions, from temperate to tropical. The crystallinity and composition of goethites vary and may reflect the environment in which they have formed.

6. Lepidocrocite, γ-FeOOH, is common in poorly drained soils (e.g., paddy soils) and in soils of humid temperate regions. The formation of this mineral in soils is favored by lower pH, lower temperature, and the absence of Fe^{3+}.

7. Ilmenite, $FeTiO_3$, does not occur commonly in soils. As a mineral resistant to weathering, it is usually inherited from igneous parent rocks.

8. Pyrite (FeS_2), ferrous sulfide (FeS), and jarosite, $KFe_3(SO_4)_2(OH)_6$, are widely distributed in submerged soils containing S (e.g., acid sulfate soils).

Both mineral and organic compounds of Fe are easily transformed in soils, and organic matter appears to have a significant influence on the formation of Fe oxides. These oxides may be amorphous, semicrystalline, or crystalline, even under the same conditions.

Transformations of Fe compounds are also affected by microorganisms. Some bacteria species (e.g., *Metallogenium* sp.) are involved in Fe cycling and are known to accumulate this metal on the surfaces of living cells.[36,809,856]

General characteristics of soil Fe are given in Chapter 4, Section II.B. Detailed descriptions of the role and behavior of Fe in soils have recently been presented by Krauskopf,[427] Norrish,[570] Schwertmann and Taylor,[699] Lindsay,[477] and Bloomfield.[81]

Many reactions are involved in the solubility of Fe in soil, but hydrolysis and complexed species appear to be most significant. Lindsay[477] reported that the mobility of Fe in soils is largely controlled by the solubility of Fe^{3+} and Fe^{2+} amorphous hydrous oxides (Figure 66). However, the formation of other Fe compounds, such as phosphates, sulfides, and carbonates, may greatly modify Fe solubilities.

The content of soluble Fe in soils is extremely low in comparison with the total Fe content. Soluble inorganic forms include Fe^{+3}, $Fe(OH)_2^+$, $FeOH^{2+}$, Fe^{2+}, $Fe(OH)_3^-$, and $Fe(OH)_4^{2-}$. In well-aerated soils, however, Fe^{2+} contributes little to the total soluble inorganic Fe, except under high soil pH conditions. The concentration of Fe in soil solutions within common soil

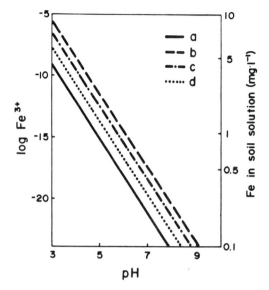

FIGURE 66.   Influence of soil pH on the activity of
Fe^{3+} maintained by Fe oxides. (a) α-FeOOH, goethite;
(b) $Fe(OH)_3$, amorphous; (c) $Fe(OH)_3$, soil constituent;
(d) Fe concentration in soil solution.[477,915]

pH levels ranges from 30 to 550 μg ℓ^{-1}, whereas in very acid soil it can exceed 2000 μg
ℓ^{-1} (Tables 11 and 12). The soluble Fe level reaches a minimum in the alkaline pH range
(Figure 66). Acid soils are therefore higher in soluble inorganic Fe than are neutral and
calcareous soils. Thus, Fe^{2+} cations when in acid anaerobic soils may become toxic, but in
alkaline well-aerated soils, the low concentration of soluble Fe species may not meet plant
requirements for this metal.

When soils are waterlogged, the reduction of Fe^{3+} to Fe^{2+} takes place and is reflected
in an increase in Fe solubility. This process of Fe reduction is strongly related to metabolism
of bacteria and can result in a high Fe^{2+} concentration in some submerged soils (e.g., paddy
soils).

Fe is known to be readily mobile in acid periodically submerged soils. The distribution
of this metal in soil profiles is highly variable and reflects several soil processes. The Fe
compounds that are produced are largely responsible for the color of soils and have been
used for the description of soil processes and for soil classification.

Like Mn compounds in soils, Fe compounds are greatly involved in the behavior of some
macronutrients and of many trace elements. The degree to which soil Fe is responsible for
trace metal solubility and availability is strongly governed by several soil factors. Conversely,
heavy metals are also known to influence the bioavailability of Fe.

Soil Fe exhibits a great affinity to form mobile organic complexes and chelates. These
compounds are largely responsible for the Fe migration between soil horizons and for Fe
leaching from soil profiles and also are important in the supply of Fe to plant roots.

The Fe content of soils is both inherited from parent rocks and the result of soil processes.
The most common range of Fe in soils is from 0.5 to 5%. Even in Fe-poor soil, there is no
absolute deficiency of Fe for plants, but only a deficiency of readily soluble amounts.

Areas of Fe deficiency in soils for certain crops are relatively widespread (Table 28), but
most occur under aridic climates and are related to calcareous, alkaline, or other specific
soils (e.g., manganiferous soils). In the humid climatic zone with a predominance of acid
soils, Fe deficiency in soils is most unlikely to occur unless anthropogenic factors disturb
the natural chemical balance.

Soil testing and the correction of the Fe deficiency have been the subjects of several studies, and all authors have stated that caution should be used in formulating methods for determining plant-available levels of Fe.[151,279,847] The chelating agents DTPA and EDTA appear to be most often recommended for measuring the availability of Fe to plants.

B. Plants

1. Absorption and Transport

The mechanisms of Fe uptake and transport by plants have received much study because they are the key processes in the supply of Fe to plants. The present knowledge on this topic has recently been extensively reviewed by Moore,[548] Chaney et al.,[128] Tiffin,[789] Mengel and Kirkby,[531] and Tinker.[798] Almost all instances of Fe deficiency in plants are considered to occur because of soil factors that govern Fe solubility. Soil conditions that render Fe unavailable are summarized in Table 28.

The Fe uptake by plants is metabolically controlled, although it may be absorbed as Fe^{3+}, Fe^{2+}, or as Fe chelates. The ability of roots to reduce Fe^{3+} to Fe^{2+} is believed to be fundamental in the absorption of this cation by most plants. At normal soil pH levels, Fe organic complexes apparently play an important role in plant nutrition. The separation of chelated Fe prior to the absorption step appears to require reduction of Fe^{3+} to Fe^{2+} at the surface of the root. Generally, roots absorb the Fe^{2+} cation.[128] In xylem exudates, Fe appears to occur unchelated; however, its transport is mediated largely by citrate chelates. In plant tissues, mobile Fe has been identified as citrates and soluble ferredoxins. Fe is not readily transported in plant tissues, and therefore its deficiency appears first in younger plant parts. As Scheffer et al.[688a] reported, the amount of Fe is relatively low in plant parts of intensive growth.

Both Fe uptake and transport between plant organs are highly affected by several plant and environmental factors, of which soil pH, concentrations of Ca and P, and ratios of several heavy metals are most pronounced. In general, a high degree of oxidation of Fe compounds, Fe precipitation on carbonates and/or phosphates, and competition of trace metal cations with Fe^{2+} for the same binding sites of chelating compounds are responsible for a low Fe uptake and for a disturbance in Fe transport within plants. Usually, the more Fe deficient, the greater the ability of plant roots to extract Fe from minerals and from chelating agents.

2. Biochemical Functions

The metabolic functions of Fe in green plants are relatively well understood, and Fe is considered the key metal in energy transformations needed for syntheses and other life processes of the cells.[83,564,630] The essential role of Fe in plant biochemistry can be summarized as follows:

1.   Fe occurs in heme and nonheme proteins and is concentrated mainly in chloroplasts.
2.   Organic Fe complexes are involved in the mechanism of photosynthetic electron transfer.
3.   Nonheme Fe proteins are involved in the reduction of nitrites and sulfates.
4.   Chlorophyll formation seems to be influenced by Fe.
5.   Fe is likely to be directly implicated in nucleic acid metabolism.
6.   Catalytic and structural roles of Fe^{2+} and Fe^{3+} are also known.

This summary is oversimplified and too generalized, but it gives some information showing that, in addition to the active Fe roles in redox reactions of chloroplasts, mitochondria, and peroxisomes, Fe also performs other functions in the plant.

Fe deficiency affects several physiological processes and therefore retards plant growth and plant yield. The deficiency of Fe is a major world-wide problem with many crops since a large number of cultivated soils are low in available content. The control of Fe deficiency often is not sufficiently effective and therefore much effort has been made in screening plants for iron efficiency. Brown[104a] reviewed the current approach to the Fe deficiency problem and stated that genotypes and plant species should be selected for their efficient absorption of Fe.

The symptoms of Fe deficiency may occur at very different Fe levels in plants, and this deficiency is highly dependent on soil, plant, nutritional, and climatic factors. The most common initial symptom of Fe deficiency is interveinal chlorosis of young leaves (Table 27). Several fruit trees and, of all the cereals, oats and rice in particular, are very susceptible to Fe chlorosis.

On soil rich in the soluble Fe fraction, an excessive Fe uptake can produce toxic effects in plants. Plant injury due to Fe toxicity is most likely to occur on strongly acid soils (ultisols, oxisols), on acid sulfate soils, and on flooded soils. This toxicity is also often associated with salinity and a low phosphorus or base status of soils. It has been reported in countries of both tropical and arid regions.[622] The concentration of about 500 ppm Fe^{2+} in the soil solution of paddy soil was reported to kill rice seedlings.[241] It should be emphasized that both an excessive Fe uptake and a low tolerance to a high concentration of Fe in plant tissues is extremely complicated by several nutritional factors.

Symptoms of Fe toxicity are not specific and usually differ among plant species and stages of growth (Table 29). Injured leaves or necrotic spots on leaves indicate an accumulation of Fe above 1000 ppm (3 to 6 times as high as the Fe content of healthy leaves). However, the most pronounced symptom is the ratio of Fe to other elements and to heavy metals in particular. The proper Fe/Mn ratio seems to be the most obligatory factor in the tolerance of plants to Fe toxicity.

Foy et al.[241] reviewed the physiology of Fe toxicity and of plant resistance. Based on this review, it may be stated that:

1. Plants high in nutrients, especially in Ca and SiO_2, can tolerate higher internal levels of Fe.
2. Rice roots are able to oxidize Fe and deposit it on root surfaces.
3. Root damage by H_2S or any other factor destroys the oxidizing power of roots and thus aggravates Fe toxicity.

Plant response to Fe toxicity, as well as to Fe deficiency, is highly variable among genotypes and plant species. Therefore, genetic manipulation through plant breeding seems to be one of the most promising lines of research on the iron problem in plant nutrition.

3. Interactions with Other Elements

Antagonistic interaction between Fe and heavy metals has been observed in several crops, and results of recent studies suggest that chlorosis brought about by heavy metal excess is apparently the result of induced Fe deficiency.[16,105,778] Excess amounts of heavy metals, and of Mn, Ni, and Co in particular, caused a reduction in absorption and translocation of Fe and resulted in a decrease of chlorophyll. On the other hand, high levels of Fe compounds in soil are known to greatly decrease trace metal uptake (Figure 16).

Pulford[633] suggested that the interaction of Fe-Zn is apparently associated with the precipitation of franklinite ($ZnFe_2O_4$) that depressed the availability of both metals. Fe/P interactions commonly occur in both plant metabolism and soil media. The affinity between Fe^{+3} and $H_2PO_4^-$ ions is known to be great, and therefore the precipitation of $FePO_4 \cdot 2H_2O$ can easily occur under favorable conditions. Furthermore, P anions compete with the plant for Fe, and thus P interferes with Fe uptake and with the internal Fe transport. The appropriate P/Fe ratio in plants is fundamental to plant health. The details of this subject are reviewed by Olsen,[581] Adams,[3] and DeKock.[174]

Table 146
MEAN LEVELS AND RANGES OF IRON IN GRASS AND CLOVER AT THE IMMATURE GROWTH STAGE FROM DIFFERENT COUNTRIES (PPM DW)

Country	Grasses		Clovers		Ref.
	Range	Mean	Range	Mean	
Australia	228—264	244	252—357	285	266
Great Britain	73—154	103	122—132[a]	—	867a
Finland	39—49	43[b]	—	—	388
East Germany	79—206	126	116—253	175	31
West Germany	110—430	187	—	—	596
Hungary	133—923	376	118—535	346[c]	803
Ireland	34—78	55	64—85	74[a]	235
Japan	55—157	106	75—229	152	770
Poland	60—140	92	76—136	117	838
U.S.	18—320	73	—	—	200
New Zealand	69—1510	320	105—1700	400	536

[a]   Alfalfa.
[b]   Timothy.
[c]   Various legumes.

Other interactions between Fe and macronutrients (Table 31) are not well understood. Thus, Fe chlorosis in plants on calcareous soils may be thought to reflect a low Fe availability in such soils because of the present insufficient evidence of Fe-Ca metabolic interferences. Interaction between Fe and S seems to be erratic in that low soil S levels may depress Fe uptake, whereas a high S content may also result in low Fe availability, depending on soil environments.

4. Concentrations in Plants

The appropriate content of Fe in plants is essential both for the health of the plant and for the nutrient supply to man and animals. The variation among plants in their ability to absorb Fe is not always consistent and is affected by changing conditions of soil and climate and by the stages of plant growth. Generally certain forbs, including legumes, are known to accumulate more Fe than are grasses. However, where Fe is easily soluble, plants may take up a very large amount of Fe. This is clearly shown by vegetation grown in soils derived from serpentine, where grass contained Fe within the range of 2127 to 3580 ppm (DW).[366]

The natural Fe content of fodder plants ranges from 18 to about 1000 ppm (DW) (Table 146). The nutritional requirement of grazing animals is usually met at the Fe concentration range from around 50 to 100 ppm (DW) in forage.

Edible parts of vegetables appear to contain fairly similar amounts of Fe, ranging from 29 to 130 ppm (DW), with lettuce being in the upper range and onion in the lower range. Fe in ash of a variety of plant species is reported to range from 220 to 1200 ppm (Table 147).

Various cereal grains do not differ much in their Fe concentrations. The common average Fe content of different cereals ranges from 25 to around 80 ppm (DW). Values above 100 ppm are reported only for a few countries (Table 148). The grand mean of 48 ppm (DW) for Fe in grains was calculated by excluding the values of 100 ppm and above.

III. COBALT

A. Soils

In the earth's crust Co has a high concentration in ultramafic rocks (100 to 220 ppm)

Table 147
MEAN IRON CONTENT OF PLANT FOODSTUFFS (PPM)

Plant	Tissue sample	FW basis Data source 574	FW basis Data source 705	DW basis Data source 705	DW basis Data source 727	AW basis Data source 381	AW basis Data source 705
Sweet corn	Grains	4.3	4.4	17	—	—	670
Bean	Pods	0.7	9.2	84	86[a]	—	1200
Cabbage	Leaves	2.4	3.3	42	—	52	450
Lettuce	Leaves	0.3	5.6	130	—	—	960
Red beet	Roots	—	—	—	71	82	—
Carrot	Roots	0.8	1.9	16	54	67	220
Onion	Bulbs	—	3.3	33	29	50	780
Potato	Tubers	2.8	3.9	21	41	58	490
Tomato	Fruits	—	3.0	58	—	—	480
Apple	Fruits	0.6	0.9	6	—	—	350
Orange	Fruits	2.2	2.0	15	—	—	430
Cucumber	Fruits	1.0	2.6	67	—	—	680

[a] Pea seeds.

Table 148
IRON CONTENT OF CEREAL GRAINS
FROM DIFFERENT COUNTRIES (PPM DW)

Country	Cereal[a]	Range	Mean	Ref.
Great Britain	Barley	—	218	112
	Oats	—	96	112
Canada	Oats	—	133	514
Egypt	Wheat	26—69	40	213
Finland	Barley	—	52	727
	Oats	—	60	727
	Rye	—	53	727
	Wheat (s)	24—50	37	508
	Wheat (w)	25—37	31	508
Norway	Barley	16—54	33	446
	Wheat	17—38	30	446
Poland	Rye	34—43	38	267,667
	Wheat	15—30	25	267,667
U.S.	Barley	53—160	88	200
	Oats	54—140	81	200
	Wheat (s)	28—100	48	200
	Wheat	25—43	38	200,492
	Rye	—	100	492
U.S.S.R.	Wheat (s)	29—37	33	586

[a] Spring, s; winter, w.

when compared to its content in acid rocks (1 to 15 ppm). The Co abundance in sedimentary rocks ranges from 0.1 to 20 ppm and seems to be associated with clay minerals or organic matter (Table 145).

There are no common rock-forming Co minerals (formed mainly with As, S, and Se), and Co is mostly hidden in various Fe minerals. In geochemical cycles Co closely resembles Fe and Mn. However, its fate in weathering processes and its distribution in sediments and in soil profile seems to be strongly determined by Mn oxide phase formation.

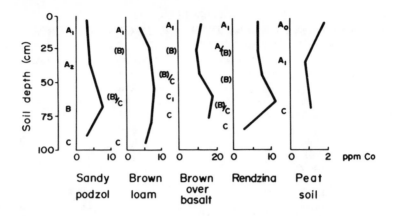

FIGURE 67.   Distribution of Co in the profiles of different soils developed under humid climate. (Letters indicate genetic soil horizons.)

In nature Co occurs in two oxidation states, Co^{2+} and Co^{3+}, and formation of the complex anion $Co(OH)_3^-$ is also possible. During weathering Co is relatively mobile in oxidizing acid environments, but due to a high sorption by Fe and Mn oxides, as well as by clay minerals, this metal does not migrate in a soluble phase.

The solubility and availability of soil Co is of great nutritional concern and therefore has received much study. McKenzie[523] has recently reviewed his research and the findings of other scientists on Co behavior in soils.

Fe oxides are known to have a great affinity for selective adsorption of Co. This has been observed in most kinds of soil and is reflected in the Co distribution in soil profiles showing a general similarity between the levels of Fe and Co in soil horizons (Figure 67). However, in certain soils enriched in Mn minerals, the association of Co with Mn dominates other factors governing Co distribution.

The sorption mechanism of Co by crystalline Mn oxides apparently differs at different pH values and generally is based on the interchange of Co^{2+} with Mn^{2+} at low pH and on the formation of hydroxy species, $Co(OH)_2$, precipitated at the oxide surface. In addition, the specific exchange sorption with bound H can occur over a broad pH range.[486] In soil the sorption of Co by Mn oxides increases greatly with pH, and the reaction seems to be rapid (Figure 68).

Soil organic matter and clay content are also important factors that govern the Co distribution and behavior. The roles of montmorillonitic and illitic clays especially have been cited by numerous investigators as being of significance because of their great sorption capacity and their relatively easy release of Co. The mobility of Co is strongly related to the kind of organic matter in soils. Organic chelates of Co are known to be easily mobile and translocated in soils (Figure 9). Although soils rich in organic matter are known to have both a low Co content and low Co availability, Co organic chelates may also be readily available to plants.[81] This is especially pronounced at higher pH and in freely drained soils. In solutions of most soils the Co concentration is fairly low and ranges from 0.3 to 87 μg ℓ^{-1} (Table 12). In general, the behavior of Co in most soil types is relatively well understood; therefore, a preliminary evaluation of a probable Co status and the availability of Co to plants can be made based on soil and geological information.

Factors contributing to Co deficiency for grazing animals are mainly associated with alkaline or calcareous soils, light leached soils, and soils with high organic matter content (Table 28). The Co content and distribution in soil profiles are also dependent on soil-forming processes and therefore differ for soils of various climatic zones. Usually higher Co contents of surface soils are observed for arid and semiarid regions, whereas exceedingly

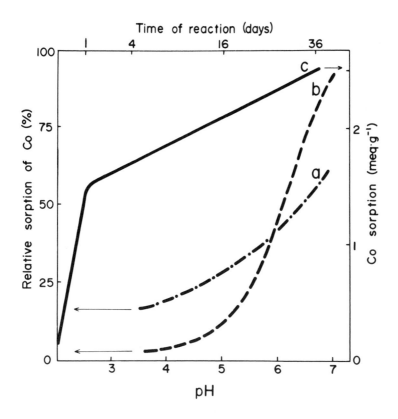

FIGURE 68.   Effects of pH on the sorption of Co by (a) Mn nodules and (b) goethite and effects of time on the sorption of Co by (c) birnessite.[523]

low soil Co has been reported for light soils of the Atlantic Coastal Plain and for soils of the glaciated region of northeastern U.S.[436]

The Co concentration in soils is inherited from parent materials. Soils over mafic rocks and soils derived from clay deposits contain the highest amounts of this metal (Tables 149 and 150). Other high Co concentrations reported for Australian ferralsols (122 ppm) and Japanese soils (116 ppm) are related either to pollution or to specific enrichment. The normal Co content of surface soils usually ranges from 1 to 40 ppm, with the highest frequency in the range of 3 to 15 ppm. The grand mean Co concentration for world-wide soils is 8.5 ppm and for the soils of the U.S. is 8.2 ppm.

Acetic acid-soluble Co in soils usually corresponds to readily available Co (Table 28). Soil application of Co as sulfate or EDTA chelate is the most common practice in controlling Co deficiency in ruminants.

Naturally high Co contents are usually observed in soils over serpentine rocks and ore deposits. Significant sources of Co pollution are related to nonferrous metal smelters, whereas coal and other fuel combustions are of considerably less importance; however, roadside soils and street dusts are known to be enriched in Co (Table 151).

B. Plants

1. Absorption and Transport

Co uptake by plants is a function of the mobile Co content of soil and of the Co concentration in solution (Figure 69). During absorption, Co behaves like other heavy metals (e.g., Fe, Mn) and is transported in forms bound to complexing organic compounds with a molecular weight in the range of 1000 to 5000, with a negative overall charge.[872] Apparently, the simultaneous transport of Co^2 (like Fe^{2+}) with citrate cannot be precluded.

Table 149

COBALT CONTENT OF SURFACE SOILS OF DIFFERENT COUNTRIES (PPM DW)

Soil	Country	Range	Mean	Ref.
Podzols and sandy soils	West Germany	0.8—6	—	689
	New Zealand	21—65	—	861
	Poland	0.1—12	2.0	382
Loess and silty soils	West Germany	4—13	—	689
	New Zealand[a]	17—24	—	861
	Poland	3—23	7.0	382
	Romania[b]	—	5.2	42
Loamy and clay soils	Bulgaria	15—23	18.1	558
	Burma	16—19	18	575
	West Germany	3—6	—	689
	New Zealand[a]	19—58	—	861
	Poland	4—29	6.0	382
	U.S.S.R.	—	5.9	346
Soils on glacial till	Denmark	—	2.1	801
	Ireland	1—17	—	236
Fluvisols	Bulgaria	—	6.4	558
	Egypt	16—21	18.3	213
	Poland	5.3—13	—	156
Rendzinas	West Germany	—	6.1	689
	Ireland	10—12	—	236
	Poland	2.2—22	4.5	156, 685
Kastanozems and brown soils	Burma	6—11	9	575
	Ireland	2—10	—	236
	U.S.S.R.	2.3—3.8	2.9	343
Ferralsols	Australia	0.2—122	—	565
	Madagascar	3—20	—	557a
Solonchaks and solonetz	Burma	12—15	14	575
	Madagascar	15—30	—	557a
	U.S.S.R.	9—14	10.4	12, 351
Chernozems	Bulgaria	9.5—18	—	558
	U.S.S.R.	0.5—50	12.0	346, 351, 419a
Paddy soils	Japan	2.4—24	9.0	395
Meadow soils	U.S.S.R.	11.7—15	—	12
Histosols and other organic soils	Bulgaria	1.7—49	—	558
	Canada	3.1—13.1	6.8	243
	Denmark	—	1.6	801
	Poland	0.2—34	3	382
Forest soils	U.S.S.R.	0.6—45	8.0	12, 419a
Various soils	Bulgaria	3.8—65	21.5	558
	Great Britain	—	17.7	818
	Canada	5—28	12.4	409
	Canada[b]	5—50	21	521
	Denmark	—	2.3	801
	West Germany	—	14.5	390
	Japan	1.3—116	10	395
	Romania	1—6.9	3.1	43

[a]   Soils derived from basalts and andesites.
[b]   Data for whole soil profiles.

Numerous studies have been made of plant uptake of Co from soils, and it has been shown that enrichment of the soil with Co has led to increased levels of this metal in plants. Co is also easily taken up by leaves through the cuticle; therefore foliar applications of Co in solution are known to be effective in the correction of Co deficiency.

Table 150
COBALT AND NICKEL CONTENTS OF SURFACE SOILS OF THE U.S.
(PPM DW)[433,706]

Soil	Co Range	Co Mean	Ni Range	Ni Mean
Podzols and sandy soils	0.4—20	3.5	<5—70	13.0
Low humic gley soils and humus groundwater podzols	0.3—3.1	1.0	—	—
Light loamy soils	3—30	7.5	5—200	22.0
Loess and soils on silt deposits	3—30	11.0	5—30	17.0
Clay and clay loamy soils	3—30	8.0	5—50	20.5
Alluvial soils	3—20	9.0	7—50	19.0
Soils over granites and gneisses	3—15	6.0	<5—50	18.5
Soils over volcanic rocks	5—50	17.0	7—150	30.0
Soils over limestones and calcareous rocks	3—20	9.5	<5—70	18.0
Soils on glacial till and drift	5—15	7.5	10—30	18.0
Light desert soils	3—20	10.0	7—150	22.0
Silty prairie soils	3—15	7.5	<5—50	16.0
Chernozems and dark prairie soils	3—15	7.5	7—70	19.5
Organic light soils	3—10	6.0	5—50	12.0
Forest soils	5—20	10.0	7—100	22.0
Various soils	3—50	10.5	<5—150	18.5

Table 151
COBALT ENRICHMENT AND CONTAMINATION IN
SURFACE SOILS (PPM DW)

Site and pollution source	Mean or range of content	Country	Ref.
Soil over serpentine rock	10—520	New Zealand	495
Mining or ore deposit	13—85	U.S.	744
Metal-processing industry	10—127	Canada	150, 333
	18	West Germany	390
	20—70	Norway	441
	42—154	U.S.	744
Sludged farmland	3.3—12.4	Holland	314
Roadside or airport area	7.9	U.S.	744
	6—14[a]	Great Britain	744

[a]   Urban street dusts.

2. Biochemical Functions

The essentiality of Co for both blue-green algae and microorganisms in fixing N_2 is now well established. It is not clear, however, whether Co is essential for higher plants, although there are some evidences of a favorable effect of Co on plant growth, as reported by Reisenauer et al.,[648] Jagodin et al.,[349] and Mengel and Kirkby.[531] Co has also been recognized as a component of a precursor of vitamin B_{12} for ruminant animals.

Smith and Carson[744] compiled information from several sources on the effects of Co on nonleguminous plants, and the final opinion is inconclusive. Although traces of Co coenzymes have been detected in nonlegumes, it is not known whether these compounds might have originated from microorganisms associated with the plant.[564] Beneficial effects of low Co concentrations on plant metabolism have not yet been fully understood. Presumably, several effects are cross-linked with interactions with other trace metals.

In legumes and alder (*Alnus* sp.) Co affects the ability of plants to fix N_2 from air. Co is chelated at the center of a porphyrin structure termed the cobamide coenzyme, which is

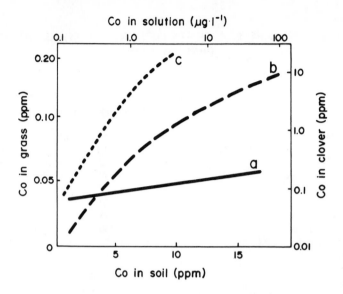

FIGURE 69.   Co uptake by plants from soil or nutrient solution. (a) Ryegrass grown on soil; (b) clover, tops, grown in culture solution; (c) clover, roots, grown in culture solution.[147,915]

effective in N_2 fixation. Nicholas reviewed all recent findings on the Co function in N_2 fixation and stated that cobamide coenzymes are involved in the migration of the H atom during the formation of the NH_3 compound by the rhizobia. Although Co is known to be essential in certain bacterial functions, it is also reported to inhibit Mg uptake by some microorganisms and to exhibit antimicrobial activity.[856]

In legumes, Co deficiency inhibits the formation of leghemoglobin and hence N_2 fixation. However, the Co requirement for this process is low. As Wilson and Reisenauer[883] have found, 10 ppb Co in the nutrient solution was adequate for the growth of alfalfa. In natural conditions Co deficiency is not known to retard the growth of either nonlegumes or legumes.

The Co deficiency in herbage has been considered mainly from the viewpoint of ruminant requirements. It has been observed that soils with a Co content of less than 5 ppm may produce herbage that is Co-deficient for the normal growth of animals. The critical Co level for ruminant diets is around 0.08 to 0.1 ppm (DW), and similar levels were found in clover from different countries (Table 151). The deficiency of Co can be controlled by the application of Co salts to the soil, and the effect of such a treatment is known to last for several years.[544,649] If the soil contains large amounts of Mn or Fe oxides capable of immobilizing Co, a shorter effective period should be expected.

When Co is taken up in excess by roots, it principally follows the transpiration stream, resulting in an enrichment of Co at the leaf margins and tips. Therefore, the common symptoms of Co toxicity are white, dead margins and tips of leaves. However, the primary reaction of plants to an excess of Co is interveinal chlorosis of new leaves, which is closely linked with Fe chlorosis. Cytological effects of Co excesses were reported to be inhibited mitosis and chromosome damage, or damage to the endoplasmic reticulum of root tips, and as disorganized phloem of the minor veins.[643,744]

In nature, although plant species range widely in their content of Co, toxicity symptoms are not often observed. When a high Co level is readily available, in polluted soil in particular, it can seriously affect plant growth and metabolic functions. Kitagishi and Yamane[395] reported that the addition of 25 and 50 ppm of Co to the soil was toxic to rice plants. Anderson et al.[23] found a Co concentration of 140 $\mu g\ \ell^{-1}$ in a soil solution from serpentine soils where oat plants were affected by Co toxicity.

Different concentrations of Co in plant tissues have been reported to produce toxicity symptoms, as follows: 43 to 142 ppm (DW) in bush beans,[841] 19 to 32 ppm (DW) in Sudan grass,[279] and 6 ppm (DW) in barley seedlings.[171]

Cereals are known as the most sensitive plants to Co excess, but Anderson et al.[23] reported that toxic effects at the Co concentrations of 10 to 20 ppm are most probably associated with an excess of Ni. Case et al.[123] proposed that the Co content of herbage should not exceed 60 ppm (DW) because of the animal health hazard.

Plants are known to accumulate large amounts of Co and to develop a mechanism of Co tolerance which is basically similar to that occurring in any metalliferous plant species. Several plant species, mainly of the families Cruciferae, Caryophyllaceae, Violaceae, Leguminosae, Boraginaceae, Myrtaceae, and Nyssaceae, are known for their high accumulation of Co and are also recommended as biogeochemical indicators. These plants growing in soils over serpentine or Cu-Co ores may contain Co as high as 2,500 to 17,700 ppm (AW).[613] Brooks[104] described *Hausmaniastrum* sp. as hyperaccumulators of Co, containing up to about 4000 ppm (DW) Co without having toxicity symptoms when grown in soil overlying a Co-Cu ore body. On the Co-deficient coastal sands of North Carolina, leaves of *Nyssa sylvatica* (black gum) contained from 1 to 216 ppm (DW) Co, where broomsedge (*Andropogon* sp.) contained only 0.05 to 0.91 ppm.[744]

3. Interactions with Other Elements

Cobalt interacts with all metals that are associated geochemically with Fe (Table 9). However, the most significant relationship has been observed between Co and Mn or Fe in the soil and between Co and Fe in the plant. Both geochemical and biochemical antagonisms between these metals have arisen from their affinity to occupy the same sites in crystalline structures and from the similarity of their metallo-organic compounds.

4. Concentrations in Plants

In addition to soil factors highly controlling the Co levels in plants, the ability of plant species to absorb Co varies considerably. Legumes are known to accumulate more Co than do grasses. The mean Co content of clover from different countries was reported to range from 0.10 to 0.57 ppm (DW), whereas this value for grass was 0.03 to 0.27 ppm (Table 152). The Co content of plant foodstuffs varies from around 8 to 100 ppm (DW), with cabbage being in the upper range and corn and apple in the lower range (Table 153).

The relatively great variation observed in Co levels given by different authors for cereal grains may reflect also analytical error (Table 154). It seems rather unlikely that grains would contain higher amounts of Co than are found in green parts. Therefore, values within the range of 4 to 80 ppb (DW) seem to be most reasonable. However, a higher content of Co may occur in grains of certain cereal genotypes, as well as in grains of plants growing in specific soil and climatic conditions.

Plants grown on Co-enriched soils such as serpentine soils or soils over Co ore bodies contain higher amounts of this metal, even when they are not known as Co accumulators. Leaves of oats from serpentine soils contained up to about 15 ppm (DW) Co,[23] cultivated grass contained up to 96 ppm (DW),[366] and native vegetation contained from 17 to 540 ppm (AW).[494]

Data is scarce concerning Co pollution in plants, although Co is known to be released into the atmosphere from coal and fuel oil burning. However, tomato plants grown in water extracts of soils collected near a Cu-Ni smelter contained from 10 to 18 ppm (DW) Co, as cited by Hutchinson and Whitby.[333]

There are no reports of Co toxicity to animals attributed to consumption of natural feedstuffs. However, in certain geochemical areas, and under the influence of man-induced pollution, the excess of Co in plants may be a health risk.

Table 152

**MEAN LEVELS AND RANGES OF COBALT IN GRASS
AND CLOVER AT THE IMMATURE GROWTH STAGE
FROM DIFFERENT COUNTRIES (PPM DW)**

	Grasses		Clovers		
Country	Range	Mean	Range	Mean	Ref.
Australia	—	—	0.07—0.53	0.19	524a
Great Britain	<0.03—1.0[a]	0.27	0.06—1.7[a]	0.57	67,544
France	0.02—0.15	0.07[b]	—	—	147
Finland	0.04—0.08	0.06[c]	—	—	829
	0.17—0.39	0.22[d]	—	—	221
East Germany	0.03—0.10	0.06	0.06—0.12	0.11	31
West Germany	0.05—0.22	0.11	0.08—0.21	0.14	576a,596
Japan	0.01—0.51	0.12[e]	0.02—0.75	0.20	770
Norway	0.02—0.04	0.03[d]	—	—	360
New Zealand	0.03—0.15	0.08[f]	—	—	863
Poland	0.01—0.24	0.08	0.05—0.26	0.10	156
Sweden	0.01—0.40	0.06	0.08—0.30	0.15	203
U.S.	<0.04—0.39	0.08	0.15—0.27	0.19	200,436

[a] For freely and poorly drained soils, respectively.
[b] Ryegrass.
[c] Hay.
[d] Timothy.
[e] Orchard grass.
[f] Sweet vernal grass.

Table 153

COBALT CONTENT OF PLANT FOODSTUFFS (PPB)

		FW basis		DW basis	AW basis
Plant	Tissue sample	Data source 547, 574	Data source 705, 710	Data source 705, 710	Data source 705, 710
Sweet corn	Grains	6.4	2.0	8.1—31	310
Snap bean	Pods	1.0	5.9	20—51	770
Cabbage	Leaves	3.6	8.0	100—160	1100
Lettuce	Leaves	6.8	1.9	46—210	360
Carrot	Roots	3.0	4.4	37—120	520
Onion	Bulbs	—	2.8	28—80	660
Potato	Tubers	2.5	6.9	37—160	860
Tomato	Fruits	—	3.2	62—200	520
Cucumber	Fruits	0.9	3.4	87—170	880
Apple	Fruits	0.3	1.2	8.3—16	460
Orange	Fruits	0.4	2.4	19—45	520

IV. NICKEL

A. Soils

There is a general similarity between the distribution of Ni, Co, and Fe in the earth's crust. Thus, Ni contents are highest in ultramafic rocks (1400 to 2000 ppm), and its concentrations decrease with increasing acidity of rocks down to 5 to 15 ppm in granites (Table 145). Sedimentary rocks contain Ni in the range of 5 to 90 ppm, with the highest range being for argillaceous rocks and the lowest for sandstones.

TABLE 154
COBALT CONTENT OF CEREAL GRAINS
FROM DIFFERENT COUNTRIES (PPB DW)

Country	Cereal[a]	Range	Mean	Ref.
Australia	Wheat	13—231	82	867
Egypt	Wheat	160—380	270	213
Finland	Barley	—	20	829
	Oats	—	10	829
	Wheat (s)	—	10	829
Norway	Barley	1—15.6	4.4	446
	Oats	<100—280[b]	—	748
	Wheat	1.5—13.7	4.7	446
U.S.	Barley	<28—44	28	200
	Oats	<24—78	41	200
	Wheat (s)	<12—48	20	200
	Wheat (w)	14—51	19	200
U.S.S.R.	Oats	170—300	237	790

[a] Spring, s; winter, w.
[b] Pot experiment.

Geochemically, Ni is siderophile and will join metallic Fe wherever such a phase occurs. Also the great affinity of Ni for S accounts for its frequent association with segregates of S bodies. In terrestrial rocks, Ni occurs primarily in sulfides and arsenides, and most of it is in ferromagnesians, replacing Fe. Ni is also associated with carbonates, phosphates, and silicates.

Ni is easily mobilized during weathering and then is coprecipitated mainly with Fe and Mn oxides. However, unlike Mn^{2+} and Fe^{2+}, Ni^{2+} is relatively stable in aqueous solutions and is capable of migration over a long distance. During weathering of Ni-rich rocks (mainly in tropical climates), the formation of garnierite, (Ni, Mg) $SiO_3 \cdot nH_2O$, which is a poorly defined mixture of clay minerals, is observed. Organic matter reveals a strong ability to absorb Ni; therefore, this metal is likely to be concentrated in coal and oil.

Although soil Ni is believed to be strongly associated with Mn and Fe oxides (Table 16), for most soils less than 15 to 30% of the total Ni is extracted with the Mn oxides.[570] A relatively high percentage of Ni extraction with EDTA from soils suggests that Ni is less strongly fixed by soil components than is Co, as described by Berrow and Mitchell.[69]

In surface soil horizons Ni appears to occur mainly in organically bound forms, a part of which may be easily soluble chelates.[81] However, Norrish[570] stated that the fraction of soil Ni carried in the oxides of Fe and Mn seems also to be the form most available for plants.

Ni distribution in soil profiles is related either to organic matter or to amorphous oxides and clay fractions, depending on soil types. Concentrations of Ni in natural solutions of surface horizons of different soils vary from 3 to 25 $\mu g \, \ell^{-1}$ at the boundary and at the center of the affected area, respectively.[23]

Information on the Ni ionic species in the soil solution is rather limited, but the Ni species described by Garrels and Christ[256] such as Ni^{2+}, $NiOH^+$, $HNiO_2^-$, and $Ni(OH)_3^-$ are likely to occur when the Ni is not completely chelated. Generally, the solubility of soil Ni is inversely related to the soil pH.

Bloomfield[81] stated that although organic matter is able to mobilize Ni from carbonates and oxides as well as to decrease Ni sorption on clays, the bonding of this metal to the organic ligands could not be particularly strong.

The Ni status in soils is highly dependent on the Ni content of parent rocks. However, the concentration of Ni in surface soils also reflects soil-forming processes and pollution.

Table 155

NICKEL CONTENT OF SURFACE SOILS OF DIFFERENT COUNTRIES
(PPM DW)

Soil	Country	Range	Mean	Ref.
Podzols and sandy soils	Austria	1—1.5	—	6
	Canada	1.3—34	8	243
	Poland	1—52	7	382
	U.S.S.R.	5—15	11	493, 580
	New Zealand[a]	6—370	—	861
Loess and silty soils	Poland	7—70	19	382
	U.S.S.R.	—	11	493
	New Zealand[a]	3.6—44	—	861
Loamy and clay soils	Austria	13—15	—	6
	Burma	27—91	50	575
	Canada	3—98	23	243
	Poland	10—104	25	382
	U.S.S.R.	—	24	493, 631, 714
	New Zealand[a]	9—110	—	861
Soils on glacial till	Denmark	—	6	801
	Ireland	5—25	—	236
Fluvisols	Austria	20—30	26	6
	U.S.S.R.	—	10	493
Gleysols	Austria	—	6	6
	Chad	3—50	—	39
	U.S.S.R.	—	36	631
Rendzinas	Austria	—	39	6
	Ireland	20—45	—	236
	Poland	7—41	21	685
Kastanozems and brown soils	Austria	6—25	18	6
	Burma	12—39	23	575
	Ireland	10—20	—	236
	U.S.S.R.	10—34	20	580, 714
Solonchaks and solonetz	Burma	32—72	55	575
	Chad	10—50	—	39
	Madagascar	30—80	—	557a
	U.S.S.R.	10—76	25	351, 580
Chernozems	U.S.S.R.	14—40	30	4, 351, 580, 714
Meadow soils	U.S.S.R.	27—75	42	631, 714
Histosols and other organic soils	Canada	6.6—119	29	243
	Denmark	1.9—5	4	1, 801
	Poland	0.2—50	9	91, 684
	U.S.S.R.	—	5	493
Forest soils	China	—	51	225
	U.S.S.R.	22—55	33	4, 631, 714
Various soils	Austria	6—38	21	6
	Great Britain	—	23	100, 818
	Canada[b]	—	20	521
	Chad	3—30	—	39
	Denmark	—	8	801
	West Germany	—	19	340
	Japan	2—660	28	395
	Madagascar	—	20	557a
	Romania	5—25	15	43

[a]   Soils derived from basalts and andesites.
[b]   Data for whole soil profiles.

Soils throughout the world contain Ni within the broad range of from 1 to around 100 ppm, while the range for soils of the U.S. is from <5 to 200 ppm (Tables 150 and 155).

Table 156
NICKEL ENRICHMENT AND CONTAMINATION IN SURFACE SOILS (PPM DW)

Site and pollution source	Mean or range in content	Country	Ref.
Soil over serpentine rocks	770	Australia	702
	1,700—5,000	New Zealand	495
	3,563—7,375	Rhodesia	566
Mine wastes	2—1,150	Great Britain	289
Metal-processing industry	206—26,000	Canada	150, 245, 775
	500—600[a]	Great Britain	38
	26—36[b]	Sweden	811
Sludged farmland	23—95	Great Britain	59
	31—101	Holland	374
	50—84	West Germany	176

[a] Soluble in HCl.
[b] Needle litter and humus layer, respectively.

The highest Ni contents are always in clay and loamy soils, in soils over basic and volcanic rocks, and in organic-rich soils. Especially peaty serpentine soils are known for high Ni levels existing in easily soluble organic complexes.[566] Also soils of arid and semiarid regions are likely to have a high Ni content. The grand mean for world soils is calculated to be 20 ppm, and 19 ppm was reported for U.S. soils.

Ni recently has become a serious pollutant that is released in the emissions from metal processing operations and from the increasing combustion of coal and oil. The application of sludges and certain phosphate fertilizers also may be important sources of Ni. Anthropogenic sources of Ni, from industrial activity in particular, have resulted in a significant increase in the Ni content of soils (Table 156). In particular, Ni in sewage sludge that is present mainly in organic chelated forms is readily available to plants and therefore may be highly phytotoxic. Soil treatments, such as additions of lime, phosphate, or organic matter, are known to decrease Ni availability to plants.

B. Plants
1. Absorption and Biochemical Functions

There is no evidence of an essential role of Ni in plant metabolism, although the reported beneficial effects of Ni on plant growth have stimulated speculation that this metal may have some function in plants.[531,542]

Welch[859] discussed recent reports that Ni is an essential component of the enzyme urease, and thus Ni may be required by nodulated legumes that transport N from roots to tops in forms of ureide compounds. The studies of the uptake and chemical behavior of Ni in plants are related mainly to its toxicity having possible implications with respect to animals and man.

Cataldo et al.[124] studied in detail the absorption, distribution, and forms of Ni in soybean plants. These authors showed that when Ni is in the soluble phase, it is readily absorbed by roots. Ni uptake by plants is positively correlated with Ni concentrations in solutions, and the mechanism is multiphasic (Figure 70).

Like other divalent cations (Co^{2+}, Cu^{2+}, and Zn^{2+}), Ni^{2+} is known to form organic compounds and complexes. Cataldo et al.[124] reported that a large portion of Ni was composed of compounds with <10,000 mol wt, whereas Wiersma and Van Goor[872] found Ni complexing by compounds with a molecular weight in the range of 1,000 to 5,000, with a

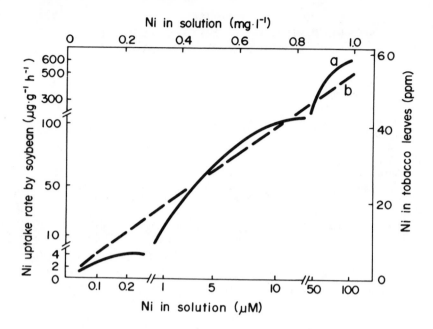

FIGURE 70.   Effects of Ni concentration in solution on Ni uptake by plants. (a) Molar Ni concentrations on the rate of Ni uptake by 21-day-old intact soybeans (whole plant); (b) Ni concentration in solution on Ni content of tobacco leaves.[124,860]

negative overall charge. Also Tiffin[789] found Ni bound to anionic organic complexes in xylem exudates. Although the transport and storage of Ni seems to be metabolically controlled, this metal is mobile in plants and is likely to be accumulated in both leaves and seeds.[176,301,860]

Ni is readily and rapidly taken up by plants from soils, and until certain Ni concentrations in plant tissues are reached, the adsorption is positively correlated with the soil Ni concentrations (Figure 71). Both plant and pedological factors affect the Ni uptake by plants, but the most pronounced factor is the influence of the soil pH. As Berrow and Burridge[67] found, increasing the soil pH from 4.5 to 6.5 decreased the Ni content of oats grain by a factor of about 8 (Figure 71).

The mechanism of Ni toxicity to plants is not well understood, although the restricted growth of plants and injuries caused by an excess of this metal have been observed for quite a long time. The most common symptom of Ni phytotoxicity is chlorosis, which seems to be Fe-induced chlorosis. Indeed, Foy et al.[241] reported a low foliar Fe level at toxic concentrations of Ni in the growth medium. With plants under Ni stress, the absorption of nutrients, root development, and metabolism are strongly retarded. Before the acute Ni toxicity symptoms are evident, elevated concentrations of this metal in plant tissues are known to inhibit photosynthesis and transpiration.[56] Also low N_2 fixation by soybean plants was reported to be caused by Ni excesses.[824]

Under natural conditions Ni toxicities are associated with serpentine or other Ni-rich soils. Anderson et al.[23] reported that oats, a Ni-sensitive crop, when affected by this metal contained Ni in leaves ranging from 24 to 308 ppm (DW). The phytotoxic Ni concentrations range widely among plant species and cultivars and have been reported for various plants to be from 40 to 246 ppm (DW).[279] Davis et al.[171] found the toxic Ni content of barley seedlings to be as low as 26 ppm (DW), and Khalid and Tinsley[389] found 50 ppm (DW) Ni in ryegrass to cause slight chlorosis.

Generally the range of excessive or toxic amounts of Ni in most plant species varies from

251

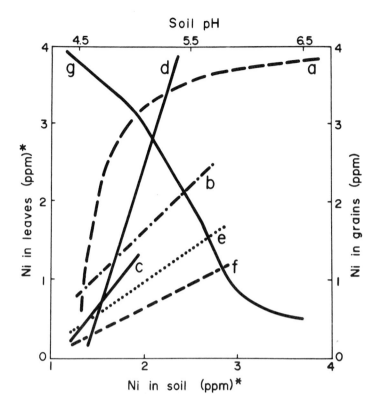

FIGURE 71.   Ni concentrations in plants as a function of Ni content of soils.
(a) leaves of *Alyssum* sp., accumulating Ni; (b) leaves of *A. montanum*, nonac-
cumulating Ni; (c) wheat leaves; (d) barley grains; (e) carrot roots; (f) radish roots.
The influence of soil pH on Ni concentration in (g) oat grains. *Ni concentrations
in soils, leaves and roots are given in powers of ten. [67,176,553,616]

10 to 100 ppm (DW) (Table 25). Several species are known for their great tolerance and
hyperaccumulation of Ni. Usually these species, mainly of the Boraginaceae, Cruciferae,
Myrtaceae, Leguminosae, and Caryophyllaceae families, are also Co accumulators. Native
vegetation of serpentine soils was reported to contain up to 19,000 ppm (AW) Ni,[494] and
Hybanthus floribundus (Violaceae family) growing on acid ferralsols accumulated 6,542
ppm (DW) Ni in leaves, 5,490 ppm in stems, and 221 ppm in root bark.[702] The pronounced
ability of some plant species to accumulate Ni when grown in soil over Ni ore bodies may
make them useful biogeochemical indicators.[417,553,899]

2. Interactions with Other Elements

The interaction between Ni and other trace metals, Fe in particular, is believed to be a
common mechanism involved in Ni toxicity. Cotaldo et al.[124] found that the absorption of
Ni by soybean roots and Ni^{2+} translocation from roots to shoots were inhibited by the
presence of Cu^{2+}, Zn^{2+}, and Fe^{2+}, whereas Wallace et al.[844] stated that Fe^{3+} (as FeEDDHA)
did not depress Ni concentrations in leaves of bush beans. Nevertheless, the excess of Ni
is believed to cause an actual Fe deficiency by inhibiting the translocation of Fe from roots
to tops. Khalid and Tinsley[389] concluded that the Ni/Fe ratio, rather than the Ni and Fe
concentrations in plants, has shown better relationships with the toxic effects of Ni. Both
antagonistic and synergistic interactions have been observed between Ni and several trace
metals (Figure 16).

Table 157
MEAN LEVELS AND RANGES OF NICKEL IN GRASS AND
CLOVER AT THE IMMATURE GROWTH STAGE FROM
DIFFERENT COUNTRIES (PPM DW)

Country	Grasses		Clovers		Ref.
	Range	Mean	Range	Mean	
Belgium	0.9—1.3	1.1	—	—	684
Great Britian	0.84—2.4[a]	—	1.3—6.2[a]	—	67
East Germany	1.3—2.5	1.7	1.0—1.3[b]	1.2	769
Finland	0.15—1.08	0.4	—	—	590
Hungary	—	—	1.4—2.4[b]	1.9	769
Ireland	0.8—2.2	1.4	—	2.7	235
Poland	0.3—1.6	0.9	0.2—2.5	1.2	684, 381
U.S.	<0.07—0.67	0.13	<0.5—5	1.5[c]	200, 710
U.S.S.R. (Lithuania)	2.4—4.8[d]	—	—	2.6[e]	502

[a] For freely and poorly drained soils, respectively.
[b] For meadow and red clover, respectively.
[c] Calculated from AW basis for alfalfa.
[d] For meadow hay and grass crop, respectively.
[e] Clover-timothy hay.

3. Concentrations in Plants

The Ni content of plants growing on uncontaminated soils may vary considerably because it is reflecting both environmental and biological factors. However, the Ni concentrations in certain foodstuffs, cereal grains, and pasture herbage from different countries do not differ widely.

The mean levels of Ni in grasses range from around 0.1 to 1.7 ppm (DW) and in clovers range from 1.2 to 2.7 ppm (DW) (Table 157). There is not much information on Ni in vegetables. Shacklette[705] showed that the Ni content of vegetables ranges from 0.2 to 3.7 ppm (DW) (Table 158).

Mean values for Ni in wheat grains ranged from around 0.2 to 0.6 ppm (DW) and seem to be relatively stable in grains from different countries (Table 159). Oat grains apparently contained somewhat higher amounts of Ni, from about 0.3 to 2.8 ppm. The Ni content of all cereals averaged 0.50 ppm (DW) if extremely high values for oats from Canada and the U.S.S.R. are not included.

Environmental Ni pollution greatly influences the concentrations of this metal in plants (Table 160). In ecosystems where Ni is an airborne pollutant, the tops of plants are likely to concentrate the most Ni, which can be washed from the leaf surfaces quite easily.[38] Sewage sludge has also been shown to be a serious source of Ni as a pollutant in plants. As Ni is easily mobile in plants, berries and grains are reported to contain elevated Ni concentrations.

The health hazards of exposure to Ni and its compounds have recently come to be recognized.[764] Therefore, a redistribution of this metal in the environment from the burning of fossil fuel, application of sludges to agricultural lands, and by industrial emissions should be of concern.

V. PLATINUM-GROUP METALS

The available analytical data are very limited for evaluating the Pt metal abundance in the earth's crust. A general estimation of the range in concentration of the Pt metals in rocks is as follows (in ppb):

Table 158
NICKEL CONTENT OF PLANT FOODSTUFFS (PPM)

Plant	Tissue sample	FW basis Data source 547, 574	FW basis Data source 705, 710	FW basis Data source 695	DW basis Data source 705,710	DW basis Data source 764	AW basis Data source 705, 710
Sweet corn	Grains	0.04	0.06	—	0.22—0.34	—	8.5
Snap bean	Pods	0.02	0.18	—	1.7—3.7	—	24.0
Kidney bean	Seeds	—	—	—	1.1[a]	2.3[b]	—
Cabbage	Leaves	0.05	0.05	0.20	0.62—0.99	3.3	6.7
Lettuce	Leaves	0.01	0.04	0.14	1.0—1.8	1.5	7.2
Carrot	Roots	0.02	0.03	—	0.26—0.98	—	3.6
Onion	Bulbs	—	0.06	—	0.59—0.84	—	13.0
Potato	Tubers	0.04	0.06	0.56	0.29—1.0	—	7.0
Tomato	Fruits	—	0.02	0.03	0.43—0.48	—	3.6
Cucumber	Fruits	0.01	0.05	—	1.3—2.0	—	13.0
Apple	Fruits	0.003	—	0.08	0.06	—	<10.0
Orange	Fruits	0.008	—	—	0.39	1.2[c]	<10.0

[a] From Reference 589.
[b] Peas.
[c] Figs.

Table 159
NICKEL CONTENT OF CEREAL GRAINS FROM DIFFERENT COUNTRIES (PPM DW)

Country	Cereal[a]	Range	Mean	Ref.
Canada	Oats	—	2.79	514
Finland	Barley	0.13—0.14	0.14	829
	Oats	0.43—0.46	0.44	829
	Wheat (s)	0.21—0.27	0.24	829
West Germany	Barley	—	0.9	176
	Rye	—	1.0	176
	Wheat	—	0.4	176
Norway	Oats	0.29—2.62	1.23[b]	748
Poland	Wheat (w)	0.2—0.5	0.4	267
Sweden	Wheat (w)	0.3—0.7	0.5	21
U.S.	Barley	0.10—0.67	0.20	200
	Oats	<0.27—0.94	0.53	200
	Wheat (s)	0.17—0.67	0.32	200
	Wheat (w)	0.18—0.47	0.25	21, 860, 906
U.S.S.R. (Lithuania)	Oat meal	—	2.55	502
	Wheat bran	—	3.60	502

[a] Spring, s; winter, w.
[b] Pot experiment.

Pd, 0.1—200
Pt, 0.1—75
Ru, 0.01—60
Os, 0.06—50
Rh, 0.01—20
Ir, 0.01—20

Table 160
NICKEL CONTENT OF PLANTS GROWN IN CONTAMINATED
SITES (PPM DW)

Site and pollution source	Plant and part	Mean or range of content	Country	Ref.
Metal-processing industry	Blueberry	4.8—6.2	Sweden	811
	Blueberry, leaves	92	Canada	866
	Celery, stalks	29	Canada	775
	Horsetail, tops	140	Canada	745
	Lettuce, leaves	2.7	Australia	57
	Lettuce, leaves	84	Canada	775
	Lettuce, leaves	11	East Germany	431
	Grass	3.9—9.0[a]	East Germany	769
	Oats, grain	1.5	East Germany	431
	Cereal, leaves	230—250	Great Britain	38
	Grass, tops	1,700—32,000	Great Britain	38
	Onion, bulbs	47	Canada	775
	Spruce, twigs	6—14[b]	Sweden	811
Oil refinery	Clover, tops	2.8—14.8	Poland	605
	Grass, tops	2.4—13.3	Poland	605
Sludged soils or urban gardens	Grass, tops	15—19[c]	Holland	297
	Collard, leaves	0.1—11.6	U.S.	127
	Lettuce	1.8—5.8[d]	Australia	831
	Lettuce	2.4—40.3[e]	Great Britain	170
	Grass	10—24	Great Britain	68
	Cereal straw	9.7—20.8	West Germany	176
	Cereal, rain	1.6—5.2	West Germany	176
	Soybean, seeds	7—26	U.S.	10
Other sources	Rice, leaves	11—20	Japan	395
	Oats, grain	60	Canada	301
	Alfalfa	44[e]	Canada	301
	Sagebrush	30[f]	U.S.	278

[a] Industrial emission and urban waste waters.
[b] Five- and one-year old, respectively.
[c] Nonwashed and washed leaves, respectively.
[d] For field and greenhouse experiments, respectively.
[e] Pot experiment.
[f] AW basis.

It should be emphasized, however, that data for the occurrence of Pt metals are extremely subject to errors derived both from sampling difficulties and analytical uncertainties.

Pt metals are concentrated mainly in ultramafic and mafic rocks and are also known to be associated with Ni-Cu ore bodies. In sedimentary and metamorphic rocks the Pt metals occur at a much lower level, with an exception of some Cu-bearing shales.

Pt-group metal deposits have been described by Page and Carlson.[595] The lodes usually are composed of all the Pt metals, with Pd and Pt by far the most abundant. Pt metals are both siderophile and chalcophile, and they are likely to be associated with oxide minerals. Most often these metals occur naturally as alloys.

Pt metals are the noble metals and do not readily unite with other elements. However, Pt is known to form some minerals, and Ru and Os may form easily volatile oxides (RuO_4, OsO_4). Some of the Pt metals apparently are readily absorbed by plants when they occur in easily soluble forms in substrata.[248]

A. Ruthenium

There are not enough data to estimate Ru abundance in soils and plants. Bowen[94] reported Ru to occur in land plants at the concentration of 5 ppb (DW). Duke[197] reported Ru in food plants from a Central American tropical forest region to range from 0.4 to <200 ppb (DW).

The radionuclides [103]Ru and [106]Ru are released during nuclear reactions. Some studies of [106]Ru behavior in soil show that this radionuclide is largely accumulated in surface soil layers.[603] In acid soils, however, [106]Ru is highly mobile and is likely to migrate down the soil profile.[17]

The availability of this nuclide is reported to be relatively high, but a large proportion of the [106]Ru is concentrated in roots.[393,698] Therefore, [106]Ru is not likely to be strongly transferred from soil to the food chain.

B. Rhodium

Data on Rh are too scarce to permit evaluation of its environmental abundance. Some authors consider Rh as the least frequent element of the Pt metals, whereas Wright and Fleischer[888] reported Rh in concentrations up to 20 ppb in black shales and as much as 7 ppm in pyrrhotites. This metal is known to be concentrated also in siderites in the order of O.OX to O.X ppm, and it may be expected to be elevated around some Fe processing industries.

C. Palladium

Pd is more abundant and more chemically reactive than the other Pt metals and is known to form minerals such as stibiopalladinite (Pd_3Sb), arsenopalladinite (Pd_3As), and others. In addition to platiniferous lodes, it is associated with Cu and Ni sulfide ores. Mn ores and Mn concentrations are also known for their capacity to accumulate Pd, as well as coals and phosphorites that may contain large amounts of this metal.

Kothny[416] found Pd in two soils to be 40 and 140 ppb, and in leaves and twigs of various trees to range from 30 to 400 ppb (AW). This author reported a great seasonal variation in Pd concretions in leaves under different climatic conditions and stages of plant growth. Fuchs and Rose[248] gave the range of Pd in surface soils to be from around 0.5 to 30 ppb, and in limber pine to range from 2 to 285 ppb (AW).

Since Pd seems to be easily taken up by plants when it occurs in soluble forms, Kothny[416] speculated that Pd^{2+} is able to replace Mn^{2+} in a metalloenzyme due to the similar ionic radii, and Fuchs and Rose[248] suggested a significant Pd mobility in the organic cycle.

Smith et al.[743] extensively reviewed information on the natural occurrence, technology, and environmental behavior of Pd. These authors concluded that losses of Pd to the environment, mainly from metal processing and use, are believed to be relatively innocuous and of little environmental hazard. The metal is only slightly phytotoxic; however, when the Pd concentration in the nutrient solution was high (from 1 to around 3 mg ℓ^{-1}), damage to the structural units of cells was observed. The inhibition of some metabolic processes has also been observed under increased Pd content in plant tissues.

D. Osmium

Like other Pt metals, Os occurs mainly in native metal alloys of variable composition. Its atypical behavior is its easy reaction with oxygen and its variable oxidation states in compounds. Smith et al.[743] summarized the available data on the geochemical, technological, and environmental behavior of Os. There is no analytical information on Os contamination, even in the vicinity of likely sources Os pollution. However, Os is known to be released to the environment as a volatile tetroxide (OsO_4) during metal processing. The authors concluded that Os is considered as an inert metal and that OsO_4 is not a serious health hazard.

E. Iridium

Generally Ir is considerably more often associated with Fe, Cu, and Ni deposits than are the other Pt metals. The environmental abundance of Ir has not been studied yet, but Bowen[94] gave the Ir content of land plants to be below 20 ppb (DW) and reported the accumulation of ^{192}Ir in the leaf margins of corn treated with that radionuclide.

F. Platinum

Pt seems to be most reluctant to enter into chemical compounds with other elements and therefore occurs mainly in alloys known as "native platinum". However, several mineral forms, such as sperrylite ($PtAs_2$) and cooperite (PtS), occur in ore bodies.

There are but few reports on the Pt occurrence in soils and plants. Fuchs and Rose[248] reported the Pt content of surface silty soils to range from <20 to 75 ppb, being the highest in the soil over weathered norite. The magnetic fraction of these soils seemed to contain most of the Pt, from <860 to <3000 ppb. Twigs of limber pine growing on these soils contained Pt concentrations ranging from 12 to 56 ppb (AW). Shacklette et al.[710] gave the range in Pt concentrations in herbage growing on metamorphic and Pt-bearing ultramafic rocks to be from 3500 to 6600 ppb (AW). Pt, as a chemically inert metal, is presumed unlikely to be of any environmental concern or health hazard.

APPENDIX

The following are common plant names used in this book, with corresponding Latin names, and families (in parentheses) to which assigned.

Alfalfa, *Medicago sativa* L. (Leguminosae)
Apple, *Pyrus malus* L. (Rosaceae)
Apricot, *Prunus armenica* L. (Rosaceae)
Asparagus, *Asparagus officinalis* L. (Liliaceae)
Avocado, *Persea americana* Mill. (Lauraceae)

Barley, *Hordeum vulgare* L. (Gramineae)
Bean, *Phaseolus* sp. (Leguminosae)
Birch, *Betula* sp. (Betulaceae)
Blueberry, *Vaccinium* sp. (Ericaceae)
Bromegrass, *Bromus* sp. (Gramineae)
Brussel sprouts, *Brassica oleraceae* var. *gemmifera* Zenker (Cruciferae)
Buckwheat, *Fagopyrum sagittatum* Gilib. (Polygonaceae)
Bush bean, *Phaseolus vulgaris* L. (Leguminosae)

Cabbage, *Brassica oleracea* var. *capitata* L. (Cruciferae)
Carnation, *Dianthus caryophyllus* L. (Caryophyllaceae)
Carrot, *Daucus carota* L. (Umbelliferae)
Cauliflower, *Brassica oleracea* var. *botrytis* L. (Cruciferae)
Celery, *Apium graveolens* var. *dulce* DC. (Umbelliferae)
Chard, *Beta vulgaris* var. *cicla* L. (Chenopodiaceae)
Cheatgrass, *Bromus secalinus* L. (Gramineae)
Cherry, *Prunus avium* L. or *P. cerasus* L. (Rosaceae)
Chinese cabbage, *Brassica pekinensis* (Lour.) Ruprecht (Cruciferae)
Chrysanthemum, *Chrysanthemum* sp. (Compositae)
Clover, *Trifolium* sp. (generally, *T. pratense* L.) (Leguminosae)
Collard, *Brassica oleracea* var. *viridis* L. (Cruciferae)
Corn, *Zea mays* L. (Gramineae)
Cotton, *Gossypium* sp. (Malvaceae)
Cucumber, *Cucumis sativus* L. (Cucurbitaceae)

Dandelion, *Taraxicum officinale* Weber (Compositae)
Douglas fir, *Pseudotsuga menziesii* (Mirb.) Franco (Pinaceae)

Flax, *Linum usitatissimum* L. (Linaceae)

Gladiolus, *Gladiolus* sp. (Iridaceae)
Grape, *Vitis vinifera* L. or *V. labruscana* Bailey (Vitaceae)

Hops, *Humulus lupulus* L. (Moraceae)
Horse bean, *Dolichos lablab* L. (Leguminosae)
Horsetail, *Equisetum* sp. (Equisetaceae)

Kidney bean, *Phaseolus vulgaris* cultivar (Leguminosae)

Labrador tea, *Ledum palustre* L. (Ericaceae)

Larch, *Larix* sp. (Pinaceae)
Lettuce, *Lactuca sativa* L. (Compositae)
Limber pine, *Pinus flexilis* James (Pinaceae)
Lucerne, *Medicago sativa* L. (Leguminosae)

Maize, *Zea mays* L. (Gramineae)
Mangel, *Beta vulgaris* var. *macrorhiza* Hort. (Chenopodiaceae)
Maple, *Acer* sp. (Aceraceae)
Mushroom, edible fleshy fungus of the Basidiomycetes
Mustard, *Brassica nigra* (L.) Koch or *B. hirta* Moench. (Cruciferae)

Nettle, *Urtica* sp. (Urticaceae)

Oak, *Quercus* sp. (Fagaceae)
Oats, *Avena sativa* L. (Gramineae)
Onion, *Allium cepa* L. (Liliaceae)
Orange, *Citrus sinensis* Osbeck (Rutaceae)
Orchardgrass, *Dactylis glomerata* L. (Gramineae)

Pea, *Pisum sativum* L. (Leguminosae)
Peanut, *Arachis hypogaea* L. (Leguminosae)

Pea, *Pisum sativum* L. (Leguminosae)
Peanut, *Arachis hypogaea* L. (Leguminosae)
Perennial rye-grass, *Lolium perenne* L. (Gramineae)
Pine, *Pinus* sp. (Pinaceae)
Potato, *Solanum tuberosum* L. (Solanaceae)
Pulses, edible seeds of various legumes

Radish, *Raphanus sativus* L. (Cruciferae)
Rape, *Brassica rapa* L. (Cruciferae)
Raya, *Brassica juncea* Coss. (Cruciferae)
Red beet, *Beta vulgaris* var. *crassa* Alef. (Chenopodiaceae)
Red clover, *Trifolium pratense* L. (Leguminosae)
Rice, *Oryza sativa* L. (Gramineae)
Rose, *Rosa* sp. (Rosaceae)
Ryegrass, *Lolium perenne* L. or *L. multiflorum* Lam. (Gramineae)

Sagebrush, *Artemisia tridentata* Nutt. (Compositae)
Sedge, various genera and species of the Cyperaceae family
Snap bean, *Phaseolus vulgaris* L. cultivar (Leguminosae)
Sorghum, *Sorghum vulgare* Pers. (Gramineae)
Soybean, *Glycine max* (L.) Merr. (Leguminosae)
Spinach, *Spinacia oleracea* L. (Chenopodiaceae)
Spruce, *Picea* sp. (Pinaceae)
Strawberry, *Fragaria ananassia* Duchesne (Rosaceae)
Sudan grass, *Sorghum sudanense* (Piper) Stapf. (Gramineae)
Sugarbeet, *Beta vulgaris* var. *saccharinum* Hort. (Chenopodiaceae)
Sunflower, *Helianthus annuus* L. (Compositae)
Sweetclover, *Melilotus officinalis* (L.) Lam. or *M. alba* Desr. (Leguminosae)
Sweet corn, *Zea mays* var. *rugosa* Bonaf. (Leguminosae)
Sweet vernal grass, *Anthoxanthum odoratum* L. (Gramineae)

Tea, *Thea sinensis* L. (Theaceae)
Timothy, *Phleum pratense* L. (Gramineae)
Tobacco, *Nicotiana tabacum* L. (Solanaceae)
Tomato, *Lycopersicum esculentum* Mill. (Solanaceae)
Triticale, an artificial hybrid of wheat and rye
Turnip, *Brassica napus* L. (Cruciferae)

Valencia orange, *Citrus sinensis* Osbeck cultivar (Rutaceae)

Wheat, *Triticum aestivum* L. (Gramineae)
Willow, *Salix* sp. (Salicaceae)
Wormwood, *Artemisia* sp. (Compositae)

REFERENCES*

1. **Aaby, B. and Jacobsen, J.,** Changes in biotic conditions and metal deposition in the last millennium as reflected in ombrotrophic peat in Draved Moses, Denmark, *Geol. Surv. Denmark, Yearb.,* 1978, 5.

2. **Abd-Elfattah, A. and Wada, K.,** Adsorption of lead, copper, zinc, cobalt and cadmium by soils that differ in cation-exchange materials, *J. Soil Sci.,* 32, 271, 1981.

3. **Adams, F.,** Interaction of phosphorus with other elements in soils and in plants, in *Proc. Symp. The Role of Phosphorus in Agriculture,* Khasawneh, F. E., Ed., Am. Soc. Agron., Madison, Wis., 1980, 655.

4. **Adierikhin, P. G., Protasova, N. A., and Shcheglov, D. J.,** Microelements in system soil-plant in Central-Chernozem regions, *Agrokhimiya,* 6, 102, 1978 (Ru).

5. **Ahlrichs, J. L.,** The soil environment, in *Organic Chemicals in the Soil Environment,* Goring, C. A. I. and Hamaker, J. W., Eds., Marcel Dekker, New York, 1972, 3.

6. **Aichberger, K.,** Schwermetallgehalte einiger Bodenprofile Oberösterrichs, *Bodenkultur,* 31, 215, 1980.

7. **Aidiniyan, N. Kh.,** Distribution of rare alkalies in colloids of soils and the participation of vegetation in this process, *Geokhimiya,* 4, 346, 1959 (Ru).

8. **Aidiniyan, N. Kh., Trioitskiy, A. I., and Balavskaya, G. A.,** Distribution of mercury in various soils of the USSR and Vietnam, *Geokhimiya,* 7, 654, 1964 (Ru).

9. **Akhtirtsev, B. P.,** Content of trace elements in grey forest soils of the central chernozem belt, *Agrokhimiya,* 9, 72, 1965 (Ru).

10. **Alexander, J., Koshut, R., Keefer, R., Singh, R., Horvarth, D. J., and Chaney, R.,** Movement of nickel from sewage sludge into soil, soybeans and voles, in *Trace Subst. Environ. Health,* Vol. 12, Hemphill, D. D., Ed., University of Missouri, Columbia, Mo., 1978, 377.

11. **Alikhanova, O. J.,** Toxic effects of boron on plants, *Agrokhimiya,* 7, 98, 1980 (Ru).

12. **Alikhanova, O. J., Zyrianova, A. N., and Cherbar, V. V.,** Minor elements in soils of some regions in the western Pamirs, *Pochvovedenie,* 11, 54, 1977 (Ru).

13. **Allaway, W. H.,** Control of the environmental levels of selenium, in *Trace Subst. Environ. Health,* Vol. 2, Hemphill, D. D., Ed., University of Missouri, Columbia, Mo., 1968, 181.

14. **Allaway, W. H. and Hodgson, J. F.,** Symposium on nutrition, forage and pastures: selenium in forages as related to the geographic distribution of muscular dystrophy in livestock, *J. Anim. Sci.,* 23, 271, 1964.

15. **Allen, V. G. and Robinson, D. L.,** Occurrence of Al and Mn in grass tetany cases and their effects on the solubility of Ca and Mg in vitro, *Agron. J.,* 72, 957, 1980.

16. **Alvarez-Tinaut, M. C., Leal, A., and Recalde-Martinez, L. R.,** Iron-manganese interaction and its relation to boron levels in tomato plants, *Plant Soil,* 55, 377, 1980.

17. **Amy, J. P.,** Contribution a l Étude de la Migration du Ruthenium-106 dans les sols, these présentée à la Faculte des Sciences de l'Universite de Nancy, France, 1970, 79.

18. **Andersson, A.,** On the influence of manure and fertilizers on the distribution and amounts of plant-available Cd in soils, *Swed. J. Agric. Res.,* 6, 27, 1976a.

19. **Andersson, A.,** On the determination of ecologically significant fractions of some heavy metals in soils, *Swed. J. Agric. Res.,* 6, 19, 1976b.

20. **Andersson, A.,** Some aspects on the significance of heavy metals in sewage sludge and related products used as fertilizers, *Swed. J. Agric. Res.,* 7, 1, 1977a.

21. **Andersson, A.,** Heavy metals in Swedish soils; on their retention, distribution and amounts, *Swed. J. Agric. Res.,* 7, 7, 1977b.

22. **Andersson, A. and Hahlin, M.,** Cadmium effects from fertilization in field experiments, *Swed. J. Agric. Res.,* 11, 3, 1981.

23. **Anderson, A. J., Meyer, D. R., and Mayer, F. K.,** Heavy metal toxicities: levels of nickel, cobalt and chromium in the soil and plants associated with visual symptoms and variation in growth of an oat crop, *Aust. J. Agric. Res.,* 24, 557, 1973.

24. **Andersson, A. and Nilsson, K. O.,** Enrichment of trace elements from sewage sludge fertilizer in soils and plants, *Ambio,* 1, 176, 1972.

25. **Andersson, A., Nilsson, K. O.,** Influence on the levels of heavy metals in soil and plant from sewage sludge used as fertilizers, *Swed. J. Agric. Res.,* 6, 151, 1976.

26. **Andersson, A. and Pettersson, O.,** Cadmium in Swedish winter wheat, *Swed. J. Agric. Res.* 11, 49, 1981.

27. **Andrzejewski, M., Baluk, A., and Czekala, J.,** Chromium content of grasslands as affected by fertilizers and cutting, *Pr. Kom. Nauk Roln. Lesn. PTPN,* 31, 11, 1971 (Po).

---

* Publications written in languages other than English, French, and German are indicated by letters in parentheses: (Bu), Bulgarian; (Ch), Chinese; (Cz), Czechoslovakian; (Du), Dutch; (Es), Spanish; (Hu) Hungarian; (Ja) Japanese; (Po), Polish; (Ro), Romanian; (Ru), Russian; (Sh), Serbo-Croatian; (Sv), Swedish; and (Uk); Ukrainian.

28. **Andrzejewski, M. and Rosikiewicz, D.,** Ability of some trace elements to form complexes with humic acids, in Trans. 7th Int. Symp. Humus et Planta, Praha, 1975, 107.

29. **Angino, E. E., Cannon, H. L., Hambidge, K. M., and Voors, A. W.,** Lithium, in *Geochemistry and the Environment*, Vol. 1, Mertz, W., Ed., N.A.S., Washington, D.C., 1974, 36.

30. **Anikina, A. P.,** Iodine in soils of geochemical landscapes of Central Baraba, *Pochvovedenie*, 1, 66, 1975 (Ru).

31. **Anke, M.,** Der Spurenelementgehalt von Grünland- und Ackerpflanzen verschiedener Böden in Thüringen, *Sonderdruck Z. Acker-u. Pflanzenbau*, 112, 113, 1961.

32. **Anke, M., Groppel, B., Lüdke, H., Grün, M., and Kleemann, J.,** Die Spurenelementversorgung der Wiederkäuer in der Deutschen Demokratischen Republik. Kupferversorgung, *Arch. Tierernaehr.*, 25, 257, 1975.

33. **Antonovics, J., Bradshaw, A. D., and Turner, R. G.,** Heavy metal tolerance in plants, *Adv. Ecol. Res.*, 7, 1, 1971.

34. **Archer, F. C.,** Trace elements in some Welch upland soils, *J. Soil Sci.*, 14, 144, 1963.

35. **Aristovskaya, T. V.,** *Microbiology of Podsols*, Izd. Nauka, Moscow, 1965, 187 (Ru).

36. **Aristovskaya, T. V. and Zykina, L. V.,** Microorganisms as indicators of R_2O_3 and manganese accumulation processes in soils, *Pochvovedenie*, 1, 88, 1979 (Ru).

37. **Arvy, M. P., Lamand, M., and Bonnemain, J. L.,** Teneuer en selenium de quelques légumineuses et graminées cultivées dans le Limousin, *C. R. Acad. Agric. Fr.*, 7, 481, 1974.

38. **Ashton, W. M.,** Nickel pollution, *Nature (London)*, 237, 46, 1972.

39. **Aubert, H. and Pinta, M.,** *Trace Elements in Soils*, Elsevier, Amsterdam, 1977, 395.

40. **Augustyn, D. and Urbaniak, H.,** Retention of cations of humic acids from Polish brown coal and some properties of metal-humic compounds, paper presented at 7th Int. Symp. Humus and Planta, Brno, Czechoslovakia, August 20, 1979.

41. **Babich, H. and Stotzky, G.,** Effects of cadmium on the biota: influence of environmental factors, *Adv. Appl. Microbiol.*, 23, 55, 1978.

41a. **Bain, C. D.,** A titanium-rich soil clay, *J. Soil Sci.*, 27, 68, 1976.

42. **Bajescu, J. and Chiriac, A.,** Some aspects of the distribution of copper, cobalt, zinc and nickel in soils of the Dobrudja, *Anal. Inst. Cent. Cerc. Agric.*, 30A, 79, 1962 (Ro).

43. **Bäjescu, J. and Chiriac, A.,** Trace element distribution in brun lessivé and lessivé soils, *Stiinta Solului*, 6, 45, 1968 (Ro).

44. **Baker, D. E.,** Acid soils, in *Proc. of Workshop on Plant Adaptation to Mineral Stress in Problem Soils*, Wright, J., Ed., Cornell University, Ithaca, No. 4, 1976, 127.

45. **Bakulin, A. A. and Mokiyenko, V. F.,** Microelement content of sands from Lower Volga Basin, *Pochvovedenie*, 4, 66, 1966 (Ru).

46. **Balks, R.,** Untersuchungen über den Bleigehalt des Bodens, *Kali-Briefe*, 1, 1, 1961.

47. **Ballantyne, D. J.,** Fluoride inhibition of the Hill reaction in bean chloroplasts, *Atmos. Environ.*, 6, 267, 1972.

48. **Baltrop, D.,** Significance of lead-contaminated soils and dusts for human populations, paper presented at Int. Symp. on Environ. Lead Research, Dubrovnik, Yugoslavia, May 14, 1975.

49. **Barrer, R. M.,** *Zeolites and Clay Minerals as Sorbents and Molecular Sieves*, Academic Press, New York, 1978, 497.

50. **Bartlett, R. J. and James, B.,** Behavior of chromium in soils. III. Oxidation, *J. Environ. Qual.*, 8, 31, 1979.

51. **Bartlett, R. J. and Kimble, J. M.,** Behavior of chromium in soils. I. Trivalent forms. II. Hexavalent forms, *J. Environ. Qual.*, 5, 379 and 383, 1976.

52. **Bartlett, R. J. and Picarelli, C. J.,** Availability of boron and phosphorus as affected by liming and acid potato soil, *Soil Sci.*, 116, 77, 1973.

53. **Baszyński, T., Król, M., and Wolińska, D.,** Photosynthetic apparatus of *Lemna minor* L. as affected by chromate treatment, in *Photosynthesis. II. Electron Transport and Photophosphorylation*, Akoyunoglou, G., Ed., Balaban Intern. Sci. Service, Philadelphia, 1981, 111.

54. **Baszyński, T., Wajda, L., Król, M., Wolińska, D., Krupa, Z., and Tukendorf, A.,** Photosynthic activities of cadmium-treated tomato plants, *Physiol. Plant.*, 4, 365, 1980.

55. **Baumeister, W. and Ernst, W.,** *Mineralstoffe und Pflanzenwachstum*, Fischer, Stuttgart, 1978, 416.

56. **Bazzaz, F. A., Carlson, R. W., and Rolfe, G. L.,** The effect of heavy metals on plants, *Environ. Pollut.*, 7, 241, 1974.

57. **Beavington, F.,** Heavy metal contamination of vegetables and soil in domestic gardens around a smelting complex, *Environ. Pollut.*, 9, 211, 1975.

58. **Beckert, W. F., Moghissi, A. A., Au, F. H. E., Bretthauer, F. W., and McFarlane, M.,** Formation of methyl mercury in a terrestrial environment, *Nature (London)*, 249, 674, 1974.

59. **Beckett, P. H. T., Davis, R. D., Brindley, P., and Chem, C.,** The disposal of sewage sludge onto farmland: the scope of the problem of toxic elements, *Water Pollut. Control*, 78, 419, 1979.

60. **Beckwith, R. S., Tiller, K. G., and Suwadji, E.,** The effects of flooding on the availability of trace metals to rice in soils of different organic matter content, in *Trace Elements in Soil-Plant-Animal Systems,* Nicholas, D. J. D. and Edan, A. R., Eds., Academic Press, New York, 1975, 135.

61. **Beer, K.,** Untersuchungen über Vorkommen und Bindungszustand des Mangans in typischen thüringer Boden, *Chem. Erde,* 25, 282, 1966.

62. **Bell, M. C. and Sneed, N. N.,** Metabolism of tungsten by sheep and swine, in *Trace Element Metabolism in Animals,* E. S. Livingston, Edinburgh, 1970, 70.

63. **Bender, J. and Szalonek, J.,** Biological activation of polluted areas and the possibilities of their rational utilization, *Arch. Ochrony Srodowiska,* 1, 99, 1980 (Po).

64. **Benes, S.,** The occurrence and migration of copper in soils from different parent rocks, *Polnohospodarstvo,* 10, 837, 1964 (Cz).

65. **Bergmann, W., Ed.,** *Mikronährstoff-Grenzwertbereiche in Pflanzen zur Diagnose des Ernährungszustandes der Pflanzen,* Institute für Pflanzenernährung, Jena-Zwätzen, East Germany, 1975.

66. **Bergmann, W. and Cumakov, A.,** *Diagnosis of Nutrient Requirement by Plants,* G. Fischer Verlag, Jena, and Priroda, Bratislava, 1977, 295 (Cz).

67. **Berrow, M. L. and Burridge, J. C.,** Sources and distribution of trace elements in soils and related crops, in *Proc. Int. Conf. on Management and Control of Heavy Metals in the Environment,* CEP Consultants Ltd., Edinburgh, U.K., 1979, 304.

68. **Berrow, M. L. and Burridge, J. C.,** Persistence of metals in available form in sewage sludge treated soils under field conditions, in *Proc. Int. Conf. Heavy Metals in the Environment,* Amsterdam, CEP Consultants Ltd., Edinburgh, U.K., 1981, 201.

69. **Berrow, M. L. and Mitchell, R. L.,** Location of trace elements in soil profile: total and extractable contents of individual horizons, *Trans. R. Soc. Edinburgh Earth Sci.,* 71, 103, 1980.

70. **Berrow, M. L. and Webber, J.,** Trace elements in sewage sludges, *J. Sci. Food Agric.,* 23, 93, 1972.

71. **Berrow, M. L., Wilson, M. J., and Reaves, G. A.,** Origin and extractable titanium and vanadium in the A horizons of Scottish podzols, *Geoderma,* 21, 89, 1978.

72. **Bhumbla, D. R. and Chhabra, R.,** Chemistry of sodic soils, in Review of Soil Research in India, in Trans. 12th Int. Congr. Soil Sci., New Delhi, India, 1982, 169.

73. **Bieliyakova, T. M.,** Fluorine in soils and plants as related to endemic fluorosis, *Pochvovedenie,* 8, 55, 1977 (Ru).

74. **Bierdnikova, A. V.,** Manganese contents in soils of Astrakhan Region, *Agrokhimiya,* 2, 128, 1978 (Ru).

75. **Bieus, A. A., Grabovskaya, L. I., and Tikhonova, N. V.,** *Geochemistry of the Surrounding Environment,* Nedra, Moscow, 1976, 247 (Ru).

76. **Bingham, F. T., Page, A. L., Mahler, R. J., and Ganje, T. J.,** Yield and cadmium accumulation of forage species in relation to cadmium content of sludge-amended soil, *J. Environ. Qual.,* 5, 57, 1976.

77. **Bingham, F. T., Page, A. L., and Strong, J. E.,** Yield and cadmium content of rice grain in relation to addition rates of cadmium, copper, nickel, and zinc with sewage sludge and liming, *Soil Sci.,* 130, 32, 1980.

78. **Biswas, T. D. and Gawande, S. P.,** Relation of manganese in genesis of catenary soils, *J. Indian Soc. Soil. Sci.,* 12, 261, 1964.

79. **Blanton, C. J., Desforges, C. E., Newland, L. W., and Ehlmann, A. J.,** A survey of mercury distributions in the Terlinqua area of Texas, in *Trace Subst. in Environ. Health,* Vol. 9, Hemphill, D. D., Ed., University of Missouri, Columbia, Mo., 1975, 139.

80. **Bloom, P. R. and McBride, M. B.,** Metal ion binding and exchange with hydrogen ions in acid-washed peat, *Soil Sci. Soc. Am. J.,* 43, 687, 1979.

81. **Bloomfield, C.,** The translocation of metals in soils, in *The Chemistry of Soil Processes,* Greenland, D. J. and Hayes, M. H. B., Eds., John Wiley & Sons, New York, 1981, 463.

82. **Bloomfield, C. and Pruden, G.,** The behavior of Cr (VI) in soil under aerobic and anaerobic conditions, *Environ. Pollut.,* 23a, 103, 1980.

83. **Boardman, N. K.,** Trace elements in photosynthesis, in *Trace Elements in Soil-Plant-Animal Systems,* Nicholas, P.J. D. and Egan, A. R., Eds., Academic Press, New York, 1975, 199.

84. **Bohn, H. L. and Seekamp, G.,** Beryllium effects on potatoes and oats in acid soil, *Water Air Soil Pollut.,* 11, 319, 1979.

85. **Bolewski, A., Fijal, J., Klapyta, Z., Manecki, A., Ptasińska, M., and Zabiński, W.,** Geochemical changes in natural environment within zones of industrial emission of fluorine compounds, *PAN Miner. Trans.,* 50, 7, 1976 (Po).

86. **Bolt, G. H. and Bruggenwert, M. G. M., Eds.,** *Soil Chemistry. A. Basic Elements,* Elsevier, Amsterdam, 1976, 281.

87. **Bolter, E., Butz, T., and Arseneau, J. F.,** Mobilization of heavy metals by organic acids in the soils of a lead mining and smelting district, in *Trace Subst. Environ. Health,* Vol. 9, Hemphill, D. D., Ed., University of Missouri, Columbia, Mo., 1975, 107.

88. **Bolton, J.,** Liming effects on the toxicity to perennial ryegrass of a sewage sludge contaminated with zinc, nickel, copper and chromium, *Environ. Pollut.,* 9, 295, 1975.

89. **Bondarenko, G. P.,** Seasonal dynamics of mobile forms of trace elements and iron in bottomland soils of the Ramenskoe widening of the Moscow river, *Nauch. Dokl. Vyssh. Shk. Biol. Nauki,* 4, 202, 1962 (Ru).

90. **Bonilla, I., Cadania, C., and Carpena, O.,** Effects of boron on nitrogen metabolism and sugar levels of sugar beet, *Plant Soil,* 57, 3, 1980.

91. **Boratyński, K., Roszyk, E., and Ziętecka, M.,** Review on research on microelements in Poland (B, Cu and Mn), *Rocz. Glebozn.,* 22, 205, 1971 (Po).

92. **Borovik-Romanova, T. F. and Bielova, I. E. A.,** About geochemistry of lithium, in *Problems of Geochemical Ecology of Organism,* Izd. Nauka, Moscow, 1974, 118 (Ru).

93. **Bowden, J. W., Nagarajah, S., Barrow, N. J., Posner, A. M., and Quirk, J. P.,** Describing the adsorption of phosphate, citrate and selenite on a variable-charge mineral surface, *Aust. J. Soil Res.,* 18, 49, 1980.

94. **Bowen, H. J. M.,** *Environmental Chemistry of the Elements,* Academic Press, New York, 1979, 333.

95. **Bowen, J. E.,** Kinetics of boron, zinc and copper uptake by barley and sugarcane, paper presented at Int. Symp. Trace Element Stress in Plants, Los Angeles, November 6, 1979, 24.

96. **Boyle, R. W.,** The geochemistry of silver and its deposits, *Geol. Surv. Can. Bull.,* 160, 264, 1968.

97. **Boyle, R. W., Alexander, W. M., and Aslin, G. E. M.,** Some observations on the solubility of gold, *Geol. Surv. Can. Pap.,* 75, 24, 1975.

98. **Boyle, R. W. and Jonasson, I. R.,** The geochemistry of arsenic and its use as an indicator element in geochemical prospecting, *J. Geochem. Explor.,* 2, 251, 1973.

99. **Bradford, G. R., Bair, F. L., and Hunsaker, V.,** Trace and major element contents of soil saturation extracts, *Soil Sci.,* 112, 225, 1971.

100. **Bradley, R. I.,** Trace elements in soils ground, Llechryd, Wales, *Geoderma,* 24, 17, 1980.

101. **Bradshaw, A. D.,** The evolution of metal tolerance and its significance for vegetation establishment on metal contaminated sites, paper presented at Int. Conf. on Heavy Metals, Toronto, October 27, 1975, 599.

102. **Braude, G. L., Nash, A. M., Wolf, W. J., Carr, R. L., and Chaney, R. L.,** Cadmium and lead content of soybean products, *J. Food Sci.,* 45, 1187, 1980.

103. **Bromfield, S. A.,** The effect of manganese-oxidizing bacteria and pH on the availability of manganeous ions and manganese oxides to oats in nutrient solutions, *Plant Soil,* 49, 23, 1978.

104. **Brooks, R. R.,** Copper and cobalt uptake by *Hausmaniastrum* species, *Plant Soil,* 48, 545, 1977.

104a. **Brown, J. C.,** Iron deficiency and boron toxicity in alkaline soils, in *Workshop on Plant Adaptation to Mineral Stress in Problem Soils,* Wright, M. J., Ed., Cornell University, Ithaca, N.Y., 1976, 83.

105. **Brown, J. C. and Devine, T. E.,** Inheritance of tolerance or resistance to manganese toxicity in soybeans, *Agron. J.,* 72, 898, 1980.

106. **Brown, B. D. and Rolston, D. E.,** Transport and transformation of methyl bromide in soils, *Soil Sci.,* 130, 68, 1980.

107. **Browne, C. L. and Fang, S. C.,** Uptake of mercury vapour by wheat. An assimilation model, *Plant Physiol.,* 61, 430, 1978.

108. **Broyer, T. C., Johnson, C. N., and Paull, R. E.,** Some aspects of lead in plant nutrition, *Plant Soil,* 36, 301, 1972.

109. **Bubliniec, E.,** Uptake of Fe, Mn, Cu, Zn, Mo and B by forest ecosystem and the possibility of meeting requirements of these elements from import by precipitation, paper presented at Int. Soil Science Conf., Prague, Czechoslovakia, August 1981, 61 (Ru).

110. **Buliński, R., Kot, A., Kutulas, K., and Szydlowska, E.,** Impact of automotive and industrial pollution on contents of cadmium, lead, mercury, zinc, and copper in cereals, *Bromatol. Chem. Toksykol.,* 10, 395, 1977 (Po).

111. **Bull, K. R., Roberts, R. D., Inskip, M. J., and Godman, G. T.,** Mercury concentrations in soils, grass, earthworms and small mammals near an industrial emission source, *Environ. Pollut.,* 12, 135, 1977.

112. **Burridge, J. C. and Scott, R. O.,** A rotating briquetted-disk method for determination of boron and other elements in plant material by emission spectroscopy, *Spectrochim. Acta,* 30, 479, 1975.

113. **Burton, M. A. S.,** Vegetation damage during an episode of selenium dioxide pollution, paper presented at Int. Symposium Trace Element Stress in Plants, Los Angeles, November 6, 1979, 67.

114. **Bussler, W.,** Microscopical possibilities for the diagnosis of trace element stress in plants, paper presented at Int. Symp. Trace Element Stress in Plants, Los Angeles, November 6, 1979, 36.

115. **Bussler, W.,** Physiological functions and utilization of copper, in *Copper in Soils and Plants,* Loneragan, J. F., Robson, A. D., and Graham, R. D., Eds., Academic Press, Sydney, 1981, 213.

116. **Byrne, A. R. and Kosta, L.,** Studies on the distribution and uptake of mercury in the area of the mercury mine at Idrija, Slovenia, Yugoslavia, *Vestn. Slov. Kem. Drus.,* 17, 5, 1970.

117. **Byrne, A. R. and Ravnik, V.,** Trace element concentrations in higher fungi, *Sci. Total Environ.,* 6, 65, 1976.

118. **Cannon, H. L.,** Lead in vegetation, in Lead in the Environment, Lovering, T. G., Ed., U.S. Geol. Surv. Prof. Pap., 957, 23, 1976.

119. **Cannon, H. L., Miesch, A. T., Welch, R. M., and Nielsen, F. H.,** Vanadium, in *Geochemistry and the Environment,* Vol. 2, Hopkins, L. L., Ed., N.A.S., Washington, D.C, 1977, 93.

120. **Cannon, H. L., Shacklette, H. T., and Bystron, H.,** Metal absorption by *Equisetum* (Horsetail), U.S. Geol. Surv. Bull., 1278a, 21, 1968.

121. **Carlisle, E. M., McKeague, J. A., Siever, R., and van Soest, P. J.,** Silicon, in *Geochemistry and the Environment,* Vol. 2, Hopps, H. C., Ed., N.A.S., Washington, D.C., 1977, 54.

122. **Cary, E. E., Allaway, W. H., and Olson, O. E.,** Control of chromium concentrations in food plants. I. Absorption and translocation of chromium in plants. II. Chemistry of chromium in soils and its availability to plants, *J. Agric. Food Chem.,* 25, I, 300; II, 305, 1977.

123. **Case, A. A., Selby, L. A., Hutcheson, D. P., Ebens, R. J., Erdman, J. A., and Feder, G. L.,** Infertility and growth suppression in beef cattle associated with abnormalities in their geochemical environment, in *Trace Subst. Environ. Health,* Vol. 6, Hemphill, D. D., Ed., University of Missouri, Columbia, Mo., 1972, 15.

124. **Cataldo, D. A., Garland, T. R., and Wildung, R. E.,** Nickel in plants, *Plant Physiol.,* 62, I, 563; II, 566, 1978.

125. **Chamel, A. and Garbec, J. P.,** Penetration of fluorine through isolated pear leaf cuticles, *Environ. Pollut.,* 12, 307, 1977.

126. **Chaney, R. L.,** Crop and food chain effects of toxic elements in sludges and effluents, in *Proc. 1st Conf. on Recycling Municipal Sludges and Effluents on Land,* National Association State University and Land Grant Colleges, Washington, D.C., 1973, 120.

127. **Chaney, R. L.,** Sludge management: risk assessment for plant and animal life, in Proc. Spring Seminar on Sludge Management, Washington, D.C., 1980, 19.

128. **Chaney, R. L., Brown, J. C., and Tiffin, L. O.,** Obligatory reduction of ferric chelates in iron uptake by soybeans, *Plant Physiol.,* 50, 208, 1972.

129. **Chaney, R. L. and Hornick, S. B.,** Accumulation and effects of cadmium on crops, paper presented at Int. Cadmium Conf., San Francisco, January 31, 1977, 125.

130. **Chaney, R. L., Hornick, S. B., and Sikora, L. J.,** Review and preliminary studies of industrial land treatment practice, in Proc. 7th Annu. Res. Symp. Land Disposal: Hazardous Waste, Cincinnati, Ohio, 1981, 201.

131. **Chapman, H. D., Ed.,** *Diagnostic Criteria for Plants and Soils,* University of California, Riverside, Calif., 1972, 793.

132. **Chattopadhyay, A. and Jervis, R. E.,** Multielement determination in market-garden soils by instrumental Photon Activation analysis, *Anal. Chem.,* 46, 1630, 1974.

133. **Chaudan, R. P. S. and Powar, S. L.,** Tolerance of wheat and pea to boron in irrigated water, *Plant Soil,* 50, 145, 1978.

134. **Chaudry, F. M., Wallace, A., and Mueller, R. T.,** Barium toxicity in plants, *Commun. Soil Sci. Plant Anal.,* 795, 1977.

135. **Cheng, B. T.,** Soil organic matter as a plant nutrient, in *Proc. Ser. Soil Organic Matter Studies,* IAEG, Vienna, 1977, 31.

136. **Cheshire, M. V., Berrow, M. L., Goodman, B., and Mundie, C. M.,** Metal distribution and nature of some Cu, Mn and V complexes in humic and fulvic acid fractions of soil organic matter, *Geochim. Cosmochim. Acta,* 41, 1131, 1977.

137. **Chhabra, R., Singh, A., and Abrol, I. P.,** Fluorine in sodic soils, *Soil Sci. Soc. Am. J.,* 44, 33, 1980.

138. **Chisholm, D.,** Lead, arsenic and copper content of crops grown on lead-arsenic-treated and untreated soils, *Can. J. Plant Sci.,* 52, 583, 1972.

139. **Chudecki, Z.,** Studies of iodine content and distribution in soils of the West Coast, *Rocz. Glebozn.,* 9d, 113, 1960 (Po).

140. **Chukhrov, F. V., Gorshkov, A. I., Tiuriukanov, A. N., Beresovskaya, V. V., and Sivtsov, A. W.,** About geochemistry and mineralogy of manganese and iron in young hypergenic products, *Izv. Akad. Nauk. SSSR Ser Geol.,* 7, 5, 1980 (Ru).

141. **Clark, R. B.,** Effect of aluminum on the growth and mineral elements of Al-tolerant and Al-intolerant corn, *Plant Soil,* 47, 653, 1977.

142. **Clarkson, D. T. and Hanson, J. B.,** The mineral nutrition of higher plants, in *Annual Review of Plant Physiology,* Vol. 31, Briggs, W. R., Green, P. W., and Jones, R. L., Eds., Ann. Reviews Inc., Palo Alto, CA, 1980, 239.

143. **Clayton, P. M. and Tiller, K. G.,** A chemical method for the determination of the heavy metal content of soils in environmental studies, *CSIRO Aust. Div. Soils Techn. Pap.,* 4, 1, 1979.

144. **Colbourn, P., Allaway, B. J., and Thornton, J.,** Arsenic and heavy metals in soils associated with regional geochemical anomalies in south-west England, *Sci. Total Environ.,* 4, 359, 1975.

145. **Connor, J. J. and Shacklette, H. T.,** Background geochemistry of some rocks, soils, plants and vegetables in the conterminous United States, *U.S. Geol. Surv. Prof. Pap.,* 574f, 168, 1975.

146. **Cooke, J. A., Johnson, M. S., Davison, A. W., and Bradshaw, A. D.,** Fluoride in plants colonising fluorspar mine waste in the Peak District and Weardale, *Environ. Pollut.,* 11, 9, 1976.

147. **Coppenet, M., Moré, E., Le Covre, L., and Le Mao, M.,** Variations de la teneur en cobalt de ray-grass étude de techniques d'enrichissement, *Ann. Agron.,* 23, 165, 1972.

148. **Cottenie, A., Verloo, M., Kiekens, L., Camerlynck, R., Velghe, G., and Dhaese, A.,** *Essential and Non Essential Trace Elements in the System Soil-Water-Plant,* I.W.O.N.L., Brussels, 1979, 75.

149. **Cox, R. M. and Hutchinson, T. C.,** Multiple metal tolerance in the grass *Deschampsia cespitosa L. Beauv.* from the Sudbury Smelting area, *New Phytol.,* 84, 631, 1980.

150. **Cox, R. M. and Hutchinson, T. C.,** Environmental factors influencing the rate of spread of the grass *Deschampsia caespitosa* invading areas around the Sudbury Nickel-Copper Smelter, *Water Air Soil Pollut.,* 16, 83, 1981.

151. **Cox, F. R. and Kamprath, E. J.,** Micronutrient soil tests, in *Micronutrients in Agriculture,* Mortvedt, J. J., Giordano, P. M., and Lindsay, W. L., Eds., Soil Science Society of America, Madison, Wis., 1972, 289.

152. **Craze, B.,** Restoration of Captains Flat mining area, *J. Soil Conserv. N.S.W.,* 33, 98, 1977.

153. **Crecelius, E. A., Johnson, C. J., and Hofer, G. C.,** Contamination of soils near a copper smelter by arsenic, antimony and lead, *Water Air Soil Pollut.,* 3, 337, 1974.

154. **Cumakov, A. and Neuberg, J.,** Bilanz der Spurenelemente in der Pflanzenproduktion der Tschechoslowakei, *Phosphorsaeure,* 28, 198, 1970.

155. **Cunningham, L. M., Collins, F. W., and Hutchinson, T. C.,** Physiological and biochemical aspects of cadmium toxicity in soybean, paper presented at Int. Conf. on Heavy Metals in the Environment, Toronto, October 27, 1975, 97.

156. **Curylo, T.,** The effect of some soil properties and of NPK fertilization levels on the uptake of cobalt by plants, *Acta Agrar. Silvestria,* 20, 57, 1981 (Po).

157. **Curzydlo, J., Kajfosz, J., and Szymczyk, S.,** Lead and bromine in plants grown along roadsides, in *Proc. Conf. No. 2, Effects of Trace Element Pollutants and Sulfur on Agric. Environmental Quality,* Kabata-Pendias, A., Ed., IUNG, Pulawy, Poland, 1980, 181 (Po).

158. **Czarnowska, K.,** The accumulation of heavy metals in soils and plants in Warsaw area, *Pol. J. Soil Sci.,* 7, 117, 1974.

159. **Czarnowska, K.,** Heavy metal contents of surface soils and of plants in urban gardens, paper presented at Symp. on Environmental Pollution, Plock, November 12, 1982 (Po).

160. **Czarnowska, K. and Jopkiewicz, K.,** Heavy metals in earthworms as an index of soil contamination, *Pol. J. Soil Sci.,* 11, 57, 1978.

161. **Dabin, P., Marafante, E., Mousny, J. M., and Myttenaere, C.,** Adsorption, distribution and binding of cadmium and zinc in irrigated rice plants, *Plant Soil,* 50, 329, 1978.

162. **Dabrowski, J., Poliborski, M., Majchrzak, J., and Polak, J.,** Studies on mercury content of cereals collected from the fields at harvest time of 1975, in Proc. 17th Natl. Semin. of the Institute of Plant Protection, Poznan, 1977, 95 (Po).

163. **Dassani, S. D., McClellan, E., and Gordon, M.,** Submicrogram level determination of mercury in seeds, grains, and food products by cold-vapor atomic absorption spectrometry, *J. Agric. Food Chem.,* 23, 671, 1975.

164. **Davey, B. G. and Wheeler, R. C.,** Some aspects of the chemistry of lithium in soils, *Plant Soil,* 5, 49, 1980.

164a. **Davies, B. E.,** Mercury content of soil in Western Britain with special reference to contamination from base metal mining, *Geoderma,* 16, 183, 1976.

165. **Davies, B. E.,** Heavy metal pollution of British agricultural soils with special reference to the role of lead and copper mining, in Proc. Int. Semin. on Soil Environment and Fertility Management in Intensive Agriculture, Tokyo, 1977, 394.

166. **Davies, B. E., Ed.,** *Applied Soil Trace Elements,* John Wiley & Sons, New York, 1980, 482.

167. **Davies, B. E., Cartwright, J. A., and Hudders, G. L.,** Heavy metals in soils and plants of urban England, in *Proc. Conf. No. 2, Effects of Trace Element Pollutants on Agric. Environ. Quality,* Kabata-Pendias, A., Ed., IUNG, Pulawy, Poland, 1978, 117.

168. **Davies, B. E. and Ginnever, R. C.,** Trace metal contamination of soils and vegetables in Shipham, Somerset, *J. Agric. Sci., Camb.,* 93, 753, 1979.

169. **Davies, T. C. and Bloxam, T. W.,** Heavy metal distribution in laterites, Southwest of Regent, Freetown Igneous Complex, Sierra Leone, *Econ. Geol.,* 74, 638, 1979.

169a. **Davies, B. E. and White, H. M.,** Trace elements in vegetable grown on soils contaminated by base metal mining, *J. Plant Nutr.,* 3, 387, 1981.

170. **Davis, R. D.,** Uptake of copper, nickel and zinc by crops growing in contaminated soils, *J. Sci. Food Agric.,* 30, 937, 1979.

171. **Davis, R. D., Beckett, P. H. T., and Wollan, E.,** Critical levels of twenty potentially toxic elements in young spring barley, *Plant Soil,* 49, 395, 1978.

172. **Davis, G. K., Jorden, R., Kubota, J., Laitinen, H. A., Matrone, G., Newberne, P. M., O'Dell, B. L., and Webb, J. S.,** Copper and molybdenum, in *Geochemistry and Environment,* Vol. 1, Davis, G. K., Ed., N.A.S., Washington, D.C., 1974, 68.

173. **Davison, A. W., Blakemore, J., and Wright, D. A.,** A re-examination of published data on the fluoride content of pastures, *Environ. Pollut.,* 10, 209, 1976.

174. **DeKock, R. C.,** Iron nutrition under conditions of stress, *J. Plant Nutr.,* 3, 513, 1981.

175. **Dev, G.,** Experiment on plant uptake of radio-strontium from contaminated soils. Effect of phosphate fetilization, *Sci. Rep. Agric. Coll. Norway,* 44, 1, 1965.

176. **Diez, Th. and Rosopulo, A.,** Schwermetallgehalte in Böden und Pflanzen nach extrem hohen Klärschlammgaben, *Sonderdruck Landw. Forsch.,* 33, 236, 1976.

177. **Dimitrov, G.,** Iodine content and migration in the Bulgarian soils, *Nauchni Tr.,* 17, 319, 1968 (Bu).

178. **Dirven, J. G. P. and Ehrencron, V. K. P.,** Trace elements in pasture soils, *Surinaamse Landbouw,* 12, 11, 1964 (Du).

179. **Dmowski, K. and Karolewski, A.,** Cumulation of zinc, cadmium and lead in invertebrates and in some vertebrates according to the degree of an area contamination, *Ekol. Pol.,* 27, 333, 1979 (Po).

180. **Dobritskaya, U. I.,** Distribution of vanadium in natural objects, *Agrokhimiya,* 3, 143, 1969 (Ru).

181. **Dobrowolski, J. W.,** Investigation of environmental conditionings of proliferative diseases with particular consideration of cattle leukaemia, *Sci. Bull. S. Staszic Univ. Min. Metall.,* 877, 139, 1981 (Po).

182. **Dobrzański, B., Dechnik, I., Gliński, J., Pondel, H., and Stawiński, J.,** Surface area of arable soils of Poland, *Rocz. Nauk Roln.* 165d, 7, 1977 (Po).

183. **Dobrzański, B., Gliński, J., and Uziak, S.,** Occurrence of some elements in soils of Rzeszów voyevodship as influenced by parent rocks and soil types, *Ann. UMCS,* 24e, 1, 1970 (Po).

184. **Doelman, P.,** Effects of lead on properties of lead-sensitive and lead-tolerant cells, in *Environmental Biogeochemistry and Geomicrobiology,* Krumbein, W. E., Ed., Ann Arbor Science, Ann Arbor, Mich., 1978, 989.

185. **Doelman, P. and Haanstra, L.,** Effect of lead on soil respiration and dehydrogenase activity, *Soil Biol. Biochem.,* 11, 475, 1979.

186. **Dolobovskaya, A. S.,** The character of biogenic accumulation of minor elements in forest litter, *Pochvovedenie,* 3, 63, 1975 (Ru).

187. **Dommerques, Y. R. and Krupa, S. V., Eds.,**., *Interactions Between Non-Pathogenic Soils Microorganisms and Plants,* Elsevier, Amsterdam, 1978, 475.

188. **Donchev, I. and Mirchev, S.,** Molybdenum in soils of Bulgaria, *Izv. Centr. Nauch. Inst. Pozvozn. Agrotekh. ''Nikola Pushkarov'',* 1, 5, 1961 (Bu).

189. **Dormaar, J. E.,** Alkaline cupric oxide oxidation of roots and alkaline-extractable organic matter on chernozemic soils, *Can. J. Soil Sci.,* 59, 27, 1979.

190. **van Dorst, S. H. and Peterson, P. J.,** Selenium speciation in the solution and its relevence to plant uptake, in *Proc. 9th Int. Plant Nutrition Colloquim,* Scaife, A., Ed., Commonwealth Agriculture, Bureaux, Bucks., 1, 134, 1982.

191. **Dossis, P. and Warren, L. J.,** Distribution of heavy metals between the minerals and organic debris in a contaminated marine sediment, in *Contaminants and Sediments,* Ann Arbor Sci., Ann Arbor, Mich., 1980, 119.

192. **Doyle, P. J. and Fletcher, K.,** Molybdenum content of bedrocks, soils and vegetation and the incidence of copper deficiency in cattle in Western Manitoba, in *Molybdenum in the Environment,* Chappell, W. R. and Peterson, K. K., Eds., Marcel Decker, New York, 1977, 371.

193. **Doyle, P., Fletcher, W. K., and Brink, V. C.,** Trace element content of soils and plants from Selwyn Mountains, Yukon and Northwest Territories, *Can. J. Bot.,* 51, 421, 1973.

194. **Dudal, R.,** Land resources for agricultural development, in Plenary Papers, 11th Congress ISSS, Edmonton, Canada, 1978, 314.

195. **Dudas, M. J. and Pawluk, S.,** Heavy metals in cultivated soils and in cereal crops in Alberta, *Can. J. Soil Sci.,* 57, 329, 1977.

196. **Duddy, I. R.,** Redistribution and fractionation of rare-earth and other elements in a weathering profile, *Chem. Geol.,* 30, 363, 1980.

197. **Duke, J. A.,** Ethnobotanical observations on the Chocó Indians, *Econ. Bot.,* 23, 344, 1970.

198. **Dvornikov, A. G. and Petrov, V. Ya.,** Some data on the mercury content in soils of the Nagolnyi Mt. Range, *Geokhimiya,* 10, 920, 1961 (Ru).

199. **Dvornikov, A. G., Ovsyannikova, L. B., and Sidenko, O. G.,** Some peculiarities of biological absorption coefficients and of biogeochemical coefficient in hydrothermal deposits of the Donbas in connection with the prognostication of hidden mercury mineralization, *Geokhimiya,* 4, 626, 1976 (Ru).

200. **Ebens, R. and Shacklette, H. T.,** Geochemistry of some rocks, mine spoils, stream sediments, soils, plants, and waters in the Western Energy Region of the conterminous United States, *U.S. Geol. Surv. Prof. Pap.,* 1237, 173, 1982.

201. **Eglinton, G. and Murphy, M. T. J., Eds.,** *Organic Geochemistry,* Springer-Verlag, Berlin, 1969, 828.

202. **Ehlig, C. F., Allaway, W. H., Cary, E. E., and Kubota, J.,** Differences among plant species in selenium accumulation from soil low in available selenium, *Agron. J.,* 60, 43, 1968.

203. **Ekman, P., Karlsson, N., and Svanberg, O.,** Investigations concerning cobalt problems in Swedish animal husbandry, *Acta Agric. Scand.,* 11, 103, 1952.

204. **El-Bassam, N., Keppel, H., and Tietjen, C.,** Arsenic transfer in soils, in Abstr. ESNA Environ. Pollut. Working Group, Cadarache, 1975a, 1.

205. **El-Bassam, N., Poelstra, P., and Frissel, M. J.,** Chrom und Quecksilber in einem seit 80 Jahren mit städtischen Abwasser berieselten Boden, *Z. Pflanzenernaehr. Bodenkd.,* 3, 309, 1975b.

206. **El-Bassam, N. and Tietjen, C.,** Municipal sludge as organic fertilizer with special reference to the heavy metals constituents, in *Soil Organic Matter Studies,* Vol. 2, IAEA, Vienna, 1977, 253.

207. **Elek, E.,** Investigation of the manganese supply in the drainage basin of the Lókos brook, *Agrokem. Talajtan.,* 15, 277, 1966 (Hu).

207a. **Ellis, B. G. and Knezek, B. D.,** Adsorption reactions of micronutrients in soils, in *Micronutrients in Agriculture,* Mortvedt, J. J., Giordano, P. M., and Lindsay, W. L., Eds., Soil Science Society of America, Madison, Wis., 1972, 59.

208. **Elfving, D. C., Haschek, W. M., Stehn, R. A., Bache, C. A., and Lisk, D. J.,** Heavy metal residues in plants cultivated on and in small mammals indigenous to old orchard soils, *Arch. Environ. Health,* 3-4, 95, 1978.

209. **Elgawhary, S. M., Malzer, G. L., and Barber, S. A.,** Calcium and strontium transport to plant roots, *Soil Sci. Am. Proc.,* 36, 794, 1972.

210. **Elinder, C. G. and Friberg, L.,** Antimony, in *Handbook on the Toxicology of Metals,* Friberg, L., Ed., Elsevier, Amsterdam, 1979, 283.

211. **El-Sheik, A. M. and Ulrich, A.,** Interactions of rubidium, sodium and potassium on the nutrition of sugar beet plants, *Plant Physiol.,* 46, 645, 1970.

212. **Elsokkary, I. H.,** Selenium distribution, chemical fractionation and adsorption in some Egyptian alluvial and lacustrin soils, *Z. Pflanzenernaehr. Bodenkd.,* 143, 74, 1980.

213. **Elsokkary, I. H. and Låg, J.,** Status of some trace elements in Egyptian soils and in wheat grains, *Jordundersokelsens Saertrykk,* 285, 35, 1980.

214. **Anon.,** Environmental Mercury and Man, report of the Department of the Environ. Central Unit on Environ. Pollut., London, 1976, 92.

215. **Erämetsä, O., Haarala, A., and Yliruokanen, I.,** Lanthanoid content in three species of *Equisetum, Suom. Kemistil.,* 46b, 234, 1973.

216. **Erämetsä, O., Viinanen, R., and Yliruokanen, I.,** Trace elements in three species of *Lycopodium, Suom. Kemistil.,* 46b, 355, 1973.

217. **Erämetsä, O. and Yliroukanen, I.,** The rare earths in lichens and mosses, *Suom. Kemistil.,* 44b, 121, 1971.

218. **Erdman, J. A., Shacklette, H. T., and Keith, J. R.,** Elemental composition of selected native plants and associated soils from major vegetation-type areas in Missouri, *U.S. Geol. Surv. Prof. Pap.,* 954c, 30, 1976a.

219. **Erdman, J. A., Shacklette, H. T., and Keith, J. R.,** Elemental composition of corn grains, soybean seeds, pasture grasses and associated soils from selected areas in Missouri, *U.S. Geol. Surv. Prof. Pap.,* 954d, 23, 1976b.

220. **Ermolaev, I.,** The effect of trace molybdenum element dressing of fodder legumes on the content of the trace element in the feed in respect to animal health, *Pochvoz. Agrokhim.,* 4, 77, 1970 (Bu).

221. **Ettala, E. and Kossila, V.,** Mineral content in heavily nitrogen fertilized grass and its silage, *Ann. Agric. Fenn.,* 18, 252, 1979.

222. **Evans, Ch. S., Asher, C. J., and Johnson, C. M.,** Isolation of dimethyl diselenide and other volatile selenium compounds from *Astragalus racemosus, Austr. J. Biol. Sci.,* 21, 13, 1968.

223. **van Faassen, G. H.,** Effects of mercury compounds on soil microbes, *Plant Soil,* 38, 485, 1973.

224. **Faber, A. and Niezgoda, J.,** Contamination of soils and plants in a vicinity of the zinc and lead smelter, *Rocz. Glebozn.,* 33, 93, 1982 (Po).

225. **Fang, C. L., Sung, T. C., and Yeh Bing,** Trace elements in the soils of north-eastern China and eastern Inner Mongolia, *Acta Pedol. Sin.,* 11, 130, 1963 (Ch).

226. **FAO/UNESCO,** *Soil Map of the World,* Vol. I, Legend, UNESCO, Paris, 1974.

227. **Farrah, H., Hatton, D., and Pickering, W. F.,** The affinity of metal ions for clay surface, *Chem. Geol.,* 28, 55, 1980.

228. **Farrah, H. and Pickering, W. F.,** The sorption of lead and cadmium species by clay minerals, *Aust. J. Chem.,* 30, 1417, 1977.

229. **Farrah, H. and Pickering, W. F.,** The sorption of mercury species by clay minerals, *Water Air Soil Pollut.,* 9, 23, 1978.

230. **Fiadotau, U. L., Bagdanau, M. P., and Brysan, M. C.,** Iodine content of plants as influenced by plant kind and stage of growth, *Vesti ANBSSR,* 4, 16, 1975 (Ru).

231. **Fiskell, J. G. A. and Brams, A.,** Root desorption analysis as a diagnostic technique for measuring soil-plant relationships, *Soil Crop Sci. Soc. Fla.* 25, 128, 1965.

232. **Fiskesjö, G.,** Mercury and selenium in a modified *Allium* test, *Hereditas,* 91, 169, 1979.

233. **Fleischer, M., Sarofim, A. F., Fassett, D. W., Hammond, P., Shacklette, H. T., Nisbet, I. C. T., and Epstein, S.,** Environmental impact of cadmium, *Environ. Health Perspect.,* 5, 253, 1974.

234. **Fleming, G. A.,** Selenium in Irish soils and plants, *Soil Sci.,* 94, 28, 1968.

235. **Fleming, G. A.,** Mineral composition of herbage, in *Chemistry and Biochemistry of Herbage,* Butler, G. W. and Bailey, R. W., Eds., Academic Press, London, 1973, 529.

236. **Fleming, G. A., Walsh, T., and Ryan, P.,** Some factors influencing the content and profile distribution of trace elements in Irish soils, in Proc. 9th Int. Congr. Soil Sci., Vol. 2, Adelaide, Australia, 341, 1968.

237. **Fletcher, K. and Brink, V. C.,** Content of certain trace elements in range forages from south central British Columbia, *Can. J. Plant Sci.,* 49, 517, 1969.

238. **Folkeson, L.,** Interspecies calibration of heavy metal concentrations in nine mosses and lichens: applicability to deposition measurements, *Water Air Soil Pollut.,* 11, 253, 1979.

239. **Folkeson, L.,** Heavy-metal accumulation in the moss *Pleurozium schreberi* in the surroundings of two peat-fired power plants in Finland, *Ann. Bot. Fenn.,* 18, 245, 1981.

240. **Forbes, E. A., Posner, A. M., and Quirk, J. P.,** The specific adsorption of divalent Cd, Co, Cu, Pb and Zn on goethite, *J. Soil Sci.,* 27, 154, 1976.

241. **Foy, C. D., Chaney, R. L., and White, M. C.,** The physiology of metal toxicity in plants, *Annu. Rev. Physiol.,* 29, 511, 1978.

242. **Förstner, U. and Müller, G.,** *Schwermetalle in Flüssen und Seen,* Springer-Verlag, Berlin, 1974, 225.

243. **Frank, R., Ishida, K., and Suda, P.,** Metals in agricultural soils of Ontario, *Can. J. Soil Sci.,* 56, 181, 1976.

244. **Frank, R., Stonefield, K. I., and Suda, P.,** Metals in agricultural soils of Ontario, *Can. J. Soil Sci.,* 59, 99, 1979.

245. **Freedman, B. and Hutchinson, T. C.,** Pollutant inputs from the atmosphere and accumulations in soils and vegetation near a nickel-copper smelter at Sudbury, Ontario, Canada, *Can. J. Bot.,* 58, 108, 1980.

246. **Frost, R. R. and Griffin, R. A.,** Effect of pH on adsorption of arsenic and selenium from landfill leachate by clay minerals, *Soil Sci. Soc. Am. J.,* 41, 53, 1977.

247. **Frøslie, A., Karlsen, J. T., and Rygge, J.,** Selenium in animal nutrition in Norway, *Acta Agric. Scand.,* 30, 17, 1980.

248. **Fuchs, W. A. and Rose, A. W.,** The geochemical behavior of platinum and palladium in weathering cycle in the Stillwater Complex, Montana, *Econ. Geol.,* 69, 332, 1974.

249. **Furr, A. K., Stoewsand, G. S., Bache, C. A., and Lisk, D. J.,** Study of guina pigs fed swiss chard grown on municipal sludge-amended soil, *Arch. Environ. Health,* 3/4, 87, 1976.

250. **Furr, A. K., Parkinson, T. F., Bache, C. A., Gutenmann, W. H., Pakkala, I., and Lisk, D. J.,** Multielement absorption by crops grown on soils amended with municipal sludges ashes, *J. Agric. Food Chem.,* 28, 660, 1980.

251. **Gadd, G. M. and Griffiths, A. J.,** Microorganisms and heavy metal toxicity, *Microb. Ecol.,* 4, 303, 1978.

252. **Gadde, R. R. and Laitinen, H. A.,** Study of the interaction of lead with corn root exudate, *Environ. Lett.,* 5, 91, 1973.

253. **Gadde, R. R. and Laitinen, H. A.,** Studies of heavy metal adsorption by hydrous iron and manganese oxides, *Anal. Chem.,* 46, 2022, 1974.

253a. **Gallego, R. and Fernandez, E.,** Trace elements in soils of the Vegas Altas plain of Guadiana, *Ann. Edafol. Agrobiol.,* 22, 307, 1963 (Es).

254. **Ganther, H. E.,** Biochemistry of selenium, in *Selenium,* Van Nostrand, New York, 1974, 546.

255. **Garbanov, S.,** Content and distribution of chromium in main soil types of Bulgaria, *Pochvozn. Agrokhim.,* 4, 98, 1975 (Bu).

256. **Garrels, R. M. and Christ, C. L.,** *Solutions, Minerals and Equilibria,* Harper & Row, New York, 1965, 450.

257. **Gartrell, J. W.,** Distribution and correction of copper deficiency in crops and pastures, in *Copper in Soils and Plants,* Loneragan, J. F., Robson, A. D., and Graham, R. D., Eds., Academic Press, New York, 1981, 313.

258. **Gartrell, J. W., Robson, A. D., and Loneragan, J. F.,** A new tissue test for accurate diagnosis of copper deficiency in cereals, *J. Agric. West Aust.,* 20, 86, 1979.

259. **Gatz, D. F., Barlett, J., and Hassett, J. J.,** Metal pollutants in agricultural soils and the St. Louis urban rainfall anomaly, *Water Air Soil Pollut.,* 15, 61, 1981.

260. **Gemmell, R. P.,** Novel revegetation techniques for toxic sites, paper presented at Int. Conf. on Heavy Metals in Environment, Toronto, October 27, 1975, 579.

261. **Gilbert, O. L.,** Effects of air pollution on landscape and land-use around Norwegian aluminium smelters, *Environ. Pollut.,* 8, 113, 1975.

262. **Gilpin, L. and Johnson, A. H.,** Fluorine in agricultural soils of Southeastern Pennsylvania, *Soil Sci. Soc. Am. J.,* 44, 255, 1980.

263. **Girling, C. A., Peterson, P. J., and Warren, H. V.,** Plants as indicators of gold mineralization at Waston Bar, British Columbia, Canada, *Econ. Geol.,* 74, 902, 1979.

264. **Gish, C. D. and Christensen, R. E.,** Cadmium, nickel, lead and zinc in earthworms from roadside soil, *Environ. Sci. Technol.,* 7, 1060, 1973.

265. **Gissel-Nielsen, G.,** Selenium concentration in Danish forage crops, *Acta Agric. Scand.,* 25, 216, 1975.

266. **Gladstones, J. S. and Loneragan, J. F.,** Nutrient elements in herbage plants in relation to soil adaptation and animal nutrition, in *Proc. 9th Int. Grassland Congr.,* University of Queensland Press, Brisbane, Australia, 1970, 350.

267. **Glinski, J. and Baran, S.,** Content of trace elements in winter wheat at different stages of development correlated with NPKCa fertilization and irrigation of soil, *Ann. UMCS,* 28/29e, 129, 1974 (Po).

268. **Glinski, J., Melke, J., and Uziak, S.,** Trace elements content in silt soils of the Polish Carpathian footland region, *Rocz. Glebozn.,* 19d, 73, 1968.

269. **Gluskoter, H. J., Ruch, R. R., Miller, W. G., Cahill, R. A., Dreher, G. B., and Kuhn, J. K.,** Trace elements in coal: occurrence and distribution, *Ill. State Geol. Surv. Circ.,* 499, 1977, 154.

270. **Godo, G. H. and Reisenauer, H. M.,** Plant effects on soil manganese availability, *Soil Sci. Soc. Am. J.,* 44, 993, 1980.

271. **Golovina, L. P., Lysenko, M. N., and Kisiel, T. J.,** Content and distribution of zinc in soils of Ukrainian Woodland (Polessie), *Pochvovedenie,* 2, 72, 1980 (Ru).

272. **Goodman, B. A. and Cheshire, M. V.,** The bonding of vanadium in complexes with humic acid, an electron paramagnetic resonance study, *Geochim. Cosmochim. Acta,* 39, 1711, 1975.

273. **Goodroad, L. L.,** Effects of P fertilizers and lime on the As, Cr, Pb and V content of soil and plants, *J. Environ. Qual.,* 8, 493, 1979.

274. **Gorbacheva, A. J.,** Distribution of molybdenum in the soil/parent rock system, *Pochvovedenie,* 6, 129, 1976 (Ru).

275. **Gorlach, E. and Gorlach, K.,** The effect of liming on the solubility on molybdenum in the soil, *Acta Agrar. Silvestria,* 16, 20, 1976 (Po).

276. **Goswani, S. C., Gulati, K. L., and Nagpaul, K. K.,** Estimation of uranium and boron contents in plants and soils by nuclear particle *etch* technique, *Plant Soil,* 48, 709, 1977.

277. **Gotoh, S., Tokudome, S., and Koga, H.,** Mercury in soil derived from igneous rock in northern Kyushu, Japan, *Soil Sci. Plant Nutr.,* 24, 391, 1978.

278. **Gough, L. P. and Severson, R. C.,** Impact of point source emission from phosphate processing on the element content of plants and soils, Soda Spring, Idaho, in *Trace Substance Environ. Health,* Vol. 10, Hemphill, D. D., Ed., University of Missouri, Columbia, Mo., 225, 1976.

279. **Gough, L. P., Shacklette, H. T., and Case, A. A.,** Element concentrations toxic to plants, animals, and man, *U.S. Geol. Surv. Bull.,* 1466, 1979, 80.

279a. **Gracey, H. I. and Stewart, J. W.,** Distribution of mercury in Saskatchawan soils and crops, *Can. J. Soil Sci.,* 54, 105, 1974.

280. **Graham, R. D.,** Absorption of copper by plant roots, in *Copper in Soils and Plants,* Loneragan, J. F., Robson, A. D., and Graham, R. D., Eds., Academic Press, New York, 1981, 141.

281. **Greenland, D. J. and Hayes, M. H. B., Eds.,** *The Chemistry of Soil Constituents,* John Wiley & Sons, New York, 1978, 469.

282. **Gregory, R. P. G. and Bradshaw, A. D.,** Heavy metal tolerance in populations of *Agrostis tenuis Sibth.* and other grasses, *New Phytol.,* 64, 131, 1965.

283. **Gribovskaya, I. F., Letunova, S. W., and Romanova, S. N.,** Microelements in the organs of legume plants, *Agrokhimiya,* 3, 81, 1968 (Ru).

284. **Griffin, R. A., Au, A. K., and Frost, R. R.,** Effect of pH on adsorption of chromium from landfill-leachate by clay minerals, *J. Environ. Sci. Health,* 12a, 431, 1977.

285. **Griffitts, W. R., Allaway, W. H., and Groth, D. H.,** Beryllium, in *Geochemistry and the Environment,* Vol. 2, Griffitts, W. R., Ed., NAS, Washington, D.C., 1977, 7.

286. **Griffitts, W. R. and Milne, D. B.,** Tin, in *Geochemistry and the Environment,* Vol. 2, Beeson, K. C., Ed., N.A.S., Washington, D.C., 1977, 88.

287. **Grim, R. E.,** *Clay Mineralogy,* McGraw-Hill, New York, 1968, 596.

288. **Groth, E.,** Fluoride pollution, *Environment,* 17, 29, 1975.

289. **Grove, J. H. and Ellis, B. G.,** Extractable chromium as related to soil pH and applied chromium, *Soil Sci. Soc. Am. J.,* 44, 238, 1980.

290. **Guliakin, I. V., Yudintseva, E. V., and Gorina, L. J.,** Accumulation of Cs-137 in yield as dependent on species peculiarities of plants, *Agrokhimiya,* 7, 121, 1975a (Ru).

291. **Guliakin, I. V., Yudintseva, E. V., and Rulevskaya, N. N.,** Fractional composition of strontium-90 and potassium in plants as influenced by the application of Ca, Mg, K and Na into soil, *Agrokhimiya,* 8, 95, 1975b (Ru).

292. **Guliakin, I. V., Yudintseva, E. V., and Levina, E. M.,** Effect of soil moisture on supply of strontium-90 and cesium-137 to plants, *Agrokhimiya,* 2, 102, 1976 (Ru).

293. **Gupta, U. C.,** Boron and molybdenum nutrition of wheat, barley and oats grown in Prince Edward Island soils, *Can. J. Soil Sci.,* 51, 415, 1971.

294. **Gupta, U. C. and Winter, K. A.,** Selenium content of soils and crops and the effects of lime and sulphur on plant selenium, *Can. J. Soil Sci.,* 55, 161, 1975.

295. **Gurievich, V. J.,** Some data on iodine accumulation by plants in region of iodine-rich waters, *Geokhim. Razviedka,* 6, 123, 1963 (Ru).

296. **Gutenmann, W. H., Bache, C. A., Youngs, W. D., and Lisk, D. J.,** Selenium in fly ash, *Science,* 191, 966, 1976.

297. **de Haan, F. A. M.,** The effects of long-term accumulation of heavy metals and selected compounds in municipal wastewater on soil, in *Wastewater Renovation and Reuse,* D'Itri, F. M., Ed., Marcel Dekker, New York, 1977, 283.

298. **Hädrich, F., Stahr, K., and Zöttl, H. W.,** Die Eignung von Al$_2$O$_3$-Keramik-platten und Ni-Sinterkerzen zur Gewinnung von Bodenlösung für die Spurenelementanalyse, *Mitt. Dtsch. Bodenkdl. Ges.,* 25, 151, 1977.

299. **Halbach, P., Borstel, D., and Gundermann, K. D.,** The uptake of uranium by organic substances in a peat bog environment on a granitic bedrock, *Chem. Geol.,* 29, 117, 1980.

300. **Hall, R. J.,** The presence and biosynthesis of carbo-fluorine compounds in tropical plants and soils, in *Trace Subst. Environ. Health,* Vol. II, Hemphill, D. D., Ed., University of Missouri, Columbia, Mo., 1977, 156.

301. **Halstead, R. L., Finn, B. J., and MacLean, A. J.,** Extractability of nickel added to soils and its concentration in plants, *Can. J. Soil Sci.,* 49, 335, 1969.

302. **Halvorsen, A. D. and Lindsay, W. L.,** The critical Zn^{2+} concentration for corn and the nonabsorption of chelated zinc, *J. Soil Sci. Soc. Am.,* 41, 531, 1977.

303. **Hanada, S., Nakano, M., Saitoh, H., and Mochizuki, T.,** Studies on the pollution of apple orchard surface soils and its improvement in relation to inorganic spray residues, I., *Bull. Fac. Agric. Hirosaki Univ.,* 25, 13, 1975 (Ja).

304. **Hansen, J. A. and Tjell, J. C.,** Guidelines and sludge utilization practice in Scandinavia, paper presented at Conf. Utilization of Sewage Sludge on Land, Oxford, April 10, 1978.

305. **Harmsen, K.,** Behaviour of Heavy Metals in Soils, Doctoral thesis, Centre for Agric. Publications and Documents, Wageningen, 1977, 170.

306. **Harmsen, K. and de Haan, F. A. M.,** Occurrence and behavior of uranium and thorium in soil and water, *Neth. J. Agric. Sci.,* 28, 40, 1980.

307. **Harter, R. D.,** Adsorption of copper and lead by Ap and B2 horizons of several Northeastern United States Soils, *Soil Sci. Am. J.,* 43, 679, 1979.

308. **Hartmans, J.,** Factors affecting the herbage iodine content, *Neth. J. Agric. Sci.,* 22, 195, 1974.

309. **Heenan, D. P. and Campbell, L. C.,** Transport and distribution of manganese in two cultivars of soybean, *Aust. J. Agric. Res.,* 31, 943, 1980.

310. **Heinrichs, H. and Mayer, R.,** Distribution and cycling of major and trace elements in two central European forest ecosystems, *J. Environ. Qual.,* 6, 402, 1977.

311. **Heinrichs, H. and Mayer, R.,** The role of forest vegetation in the biogeochemical cycle of heavy metals, *J. Environ. Qual.,* 9, 111, 1980.

312. **Hem, J. D.,** Chemical behavior of mercury in aqueous media, *U.S. Geol. Surv. Prof. Pap.,* 713, 19, 1970.

313. **Hem, J. D.,** Redox processes at surface of manganese oxide and their effects on aquaeous metal ions, *Chem. Geol.,* 21, 199, 1978.

314. **Hemkes, O. J., Kemp, A., and Broekhoven, L. W.,** Accumulation of heavy metals in the soil due to annual dressing with sewage sludge, *Neth. J. Agric. Sci.,* 28, 228, 1980.

315. **Hendrix, D. L. and Higinbotham, N.,** Heavy metals and sulphhydryl reagents as probes of ion uptake in pea stem, in *Membrane Transport in Plants,* Spring-Verlag, Berlin, 1974, 412.

316. **Hennig, A.,** *Mineralstoffe, Vitamine, Ergotropika,* DDR-VEB, Berlin, 1972, 263.

317. **Hewitt, E. J.,** Sand and Water Culture Methods Used in the Study of Plant Nutrition, Commonwealth Agriculture Bureaux, Bucks., U.K., 1966, 547.

318. **Hildebrand, E. E.,** Die Bindung von Immissionsblei in Böden, *Freiburger Bodenkundliche Abhandlungen,* 4, 1, 1974.

319. **Hildebrand, E. E. and Blume, W. E.,** Lead fixation by clay minerals, *Naturwissenschaften,* 61, 169, 1974.

320. **Hodgson, J. F., Geering, H. R., and Norvell, W. A.,** Micronutrient cation complexes in soil solution, *Soil Sci. Soc. Am. Proc.,* I, 29, 665, 1965; II, 30, 723, 1966.

321. **Hogg, T. J., Stewart, J. W. B., and Bettany, J. R.,** Influence of the chemical form of mercury on its adsorption and ability to leach through soils, *J. Environ. Qual.,* 7, 440, 1978.

322. **Hondenberg, A. and Finck, A.,** Ermittlung von Toxizitäts-Grenzwerter für Zink, Kupfer und Blei in Hafer und Rotklee, *Z. Pflanzenernaehr. Bodenkd.,* 4/5, 489, 1975.

323. **Horak, O.,** Untersuchung der Schwermetallbelastung von Pflanzen durch Klärschlammgaben im Gefäss-versuch, *Sonderdruck. Bodenkultur,* 31, 172, 1980.

324. **Hornick, S. B., Baker, D. E., and Guss, S. B.,** Crop production and animal health problems associated with high molybdenum soils, paper presented at Symp. on Molybdenum in the Environment, Denver, June 16, 1975, 12.

325. **Horowitz, C. T., Schock, H. H., and Horowitz-Kisimova, L. A.,** The content of scandium, thorium, silver, and other trace elements in different plant species, *Plant Soil,* 40, 397, 1974.

326. **Horst, W. J. and Marschner, H.,** Effect of silicon on manganese tolerance of bean plants (*Phaseolus vulgaris* L.), *Plant Soil,* 50, 287, 1978.

327. **Horváth, A. and Möller, F.,** Der Arsengehalt des Bodens, in *Proc. Arsen Symp.,* Anke, M., Schneider, H. J., and Brückner, Chr., Eds., Friedrich-Schiller University, Jena, E. Germany, 1980, 95.

328. **Horváth, A., Bozsai, G., Szabados, M., Károlyi, E., and Szaló, M.,** Study of heavy metal soil pollution in the environment of lead-works, *Magy. Kem. Lapja,* 35, 135, 1980 (Hu).

329. **Huang, P. M.,** Retention of arsenic by hydroxy-aluminum on surface of micaceous mineral colloids, *Soil Sci. Soc. Am. Proc.,* 39, 271, 1975.

330. **Huff, L. C.,** Migration of lead during oxidation and weathering of lead deposits, in *Lead in the Environment,* Lovering, T. G., Ed., *Geol. Survey Prof. Pap.,* 957, 21, 1976.

330a. **Huffman, E. W. and Hodgson, J. F.,** Distribution of cadmium and zinc/cadmium ratios in crops from 19 states east of the Rocky Mountains, *J. Environ. Qual.,* 2, 289, 1973.

331. **Hughes, M. K., Lepp, N. W., and Phipps, D. A.,** Aerial heavy metal pollution and terrestrial ecosystems, *Adv. Ecol. Res.,* 11, 217, 1980.

332. **Hutchinson, T. C., Czuba, M., and Cunningham, L.,** Lead, cadmium, zinc, copper and nickel distributions in vegetables and soils of an intensely cultivated area and levels of copper, lead and zinc in the growers, in *Trace Subst. Environ. Health,* Vol. 8, Hemphill, D. D., Ed., University of Missouri, Columbia, Mo., 1974, 81.

333. **Hutchinson, T. C. and Whitby, L. M.,** A study of airborne contamination of vegetation and soils by heavy metals from the Sudbury, Ontario, copper-nickel smelters, in *Trace Subst. Environ. Health,* Vol. 7, Hemphill, D. D., Ed., University of Missouri, Columbia, Mo., 1973, 179.

334. **Hutton, J. T.,** Titanium and zirconium minerals, in *Minerals in Soil Environments,* Dixon, J. B. and Weed, S. B., Eds., Soil Science Society of America, Madison, Wis., 1977, 673.

335. **Ignatowicz, I. and Żmigrodzka, T.,** Microelements (Mn, Cu, Mo) in grains of winter cereals from area of Zielonagóra voyevodship, *Rocz. Glebozn.,* 23, 113, 1972 (Po).

336. **Iimura, K., Ito, H., Chino, M., Morishita, T., and Hirata, H.,** Behavior of contaminant heavy metals in soil-plant system, in Proc. Inst. Sem. SEFMIA, Tokyo, 1977, 357.

337. **Ilyin, V. B.,** *Biogeochemistry and Agrochemistry of the Trace Elements (Mn, Cu, Mo, B) in the South of Western Siberia,* Izd. Nauka, Novosibirsk, 1973, 389 (Ru).

338. **Ilyin, V. B. and Stiepanova, M. D.,** Distribution of lead and cadmium in wheat plants growing on soils contaminated with these metals, *Agrokhimiya,* 5, 114, 1980 (Ru).

339. **Ireland, M. P.,** Metal accumulation by the earthworms *Lumbricus rubellus, Dendrobaena veneta* and *Eiseniella tetraedra* living in heavy metal polluted sites, *Environ. Pollut.,* 13, 205, 1979.

340. **Isermann, K.,** Method to reduce contamination and uptake of lead by plants from car exhaust gases, *Environ. Pollut.,* 12, 199, 1977.

341. **Itoh, S., Tokumaga, Y., and Yumura, Y.,** Concentration of heavy metals contained in the soil solution and the contamination of vegetable crops by the excessive absorption of heavy metals, *Bull. Veg. Ornamental Crops Res. Stn.,* 5a, 145, 1979 (Ja).

342. **Itoh, S. and Yumura, Y.,** Studies on the contamination of vegetable crops by excessive absorption of heavy metals, *Bull. Veg. Ornamental Crops Res. Stn.,* 6a, 123, 1979 (Ja).

343. **Ivanov, G. M. and Cybzhitov, C. H.,** Microelements in steppe facies of Southern Zabaykalie, *Agrokhimiya,* 2, 103, 1979 (Ru).

344. **Iverson, W. P. and Brinckman, E. F.,** Microbial transformation of heavy metals, in *Water Pollution Microbiology,* Vol. 2, Mitchell, R., Ed., Wiley Interscience, New York, 1978, 201.

345. **Ivlev, A. M., Teno-Chak-Mun, and Zbruyeva, A. I.,** On biogeochemistry of manganese in southern part of Sakhalin, in *Biogeokhimiya Zony Gipergeneza,* NEDRA, Moscow, 1971, 92 (Ru).

346. **Izerskaya, L. A. and Pashnieva, G. E.,** Manganese, copper and cobalt in soils of the Tomsk Region, *Agrokhimiya,* 5, 94, 1977 (Ru).

347. **Jacks, G.,** Vanadium in an area just outside Stockholm, *Environ. Pollut.,* 11, 289, 1976.

348. **Jackson, J. F. and Chapman, K. S. R.,** The role of boron in plants, in *Trace Elements in Soil-Plant-Animal Systems,* Nicholas, D. J. D. and Egan, A. R., Eds., Academic Press, New York, 1975, 213.

349. **Jagodin, B. A., Troykaya, G. N., Genepozoba, I. P., Sawich, M. S., and Ovcharenko, G. A.,** Cobalt in plant metabolism, in *Biologicheskaya Rol Mikroelementov i ich Primienieniye v Selskom Khozaystvie i Medicinie,* Nauka, Moscow, 1974, 329 (Ru).

350. **Jakubick, A. T.,** Geochemistry and physics of plutonium migration, in *Origin and Distribution of the Elements,* Vol. 11, Ahrens, L. H., Ed., Pergamon Press, Oxford, 1979, 775.

351. **Jakushevskaya, I. V. and Martynienko, A. G.,** Minor elements in the landscapes of separated forest stand steppe, *Pochvovedenie,* 4, 44, 1972 (Ru).

352. **James, R. O. and Barrow, N. J.,** Copper reactions with inorganic components of soils including uptake by oxide and silicate minerals, in *Copper in Soils and Plants,* Loneragan, J. F., Robson, A. D., and Graham, R. D., Eds., Academic Press, New York, 1981, 47.

353. **Jarvis, S. C. and Jones, L. H. P.,** The contents and sorption of cadium in some agricultural soils of England and Wales, *J. Soil Sci.,* 31, 469, 1980.

354. **Jasiewicz, C.,** The effect of copper and fertilization with various forms of nitrogen on some physiological indices in maize, *Acta Agrar. Silvestria,* 20, 95, 1981 (Po).

355. **Jaworowski, Z. and Grzybowska, D.,** Natural radionuclides in industrial and rural soils, *Sci. Total Environ.,* 7, 45, 1977.

356. **Jenne, E. A.,** Trace element sorption by sediments and soils-sites and processes, in *Molybdenum in the Environment,* Vol. 2, Chappell, W. R. and Petersen, K. K., Eds., Marcel Dekker, New York, 1977, 555.

357. **Jennett, J. Ch., Wixon, B. G., Lowsley, I. H., and Purushothaman, K.,** Transport and distribution from mining, milling, and smelting operations in a forest ecosystem, in *Lead in the Environment,* Bogges, W. R. and Wixon, B. G., Eds., Report for NSF and RANN, Washington, D.C., 1977, 135.

358. **Jermakov, V. V. and Kovalskiy, V. V.,** Geochemical ecology of organisms at elevated concentrations of selenium in environment, *Tr. Biogeokhim. Lab.,* 12, 204, 1968 (Ru).

359. **Jernelöv, A.,** Microbial alkylation of metals, paper presented at Int. Conf. on Heavy Metals in the Environment, Toronto, October 27, 1975, 845.

360. **Johansen, O. and Steinnes, E.,** Routine determination of traces of cobalt in soils and plant tissue by instrumental neutron activation analysis, *Acta Agric. Scand.,* 22, 103, 1972.

361. **John, M. K.,** Cadmium adsorption maxima of soil as measured by the Langmuir isotherm, *Can. J. Soil. Sci.,* 52, 343, 1972.

362. **John, M. K., Chuah, H. H., and van Laerhoven, C. J.,** Boron response and toxicity as affected by soil properties and rates of boron, *Soil Sci.,* 124, 34, 1977.

363. **John, M. K., van Laerhoven, C. J., and Cross, Ch. H.,** Cadmium, lead and zinc accumulation in soil near a smelter complex, *Environ. Lett.,* 10, 25, 1975.

364. **Johnson, C. M.,** Selenium in soils and plants: contrasts in conditions providing safe but adequate amounts of selenium in the food chain, in *Trace Elements in Soil-Plant-Animal Systems,* Nicholas, D. J. D. and Egan, A. R., Eds., Acaemic Press, New York, 1975, 165.

365. **Johnson, R. D., Miller, R. E., Williams, R. E., Wai, C. M., Wiese, A. C., and Mitchell, J. E.,** The heavy metal problem of Silver Valley, Northern Idaho, paper presented at Int. Conf. Heavy Metals in Environment, Toronto, October 27, 1975, 465.

366. **Johnston, W. R. and Proctor, J.,** Metal concentrations in plants and soils from two British serpentine sites, *Plant Soil,* 46, 275, 1977.

367. **Jonasson, I. R.,** Mercury in the natural environment: a review of recent work, *Geol. Surv. Can.,* 70/57, 37, 1970.

368. **Jonasson, I. R. and Boyle, R. W.,** Geochemistry of mercury and origins of natural contamination of the environment, *Can. Min. Metall. Bull.,* 1, 1, 1972.

369. **Jones, J. B.,** Plant tissue analysis for micronutrients, in *Micronutrients in Agriculture,* Mortvedt, J. J., Giordano, P. M., and Lindsay, W. L., Eds., Soil Science Society of America, Madison, Wis., 1972, 319.

369a. **Jones, L. H. P. and Clement, C. R.,** Lead uptake by plants and its significance for animals, in *Lead in the Environment,* Hepple, P., Ed., Institute of Petroleum, London, 1972, 29.

370. **Jones, L. H. P., Jarvis, S. C., and Cowling, D. W.,** Lead uptake from soils by perennial ryegrass and its relation to the supply of an essential element (sulphur), *Plant Soil,* 38, 605, 1973.

371. **Jönson, G. and Pehrson, B.,** Selenium — a trace element of great significance for the health of livestock, in *Geochemical Aspects in Present and Future Research,* Låg, J., Ed., Universitesforlaget, Oslo, 1980, 103.

372. **von Jung, J., Isermann, K., and Henjes, G.,** Einfluss von cadmiumhaltigen Düngerphosphaten auf die Cadmiumanreicherung von Kulturböden und Nutzpflanzen, *Landwirtsch. Forsch.,* 32, 262, 1979.

373. **Juszkiewicz, T. and Szprengier, T.,** Mercury content of feed cereal grains, *Med. Weterynaryjna,* 32, 415, 1976 (Po).

373a. **Kabata-Pendias, A.,** The sorption of trace elements by soil-forming minerals, *Rocz. Glebozn.*, 19d, 55, 1968.

374. **Kabata-Pendias, A.,** Leaching of micro- and macro-elements in columns with soil derived from granite, *Pamiet. Pulawski*, 38, 111, 1969 (Po).

375. **Kabata-Pendias, A.,** Chemical composition of soil solutions, *Rocz. Glebozn.*, 23, 3, 1972 (Po).

376. **Kabata-Pendias, A.,** Effects of inorganic air pollutants on the chemical balance of agricultural ecosystems, paper presented at United Nations – ECE Symp. on Effects of Air-borne Pollution on Vegetation, Warsaw, August 20, 1979, 134.

376a. **Kabata-Pendias, A.,** Current problems in chemical degradation of soils, paper presented at Conf. on Soil and Plant Analyses in Environmental Protection, Falenty/Warsaw, October 29, 1979, 7 (Po).

377. **Kabata-Pendias, A.,** Heavy metal sorption by clay minerals and oxides of iron and manganese, *Mineral. Pol.*, 11, 3, 1980.

378. **Kabata-Pendias, A.,** Heavy metal concentrations in arable soils of Poland, *Pamiet. Pulawski*, 74, 101, 1981 (Po).

379. **Kabata-Pendias, A. and Gondek, B.,** Bioavailability of heavy metals in the vicinity of a copper smelter, in *Trace Subst. Environ. Health*, Vol. 12, Hemphill, D. D., Ed., University of Missouri, Columbia, Mo., 1978, 523.

380. **Kabata-Pendias, A., Bolibrzuch, E., and Tarlowski, P.,** Impact of a copper smelter on agricultural environments, *Rocz. Glebozn.*, 32, 207, 1981.

381. **Kabata-Pendias, A. and Pendias, H.,** *Trace Elements in the Biological Environment*, Wyd. Geol., Warsaw, 1979, 300 (Po).

382. **Kabata-Pendias, A. and Piotrowska, M.,** Total contents of trace elements in soils of Poland, *Materialy IUNG*, Pulawy, Poland, 8s, 7, 1971, (Po).

383. **Kaniuga, Z., Ząbek, J., and Sochanowicz, B.,** Photosynthetic apparatus in chilling-sensitive plants, *Planta*, 144, 49, 1978.

384. **Karlsson, N.,** On molybdenum in Swedish soil and vegetation and some related questions, *Statens Lantbrukskem. Kontrollanst. Medd.*, 23, 1961, 243 (Sv).

385. **Karoń, B.,** Vanadium content of cultivated plants, *Z. Prob. Post. Nauk Roln.*, 179, 361, 1976 (Po).

386. **Kashin, V. K., Osipov, K. J., Utinova-Ivanowa, L. P., and Jefimov, M. V.,** Iodine in sward of haylands of Mujsk valley of Buratia, *Angrokhimiya*, 6, 99, 1980 (Ru).

387. **Kaufman, P. B., Bigelow, W. C., Petering, L. B., and Drogosz, F. B.,** Silica in developing epidermal cells of *Avena* internodes, *Science*, 166, 1015, 1969.

388. **Kähäri, J. and Nissinen, H.,** The mineral element contents of timothy (*Phleum pratense* L.) in Finland, *Acta Agric. Scand. Suppl.*, 20, 26, 1978.

389. **Khalid, B. Y. and Tinsley, J.,** Some effects of nickel toxicity on ryegrass, *Plant Soil*, 55, 139, 1980.

390. **Kick, H., Bürger, H., and Sommer, K.,** Gesamthalte an Pb, Zn, Sn, As, Cd, Hg, Cu, Ni, Cr, und Co in landwirtschaftlich und gärtnerisch genutzten Böden Nordrhein-Westfalens, *Landwirtsch. Forsch.*, 33, 12, 1980.

391. **King, P. M. and Alston, A. M.,** Diagnosis of trace element deficiencies in wheat on Eyre Peninsula, South Australia, in *Trace Elements in Soil-Plant-Animal Systems*, Nicholas, D. J. D. and Egan, A. R., Eds., Academic Press, New York, 1975, 339.

392. **Kinniburgh, D. G., Jackson, M. L., and Syers, J. K.,** Adsorption of alkaline earth, transition, and heavy metal cations by hydrous oxide gels of iron and aluminum, *Soil Sci. Soc. Am. J.*, 40, 796, 1976.

393. **Kirchmann, R. and D'Souza, T. J.,** Behaviour of ruthenium in an established pasture soil and its uptake by grasses, in *Isotopes and Radiation in Soil-Plant Relationship*, IAEA, Vienna, 1972, 587.

394. **Kiriluk, V. P.,** Accumulation of copper and silver in chernozems of vineyards, in *Microelements in Environment*, Vlasyuk, P. A., Ed., Naukova Dumka, Kiyev, 1980, 76 (Ru).

395. **Kitagishi, K. and Yamane, I., Eds.,** *Heavy Metal Pollution in Soils of Japan*, Japan Science Society Press, Tokyo, 1981, 302.

396. **Klein, D. H.,** Mercury and other metals in urban soils, *Environ. Sci. Technol.*, 6, 560, 1972.

397. **Kloke, A.,** Blei-Zink-Cadmium Anreicherung in Böden und Pflanzen, *Staub Reinhalt. Luft*, 34, 18, 1974.

398. **Kloke, A.,** Content of arsenic, cadmium chromium, fluorine, lead, mercury and nickel in plants grown on contaminated soil, paper presented at United Nations-ECE Symp. on Effects of Air-borne Pollution on Vegetation, Warsaw, August 20, 1979, 192.

399. **Kloke, A.,** Der Einfluss von Phosphatdüngern auf den Cadmiumgehalt in Pflanzen, *Gesunde Pflanz.*, 32, 261, 1980a.

400. **Kloke, A.,** Materialien zur Risikoeinschätzung des Quecksilberproblems in der Bundesrepublik Deutschland, *Nachrichtenbl. Dtsch. Pflanzenschutz.*, 32, 120, 1980b.

401. **Kluczyński, B.,** The influence of fluorine and its compounds on plants, *Arbor. Kornickie*, 21, 401, 1976 (Po).

402. **Knälmann, M.,** Sorption of iodine in soils, paper presented at Annu. Meeting of ESNA, Budapest, September 26, 1972.

403. **Kobayashi, J.,** Air and water pollution by cadmium, lead and zinc attributed to the largest zinc refinery in Japan, in *Trace Subst. Environ. Health,* Vol. 5, Hemphill, D. D., Ed., University of Missouri, Columbia, Mo., 1971, 117.

404. **Kobayashi, J.,** Effect of cadmium on calcium metabolism of rats, in *Trace Subst. Environ. Health,* Vol. 7, Hemphill, D. D., Ed., University of Missouri, Columbia, Mo., 1973, 295.

405. **Kobayashi, J., Morii, F., and Muramoto, S.,** Removal of cadmium from polluted soil with the chelating agent, EDTA, in *Trace Subst. Environ. Health,* Vol. 8, Hemphill, D. D., Ed., University of Missouri, Columbia, Mo., 1974, 179.

406. **Kodama, H. and Schnitzer, M.,** Effect of fulvic acid on the crystallization of aluminum hydroxides, *Geoderma,* 24, 195, 1980.

407. **Kokke, R.,** Radioisotopes applied to environmental toxicity research with microbes, in *Radiotracer Studies of Chemical Residues in Food and Agriculture,* IAEA, Vienna, 1972, 15.

408. **Koljonen, T.,** The availability of selenium as nutrient in different geological environments, with special reference to Finland and Iceland, *Ambio,* 7, 169, 1978.

409. **Koons, R. D. and Helmke, P. A.,** Neutron activation analysis of standard soils, *Soil Sci. Soc. Am. J.,* 42, 237, 1978.

410. **de Koning, H. W.,** Lead and cadmium contamination in the area immediately surrounding a lead smelter, *Water Air Soil Pollut.,* 3, 63, 1974.

411. **Korkman, J.,** The effect of selenium fertilizers on the selenium content of barley, spring wheat and potatoes, *J. Sci. Agric. Soc. Finland,* 52, 495, 1980.

412. **Kosanovic, V. and Halasi, R.,** Boron in soils of Voyvodine, *Letop. Nauch. Radova,* 6, 167, 1962 (Sh).

413. **Kosta, L., Byrne, A. R., Zelenko, V., Stegnar, P., Dermelj, M., and Ravnik, V.,** Studies on the uptake and transformations of mercury in living organisms in the Idrija region and comparative area, *Vestn. Slov. Kem. Drus.,* 21, 49, 1974.

414. **Kosta, L., Zelenko, V., and Ravnik, V.,** Trace elements in human thyroid with special reference to the observed accumulation of mercury following long-term exposure, in *Comp. Studies of Food and Environ. Contam.,* IAEA, Vienna, 1974, 541.

415. **Kosta, L., Zelenko, V., Stegnar, P., Ravnik, V., Dermelj, M., and Byrne, A. R.,** Fate and significance of mercury residues in an agricultural ecosystem, paper presented at Meeting Isotope Tracer Studies of Chemical Residues in Food and in the Agricultural Environment, Vienna, October 30, 1972, 87.

416. **Kothny, E. L.,** Palladium in plant ash, *Plant Soil,* 53, 547, 1979.

417. **Kovalevskiy, A. L.,** *Biogeochemical Exploration for Mineral Deposits,* published for the USDI and the NSF, Amerind Publ. Co. Pvt. Ltd., New Delhi, 1979, 136.

418. **Kovalskiy, V. V.,** *Geochemical Ecology,* Izd. Nauka, Moscow, 1974a, 298 (Ru).

419. **Kovalskiy, V. V.,** Geochemical environment, health and diseases, in *Trace Subst. Environ. Health,* Vol. 8, Hemphill, D. D., Ed., University of Missouri, Columbia, Mo., 137, 1974.

419a. **Kovalskiy, V. V. and Andryanova, G. A.,** *Trace Elements (Cu, Co, Zn, Mo, Mn, B, I, Sr) in soils of USSR,* Buryatskoye Knigi Izd., Ulan-Ude, 1968, 56 (Ru).

420. **Kovalskiy, V. V. and Yarovaya, G. A.,** Biogeochemical province enriched in molybdenum, *Agrokhimiya,* 8, 68, 1966 (Ru).

421. **Kovalskiy, V. V. and Letunova, S. V.,** Geochemical ecology of microorganisms, *Tr. Biogeochim. Lab.,* 13, 3, 1974 (Ru).

422. **Kovalskiy, V. V., Letunova, S. V., and Aleksieyeva, S. A.,** Accumulation of nickel and other elements in the microbiota of a soil in South Ural Subregion of the biosphere, in *Proc. Nickel Symp.,* Anke, M., Schneider, H. J., and Brückner, Chr., Eds., Friedrich-Schiller University, Jena, E. Germany, 1980, 163.

423. **Kovalskiy, V. V., Vorotnickaya, I. J., Lekarev, V. S., and Nikitina, V. J.,** Biogeochemical changes in uranium levels in pasture of Issyk-kul valey, *Tr. Biogeochim. Lab.,* 12, 5, 1968 (Ru).

424. **Kozak, L. and Tarkowski, C.,** The Cu, Zn, Mn, Fe and Mg contents at different growth stages of triticale, wheat and rye, *Rocz. Nauk Roln. Ser. A,* 104, 113, 1979 (Po).

425. **Krampitz, G.,** Die biologische Bedeutung von Beryllium-Verbindungen, in *Proc. Arsen Symposium,* Anke, M., Schneider, H. J., Brückner, Chr., Eds., Friedrich-Schiller University, Jena, E. Germany, 1980, 245.

426. **Krasinskaya, N. P. and Letunova, S. V.,** Accumulation of Zn, Mo and B by biomass of soil microflora with changing level of content of these elements in psuedo-podzolized brown soils of subtropical zone of Abkhazia, *Agrokhimiya,* 6, 111, 1981 (Ru).

427. **Krauskopf, K. B.,** Geochemistry of micronutrients, in *Micronutrients in Agriculture,* Mortvedt, J. J., Giordano, P. M., and Lindsay, W. L., Eds., Soil Science Society of America, Madison, Wis., 1972, 7.

428. **Krähmer, R. and Bergmann, W.,** Verteilung verschiedener Kupfer-Fraktionen in Acker- und Grünland-profilen sowie deren Bezichung zum salpetersäurelöslichen Kupfer nach Westerhoff, *Arc. Acker-u. Pflan-zenbau-u. Bodenk.,* 22, 405, 1978.

429. **Kristensen, K. K. and Bonde, G. J.,** The current status of bacterial and other pathogenic organisms in municipal wastewater and their potential health hazards with regard to agricultural irrigation, in *Wastewater Renovation and Reuse,* D'Itri, F. M., Ed., Marcel Dekker, New York, 1977, 705.

430. **Kronemann, H., Anke, M., Grün, M., and Partschefeld, M.,** The cadmium concentration of feedstuffs, foodstuffs and of water in the GDR and in an area with nonferrous metal industry, in *Proc. Kadmium Symp.,* Anke, M. and Schneider, H. J., Eds., Friedrich-Schiller University, Jena, E. Germany, 1979, 230.

431. **Kronemann, H., Anke, M., Thomas, S., and Riedel, E.,** The nickel concentration of different food- and feed-stuffs from areas with and without nickel exposure, in *Proc. Nickel Symp.,* Anke, M., Schneider, H. J., and Brückner, Chr., Eds., Friedrich-Schiller University, Jena, E. Germany, 1980, 221.

432. **Krupskiy, N. K., Golovina, L. P., Aleksandrova, A. M., and Kisiel, T. I.,** Distribution of manganese in soils of Ukrainian forest-steppe zone, *Pochvovedenie,* 11, 41, 1978 (Ru).

433. **Kubota, J.,** Distribution of cobalt deficiency in grazing animals in relation to soils and forage plants of the United States, *Soil Sci.,* 106, 122, 1968.

434. **Kubota, J.,** Areas of molybdenum toxicity to grazing animals in the Western States, *J. Range Manage.,* 28, 252, 1975.

435. **Kubota, J.,** Molybdenum status of United States soils and plants, in *Molybdenum in the Environment,* Chappell, W. R. and Peterson, K. K., Eds., Marcel Dekker, New York, 1977, 555.

436. **Kubota, J. and Allaway, W. H.,** Geographic distribution of trace element problems in *Micronutrients in Agriculture,* Mortvedt, J. J., Giordano, P. M., and Lindsay, W. L., Eds., Soil Science Society of America, Madison, Wis., 1972, 525.

436a. **Kubota, J., Cary, E. E., and Gissel-Nielsen, G.,** Selenium in rainwater of the United States and Denmark, in *Trace Subst. Environ. Health,* Vol. 9, Hemphill, D. D., Ed., University of Missouri, Columbia, Mo., 1975, 123.

437. **Kukurenda, H. and Lipski, R.,** Solubility of manganese in different soils and its availability to plants, *Pamiet. Pulawski,* 76, 172, 1982 (Po).

438. **Kulikova, A. Kh. and Nurgaleyeva, G. M.,** Dynamics of mercury organic compounds in leaching chernozems and their supply to agricultural crops, *Pochvovedenie,* 12, 142, 1979 (Ru).

439. **Kunishi, H. M. and Taylor, A. W.,** Immobilization of radio-strontium in soil by phosphate addition, *Soil Sci.,* 113, 1, 1972.

440. **Kuo, S. and Mikkelsen, D. S.,** Zinc adsorption by two alkaline soils, *Soil Sci.,* 128, 274, 1979.

441. **Låg, J.,** Soil science and geomedicine, *Acta Agric. Scand.,* 22, 150, 1972.

442. **Låg, J.,** Soil selenium in relation to precipitation, *Ambio,* 3, 237, 1974.

443. **Låg, J. and Dev, G.,** Distribution of exchangeable manganese in some Norwegian podzol profiles, *J. Indian Soc. Soil Sci.,* 12, 215, 1964.

444. **Låg, J. and Steinnes, E.,** Study of mercury and iodine distribution in Norwegian forest soils by neutron activation analysis, in *Nuclear Techniques in Environmental Pollution,* IAEA, Vienna, 1971, 429.

445. **Låg, J. and Steinnes, E.,** Regional distribution of halogens in Norwegian forest soils, *Geoderma,* 16, 317, 1976.

446. **Låg, J. and Steinnes, E.,** Contents of some trace elements in barley and wheat grown in Norway, *Meld. Nor. Landrukshoegsk,* 57, 1, 1978a.

447. **Låg, J. and Steinnes, E.,** Regional distribution of mercury in humus layers of Norwegian forest soils, *Acta Agric. Scand.,* 28, 393, 1978b.

447a. **Lagerwerff, J. V.,** Lead, mercury and cadmium in environmental contaminants, in *Micronutrients in Agriculture,* Mortvedt, J. J., Giordano, P. P., and Lindsay, W. L., Eds., Soil Science Society of America, Madison, Wis., 1972, 593.

448. **Lagerwerff, J. V., Armiger, W. H., and Specht, A. W.,** Uptake of lead by alfalfa and corn from soil and air, *Soil Sci.,* 115, 455, 1973.

449. **Lagerwerff, J. V. and Biersdorff, G. T.,** Interaction of zinc with uptake and translocation of cadmium in radish, in *Trace Subst. Environ. Health,* Vol. 5, Hemphill, D. D., Ed., University of Missouri, Columbia, Mo., 1972, 515.

450. **Lagerwerff, J. V. and Kemper, W. D.,** Reclamation of soils contaminated with radioactive strontium, *Soil Sci. Soc. Am. Proc.,* 39, 1077, 1975.

451. **Langerwerff, J. V. and Milberg, R. P.,** Sign-of-charge of species of Cu, Cd and Zn extracted from sewage sludge, and effect of plants, *Plant and Soil,* 49, 117, 1978.

452. **Lahann, R. W.,** Molybdenum hazard in land disposal of sewage sludge, *Water Air and Soil Pollut.,* 6, 3, 1976.

453. **Lakin, H. W., Curtin, C. G., and Hubert, A. E.,** Geochemistry of gold in the weathering cycle, *U.S. Geol. Surv. Bull.,* 1330, 80, 1974.

454. **Lakin, H. W. and Dawidson, D. F.,** The relation of the geochemistry of selenium to its occurrence in soil, in Proc. Selenium in Biomedicine, Westport, Conn., 1967, 27.

455. **Lal, F. and Biswas, T. D.,** Factors affecting the distribution and availability of micronutrient elements in major soil groups of Rajasthan, *J. Indian Soc. Soil Sci.,* 22, 333, 1974.

456. **Lambert, D. H., Baker, D. E., and Cole, H., Jr.,** The role of mycorrhizae in the interactions of P with Zn, Cu and other elements, *Soil Sci. Soc. Am. J.,* 43, 976, 1979.

457. **Lambert, D. H., Cole. H., and Baker, D. E.,** The role of boron in plant response to mycorrhizal infection, *Plant Soil,* 57, 431, 1980.

458. **Landa, E. R.,** The retention of metallic mercury vapor by soils, *Geochim. Cosmochim. Acta,* 42, 1407, 1978.

459. **Landa, E. R.,** The volatile loss of mercury from soils amended with methyl-mercury chloride, *Soil Sci.,* 128, 9, 1979.

460. **Lane, S. D., Martin, E. S., and Garrod, J. F.,** Lead toxicity effects on indole-3-acetic acid-induced cell elongation, *Planta.,* 144, 79, 1978.

461. **Larsen, S. and Widdowson, A. E.,** Soil fluorine, *J. Soil Sci.,* 22, 210, 1971.

462. **Laul, J. C., Weimer, W. C., and Rancitelli, L. A.,** Biogeochemical distribution of rare earths and other trace elements in plants and soils, in *Origin and Distribution of the Elements,* Vol. 11, Ahrens, L. H., Ed., Pergamon Press, Oxford, 1979, 819.

463. **Leal, A., Gomez, M., Sanchez-Raya, J. A., Yanēz, J., and Recalde, L.,** Effect of boron absorption on accumulation and distribution of phosphate, paper presented at 3rd Coll. Le Contrôle de l'adimentation des Plantes Cultivées, Budapest, September 4, 1972, 763.

464. **Letunova, S. V.,** Geochemical ecology of soil microorganisms, in *Trace Element Metabolism in Animals,* Mills, C. F., Ed., Churchill Livingston, Edinburgh, 1970, 549.

465. **Letunova, S. V. and Gribovskaya, I. F.,** Influence of soil microflora on biogenic migration of copper, molybdenum and lead in some biogeochemical provinces of Armenia, *Agrokhimiya,* 3, 123, 1975 (Ru).

466. **Letunova, S. V., Kovalskiy, V. V., and Bochkova, L. P.,** Importance of soil microflora in biogenic migration of manganese in manganese biochemical province of Georgia, *Agrokhimiya,* 12, 88, 1976 (Ru).

467. **Letunova, S. V. and Krivitskiy, V. A.,** Concentration of zinc in biomass of soil microflora in South-Urals copper-zinc subregion of biosphere, *Agrokhimiya,* 6, 104, 1979 (Ru).

468. **Lévesque, M. and Vendette, E. D.,** Selenium determination in soil and plant materials, *Can. J. Soil Sci.,* 51, 142, 1971.

469. **Lewin, V. H. and Beckett, P. H. T.,** Monitoring heavy metal accumulation in agricultural soils treated with sewage sludge, *Effluent Water Treat. J.,* May, p. 217, 1980.

470. **Lexmond, Th. M., de Haan, F. A. M., and Frissel, M. J.,** On the methylation of inorganic mercury and the decomposition of organomercury compounds — a review, *Neth. Agric. Sci.,* 24, 79, 1976.

471. **Liang, C. N. and Tabatabai, M. A.,** Effects of trace elements on nitrogen mineralisation in soils, *Environ. Pollut.,* 12, 141, 1977.

472. **Lin, H. C.,** Problems of soil arsenic, *Mem. Coll. Agric. Natl. Taiwan Univ.,* 11, 1, 1977.

473. **Lindberg, P. and Bingefors, S.,** Selenium levels of forages and soils in different regions of Sweden, *Acta Agric. Scand.,* 20, 133, 1970.

474. **Lindberg, P. and Lannek, N.,** Amounts of selenium in Swedish forages, soils and animal tissues, in *Trace Element Metabolism in Animals,* Mills, C. F., Ed., Churchill, Livingston, Edinburgh, 1970, 421.

475. **Lindsay, W. L.,** Inorganic phase equilibria of micronutrients in soils, in *Micronutrients in Agriculture,* Mortvedt, J. J., Giordano, P. M., and Lindsay, W. L., Eds., Soil Science Society of America, Madison, Wis., 1972a, 41.

476. **Lindsay, W. L.,** Zinc in soils and plant nutrition, *Adv. Agron.,* 24, 147, 1972b.

477. **Lindsay, W. L.,** *Chemical Equilibria in Soils,* Wiley-Interscience, New York, 1979, 449.

478. **Linzon, S. N.,** Fluoride effects on vegetation in Ontario, paper presented at 2nd Int. Clean Air Congr., Washington, D.C., December 6, 1970.

479. **Linzon, S. N.,** Phytotoxicology Excessive Levels for Contaminants in Soil and Vegetation, report of Ministry of the Environment, Ontario, Canada, 1978.

480. **Linzon, S. N., Chai, B. L., Temple, P. J., Pearson, R. G., and Smith, M. L.,** Lead contamination of urban soils and vegetation by emissions from secondary lead industries, *APCA J.,* 26, 651, 1976.

481. **Lipsey, R. L.,** Accumulation and physiological effects of methyl mercury hydroxide on maize seedlings, *Environ. Pollut.,* 8, 149, 1975.

482. **Little, P. and Martin, M. H.,** A survey of zinc, lead and cadmium in soil and natural vegetation around a smelting complex, *Environ. Pollut.,* 3, 241, 1972.

483. **Liu, D. J., Pomeranz, Y., and Robins, G. S.,** Mineral content of developing and malted barley, *Cereal Chem.,* 52, 678, 1975.

484. **Liu, D. J., and Robbins, G. S., and Pomeranz, Y.,** Composition and utilization of milled barley products, *Cereal Chem.,* 51, 309, 1974

485. **Lodenius, M.,** Regional distribution of mercury in *Hypogumnia physodes* in Finland, *Ambio,* 10, 183, 1981.

486. **Loganathan, P., Burau, R. G., and Fuerstenau, D. W.,** Influence of pH on the sorption of Co^{2+}, Zn^{2+} and Ca by a hydrous manganese oxide, *Soil Sci. Soc. Am. J.,* 41, 51, 1977.

487. **Lognay, G.,** Spectrofluorometric determination of selenium in plants, *Bull. Rech. Agron. Gembloux,* 15, 71, 1980.

488. **Lokay, D. and Pavel, J.,** Das Anzeigevermögen der Kadmium-belastung durch Ackerrotklee, Weizen und Roggen, in *Proc. Kadmium Symposium,* Anke, M. and Schneider, H. J., Eds., Friedrich-Schiller University, Jena, E. Germany, 1979, 221.

488a. **van Loon, J. C.,** Agricultural use of sewage treatment plant sludges, a potential source of mercury contamination, *Environ. Lett.,* 4, 529, 1973.

489. **Loneragan, J. F.,** The availability and absorption of trace elements in soil-plant systems and their relation to movement and concentration of trace elements in plants, in *Trace Elements in Soil-Plant-Animal Systems,* Nicholas, D. J. D. and Egan, A. R., Eds., Academic Press, New York, 1975, 109.

490. **Loneragan, J. F.,** Distribution and movement of copper in plants, in *Copper in Soils and Plants,* Loneragan, J. F., Robson, A. D., and Graham, R. D., Eds., Academic Press, New York, 1981, 165.

491. **Loneragan, J. F., Grove, T. S., Robson, A. D., and Snowball, K.,** Phosphorus toxicity as a factor in zinc-phosphorus interactions in plants, *Soil Sci. Soc. Am. J.,* 43, 966, 1979.

492. **Lorenz, K., Reuter, F. W. and Sizer, C.,** The mineral composition of triticales and triticale milling fractions by X-ray fluorescence and atomic absorption, *Cereal Chem.,* 51, 534, 1974.

493. **Lukashev, K. J. and Pietukhova, N. N.,** Minor elements in landscapes of Byelorussian S.S.R., *Pochvovedenie,* 8, 47, 1974 (Ru).

494. **Lyon, G. L., Brooke, R. R., Peterson, P. J., and Butler, G. W.,** Trace elements in a New Zealand serpentine flora, *Plant Soil,* 29, 225, 1968.

495. **Lyon, G. L., Brooks, R. R. Peterson, P. J., and Butler, G. W.,** Some trace elements in plants from serpentine soils, *N. Z. J. Sci.,* 13, 133, 1970.

496. **MacLean, A. J.,** Mercury in plants and retention of mercury by soils in relation to properties and added sulfur, *Can. J. Soil Sci.,* 54, 287, 1974.

497. **MacLean, A. J.,** Cadmium in different plant species and its availability in soils as influenced by organic matter and addition of lime, P, Cd, and Zn, *Can. J. Soil Sci.,* 56, 129, 1976.

498. **MacLean, A. J., Stone, B., and Cordukes, W. E.,** Amounts of mercury of some golf course sites, *Can. J. Soil Sci.,* 53, 130, 1973.

499. **MacLean, D. C. and Schneider, R. E.,** Effect of gaseous hydrogen fluoride on the yield of field-grown wheat, *Environ. Pollut.,* 24a, 39, 1981.

500. **Macuch, P., Hluchan, P., Mayer, J., and Abel, E.,** Air pollution by fluoride compounds near an aluminum factory, *Fluoride Q. Rep.,* 2, 28, 1969.

501. **Makarov, V. A.,** Output of trace elements with yield of agricultural crops, *Zap. Leningr. Skh. Inst.,* 160, 23, 1971 (Ru).

502. **Malaiškaite, B. S. and Radišauskas, J. G.,** Nickel supplies to bulls in the Lithuanian SSR, in *Proc. Nickel Symp.,* Anke, M., Schneider, H. J., and Brückner, Chr., Eds., Friedrich-Schiller University, Jena, E. Germany, 1980, 253.

503. **Malgin, M. A.,** Iodine in Altai soils, *Pochvovedenie,* 8, 74, 1980 (Ru).

504. **Maliszewska, W.,** Influence de certains oligo-éléments sur l'activité de quelques processus microbiologiques du sol, *Rev. Ecol. Biol. Sol.,* 9, 505, 1972.

505. **Maliszewska, W. and Wierzbicka, H.,** The influence of lead, zinc, and copper on the development and activity of microorganisms in soils, in *Proc. Effects of Trace Element Pollutants on Agric. Environ. Quality,* Volume I; Kabata-Pendias, A., Ed., IUNG, Pulawy, Poland, 1978, 135 (Po).

506. **Malone, G., Koeppe, D. E., and Miller, R. J.,** Localization of lead accumulated by corn plants, *Plant Physiol.,* 53, 388, 1974.

507. **Manecki, A., Klapyta, Z., Schejbal-Chwastek, M., Skowroński, A., Tarkowski, J., and Tokarz, M.,** The effect of industrial pollutants of the atmosphere on the geochemistry of natural environment of the Niepolomice forest, *PAN Miner. Trans.,* 71, 7, 1981 (Po).

508. **Marjanen, H.,** On the relationship between the contents of trace elements in soils and plants and the cancer incidence in Finland, in *Geochemical Aspects in Present and Future Research,* Låg, Ed., Universitetsforlaget, Oslo, 1980, 149.

509. **Martell, E. A.,** Tobacco radioactivity and cancer in smokers, *Am. Sci.,* 7/8, 404, 1975.

510. **Martin, H., Ed.,** *Pesticide Manual,* British Crop Protection Council, 1968, 464.

510a. **Martin, J. P.,** Side effects of organic chemicals, in *Organic Chemicals in the Soil Environment,* Goring, C. I., and Hamaker, J. W., Eds., Marcel Dekker, New York, 1972, 733.

511. **Marutian, S. A.,** Activity of micro- and macroelements in vine shoots during nongrowing season, paper presented at 3rd Coll. Le Contrôle de l'Alimentation des Plantes Cultivées, Budapest, September 4, 1972, 763.

512. **Maszner, P.,** The influence of industrial dusts and gases on soils of Luboń region, *Rocz. Glebozn.,* 30, 199, 1979 (Po).

513. **Mathews, H. and Thornton, I.,** Agricultural implication of Zn and Cd contaminated land at Shipham, Somerset, in *Trace Subst. Environ. Health,* Vol. 14, Hemphill, D. D., Ed., University of Missouri, Columbia, Mo., 1980, 478.

514. **Mathur, S. P., Hamilton, H. A. and Preston, C. M.,** The influence of variation in copper content of an organic soil on the mineral nutrition of oats grown *in situ, Commun. Soil Sci. Plant Anal.,* 10, 1399, 1979.

515. **Mathur, S. P., Macdougall, J. I., and McGrath, M.,** Levels of activities of some carbohydrates, proteases, lipase and phosphatase in organic soils of differing copper content, *Soil Sci.,* 129, 376, 1980.

516. **Matsumoto, H., Syo, S., and Takahashi, E.,** Translocation and some forms of germanium in rice plants, *Soil Sci. Plant Nutr.,* 21, 273, 1975.

517. **McBride, M. B.,** Retention of Cu^{2+}, Ca^{2+}, Mg^{2+} and Mn^{2+} by amorphous alumina, *Soil Sci. Soc. Am. J.,* 42, 27, 1978.

518. **McBride, M. B.,** Forms and distribution of copper in solid and solution phases of soil, in *Copper in Soils and Plants,* Loeragan, J. F., Robson, A. D., and Graham, R. D., Eds., Academic Press, New York, 1981, 25.

519. **McBride, M. B. and Blasiak, J. J.,** Zinc and copper solubility as a function of pH in an acid soil, *Soil Sci. Soc. Am. J.,* 43, 866, 1979.

520. **McKeague, J. A. and Kloosterman, B.,** Mercury in horizons of some profiles in Canada, *Can. J. Soil Sci.,* 54, 503, 1974.

521. **McKeague, J. A. and Wolynetz, M. S.,** Background levels of minor elements in some Canadian soils, *Geoderma,* 24, 299, 1980.

522. **McKenzie, R. M.,** Trace elements in some South Australian terra rossa and rendzina soils, *Aust. J. Agric. Res.,* 10, 52, 1959.

523. **McKenzie, R. M.,** The mineralogy and chemistry of soil cobalt, in *Trace Elements in Soil-Plant-Animal systems,* Nicholas, D. J. D. and Egan, A. R., Eds., Academic Press, New York, 1975, 83.

524. **McKenzie, R. M.,** Manganese oxides and hydroxides, in *Minerals in Soil Environment,* Dixon, J. B. and Weed, S. B., Eds., Soil Science Society of America, Madison, Wis., 1977, 181.

524a. **McKenzie, R. M.,** The effect of two manganese dioxides on the uptake of lead, cobalt, nickel, copper and zinc by subterraneum clover, *Austr. J. Soil Res.,* 16, 209, 1978.

525. **McKenzie, R. M.,** The adsorption of lead and other heavy metals on oxides of manganese and iron, *Aust. J. Soil Res.,* 18, 61, 1980a.

526. **McKenzie, R. M.,** The manganese oxides in soils, in *Geology and Geochemistry of Manganese,* Varentsov, I. M. and Grasselly, G., Eds., Akadémiai Kiadó, Budapest, 1980b, 259.

527. **McLaren, R. G. and Crawford, D. V.,** Studies on soil copper, *J. Soil Sci.,* I, 24, 172; II,24,443; III, 25, 111, 1973.

528. **McLean, E. D.,** Chemistry of soil aluminum, *Commun. Soil Sci. Plant. Anal.,* 7, 619, 1976.

529. **Megumi, K. and Mamuro, T.,** Concentration of uranium series nuclides in soil particles in relation to their size, *J. Geophys. Res.,* 82, 353, 1977.

530. **Mekaru, T. and Uehara, G.,** Anion adsorption in ferruginous tropical soils, *Soil Sci. Soc. Am. Proc.,* 36, 296, 1972.

531. **Mengel, K. and Kirkby, E. A.,** *Principles of Plant Nutrition,* International Potash. Institute, Worblaufen-Bern, 1978, 593.

532. **Menzies, J. D. and Chaney, R. L.,** Waste characteristics, in Factors Involved in Land Application of Agricultural and Municipal Wastes, National Program Staff, U.S. Department of Agriculture, Beltsville, Md., 1974, 18.

533. **Merry, R. H. and Tiller, K. G.,** The contamination of pasture by a lead smelter in semi-arid environment, *Austr. J. Exp. Agric. Anim. Husb.,* 18, 89, 1978.

534. **Mertz, W.,** Chromium occurrence and function in biological systems, *Physiol. Rev.,* 49, 163, 1969.

535. **Mertz, W., Angino, E. E., Cannon, H. L., Hambidge, K. M., and Voors, A. W.,** Chromium, in *Geochemistry and the Environment,* Vol. 1, Mertz, W., Ed., N.A.S., Washington, D.C., 1974, 29.

536. **Metson, A. J., Gibson, E. J., Hunt, J. L., and Sauders, W. M. H.,** Seasonal variations in chemical composition of pasture, *N.Z. J. Agric. Res.,* 22, 309, 1979.

537. **Mickievich, B. F., Sushchyk, J. J., Yermolenko, W. I., Babak, K. A., and Kornienko, T. G.,** *Berylium in the Hipergenic Zone,* Naukova Dumka, Kiyev, 1977, 167 (Ru).

538. **Miesch, A. T. and Huffman, C.,** Abundance and distribution of lead, zinc, cadmium, and arsenic in soils, in Helena Valley, Montana, Area Environmental Pollution Study, Environmental Protection Agency, Research Triangle Park, N.C., 1972, 65.

539. **Miller, G. W., Yu, M. H., and Psenak, M.,** Presence of fluoro-organic compounds in higher plants, *Fluoride,* 6, 203, 1973.

540. **Mills, J. G. and Zwarich, M. A.,** Heavy metal content of agricultural soils in Manitoba, *Can. J. Soil Sci.,* 55, 295, 1975.

541. **Mirchev, S.,** Zinc and copper compounding forms in chernozems and forest soils, *Pochvozn. Agrokhim.,* 13, 84, 1978 (Bu).

542. **Mishra, D. and Kar, M.,** Nickel in plant growth and metabolism, *Bot. Rev.,* 40, 395, 1974.

543. **Mitchell, R. L.,** Trace elements in some constituent species of moorland grazing, *J. Br.Grass. Soc.,* 9, 301, 1954.

544. **Mitchell, R. L.,** Cobalt in soil and its uptake by plants, paper presented at 9th Simposio Int. di Agrochimica, Punta Ala, Argentina, October 2, 1972, 521.

545. **Mitev, Kh. and Gyurov, G.,** Manganese in Bulgarian pseudopodsols, *Nauchni Tr.,* 22, 63, 1973 (Bu).

546. **Miyake, Y. and Takahashi, E.,** Silicon deficency of tomato plant, *Soil Sci. Plant Nutr.,* 24, 175, 1978.

547. **Montford, M. A., Shank, K. E., Hendricks, C., and Oakes, T. W.,** Elemental concentrations in food products, in *Trace Subst. in Environ. Health,* Vol. 14, Hemphill, D. D., Ed., University of Missouri, Columbia, Mo., 1980, 155.

548. **Moore, D. P.,** Mechanisms of micronutrient uptake by plants, in *Micronutrients in Agriculture,* Mortvedt, J. J., Giordano, P. M., and Lindsay, W. L., Eds., Soil Science Society of America, Madison, Wis., 1972, 17.

549. **Moore, H. E., Martell, E. A., and Poet, S. E.,** Sources of polonium-210 in atmosphere, *Environ. Sci. Technol.,* 10, 586, 1976.

550. **Moore, H. E. and Poet, S. E.,** Background levels of Ra-226 in the lower troposphere, *Atmos. Environ.,* 10, 381, 1976.

551. **Moré, E. and Coppnet, M.,** Teneurs en sélénium des plantes fourragères influence de la fertilisation et des apports de sélénite, *Ann. Agron.,* 31, 297, 1980.

552. **Morris, D. F. C. and Short, E. L.,** Minerals of rhenium, *Mineral. Mag.,* 35, 871, 1966.

553. **Morrison, R. S., Brooks, R. R., and Reeves, R. D.,** Nickel uptake by *Alyssum* species, *Plant Sci. Lett.,* 17, 451, 1980.

554. **Mortvedt, J. J.,** Soil reactions of Cd contaminants in P fertilizers, *Agron. Abstr.,* Dec. 3, 1978.

555. **Moxon, A. L. and Olson, O. E.,** Selenium in agriculture, in *Selenium,* Van Nostrand, New York, 1974, 675.

556. **Murphy, L. S. and Walsh, L. M.,** Correction of micronutrient deficiencies with fertilizers, in *Micronutrients in Agriculture,* Mortvedt, J. J., Giordano, P. M., and Lindsay, W. L., Eds., Soil Science Society of America, Madison, Wis., 1972, 347.

557. **Murray, F.,** Effects of fluorides on plant communities around an aluminium smelter, *Environ. Pollut.,* 24, 45, 1981.

557a. **Nalovic, L. and Pinta, M.,** Recherches sur les éléments traces dans les sols tropicaux; étude de quelques sols de Madagascar, *Geoderma,* 3, 117, 1970.

558. **Naidenov, M. and Travesi, A.,** Nondestructive neutron activation analysis of Bulgarian soils, *Soil Sci.,* 124, 152, 1977.

559. **Nakos, G.,** Lead pollution. Fate of lead in the soil and its effects on *Pinus halepensis, Plant Soil,* 53, 427, 1979.

560. **Nambiar, K. K. M. and Motiramani, D. P.,** Tissue Fe/Zn ratio as a diagnostic tool for prediction of Zn deficiency in crop plants, *Plant Soil,* 60, 357, 1981.

561. **Nasseem, M. G. and Roszyk, E.,** Studies on copper and zinc forms in some Polish and Egyptian soils, *Pol. J. Soil Sci.,* 10, 25, 1977.

562. **Neuberg, J., Zelený, F., and Zavadilová, L.,** Investigation on manganese content in the chief plants of crop rotation, *Rostl. Vyroba,* 24, 347, 1978a (Cz).

563. **Neuberg, J., Zelený, F., Hovorková-Zavadilová, L., and Hrozinková, A.,** Molybdenum content of main crops of the field crop rotation, *Rostl. Vyroba,* 24, 567, 1978b (Cz).

564. **Nicholas, D. J. D.,** The functions of trace elements in *Trace Elements in Soil-Plant-Animal Systems,* Nicholas, D. J. D. and Egan, A. R., Eds., Academic Press, New York, 1975, 181.

565. **Nicolls, K. D. and Honeysett, J. L.,** The cobalt status of Tasmanian soils, *Austr. J. Agric. Res.,* 13, 368, 1964.

566. **Nielsen, F. H., Reno, H. T., Tiffin, L. O., and Welch, R. M.,** Nickel, in *Geochemistry and the Environment,* Vol. 2, Nielsen, F. H., Ed., N.A.S., Washington, D.C., 1977, 40.

567. **Niyazova, G. A. and Letunova, S. V.,** Microelements accumulation by soil microflora at the conditions of the Sumsarsky lead-zinc biogeochemical province in Kirghizya, *Ekologiya,* 5, 89, 1981 (Ru).

568. **Nömmik, H.,** *Fluorine in Swedish Agrcultural Products, Soil and Drinking Water,* Esselte Aktiebolag, Stockholm, 1953, 121.

569. **Nordberg, G. F.,** Health hazards of environmental cadmium pollution, *Ambio,* 3, 55, 1974.

570. **Norrish, K.,** The geochemistry and mineralogy of trace elements, in *Trace Elements in Soil-Plant-Animal Systems,* Nicholas, D. J. D. and Egan, A. R., Eds., Academic Press, New York, 1975, 55.

571. **Norvell, W. A.,** Equilibria of metal chelates in soil sollution, in *Micronutrients in Agriculture,* Mortvedt, J. J., Giordano, P. M., and Lindsay, W. L., Eds., Soil Science Society of America, Madison, Wis., 1972, 115.

572. **Nowosielski, O.,** The use of simplified *Aspergillus niger* method for chemical analysis of agricultural materials, *Rocz. Nauk Roln.,* 87a, 201, 1963 (Po).

573. **Nwankwo, J. N. and Elinder, C. G.,** Cadmium, lead and zinc concentratons in soils and food grown near a zinc and lead smelter in Zambia, *Bull. Environ. Contam. Toxicol.,* 22, 265, 1979.

574. **Oakes, T. W., Shank, K. E., Easterly, C. E., and Quintana, L. R.,** Concentrations of radionuclides and selected stable elements in fruits and vegetables, in *Trace Subst. Environ. Health,* Vol II, Hemphill, D. D., Ed., University of Missouri, Columbia, Mo., 1977, 123.

575. **Obukhov, A. I.,** Minor element content and distribution in soils of the aridic tropical zone in Burma, *Pochvovedenie,* 2, 93, 1968 (Ru).

576. **Oelschläger, W.,** Zusammensetzung staubförmiger Kraftfahrzeug-Emissionen, *Sonderdruck Staub Reinhalt. Luft.,* 6, 3, 1972.

576a. **Oelschläger, W.,** Die Problematik bei der Probeentnahme von Futtermitteln in und ausserhalb von Emissionsgebieten, *Sonderdruck Landwirtsch. Forsch.,* 26, 89, 1973.

577. **Oelschläger, W.,** Über die Kontamination von Futtermitteln und Nahrungsmitteln mit Cadmium, *Landwirtsch. Forsch.,* 27, 247, 1974.

578. **Oelschläger, W. and Menke, K. H.,** Über Selengehalte pflanzlicher, tierischer und anderer Stoffe, *Sonderdruck Z. Ernaehrungswiss.,* 9, 208, 216, 1969.

579. **Oelschläger, W., Wöhlbier, W., and Menke, K. W.,** Über Fluor-Gehalte pflanzlicher tierischer und anderer Stoffe aus Gebieten ohne und mit Fluoremissionen, *Landwirtsch. Forsch.,* 20, 199, 1968.

580. **Ogoleva, V. P. and Tcherdakoa, L. N.,** Nickel in soils of the Volgograd region, *Agrokhimiya,* 9, 105, 1980 (Ru).

581. **Olsen, S. R.,** Micronutrient interactions, in *Micronutrients in Agriculture,* Mortvedt, J. J., Giordano, P. M., and Lindsay, W. L., Eds., Soil Science Society of America, Madison, Wis., 1972, 243.

582. **Olson, K. W. and Skogerboe, R. K.,** Identification of soil lead compounds from automotive sources, *Environ. Sci. Technol.,* 9, 277, 1975.

583. **Omueti, J. A. I. and Jones, R. L.,** Fluoride adsorption by Illinois soils, *J. Soil Sci.,* 28, 564, 1977.

584. **Omueti, J. A. I. and Jones, R. L.,** Fluorine distribution with depth in relation to profile development in Illinois, *Soil Sci. Soc. Am. J.,* 44, 247, 1980.

585. **Ormrod, D. P.,** *Pollution in Horticulture,* Elsevier, Amsterdam, 1978, 280.

586. **Orlova, E. D.,** Influence of microfertiizers on microelement uptake by leaves, grain and straw of spring wheat, in *Mikroelementy v Pochvakh, Rastieniyakh i Vodakh Yuzhnoy Chasti Zapadnoy Sibiriy,* Kovalev, R. V., Ed., Izd. Nauka Novosibirsk, 1971, 98 (Ru).

587. **Ovcharenko, F. D., Gordienko, S. A., Glushchenko, T. F., and Gavrish, I. N.,** Methods and results of the complex formation of humic acids from peat, in Trans. 6th Int. Symp. Humus Planta, Praha, 1975, 137.

588. **Ozoliniya, G. R. and Kiunke, L. M.,** Content of little investigated elements in parts of flax, barley and lettuce, in *Fizyologo-Biokhimicheskiye Issledovaniya Rasteniy,* Zinante, Riga, 1978, 111 (Ru).

589. **Ozoliniya, G. R. and Zariniya, V.,** Accumulation of copper, molybdenum, cobalt and nickel in seeds of kidney beans, in *Mikroelementy w Kompleksie Mineralnogo Pitaniya Rasteniy,* Zinante, Riga, 1975, 117 (Ru).

590. **Paasikallio, A.,** The mineral element contents of timothy (*Phleum pratense* L.) in Finland, *Acta Agric. Scand. Suppl.,* 20, 40, 1978.

591. **Paasikallio, A.,** The effect of soil pH and Fe on the availability of Se-75 in sphagnum peat soil, *Ann. Agric. Fenn.,* 20, 15, 1981.

592. **Padzik, A. and Wlodek, S.,** Beryllium contamination in the rural and industrial areas in Poland, *Rocz. Panstw. Zakl. Hig.,* 30, 397, 1979 (Po).

593. **Page, A. L.,** Fate and Effects of Trace Elements in Sewage Sludge when Applied to Agricultural Lands, U.S. Environmental Protection Agency Report, Cincinnati, Ohio, 1974, 97.

594. **Page, A. L., Ganje, T. J., and Joshi, M. S.,** Lead quantities in plants, soil and air near some major highways in Southern California, *Hilgardia,* 41, 1, 1971.

595. **Page, N. J. and Carlson, R. R.,** Review of platinum-group metal chemistry and the major occurrences in the world, *U.S. Geol. Surv. Open-File Rep.,* 80-90, 21, 1980.

596. **Pahl, E., Voigländer, G., and Kirchgessner, M.,** Untersuchungen über Spurenelementgehalt des Weidefutters einer mehrfach genutzten Weidelgrass — Weissklee weide wärent zweier Vegetationsperioden, *Z. Acker Pflanzenbau,* 131, 70, 1970.

597. **Pais, I., Fehér, M., Farkas, E., Szabó, Z., and Cornides, I.,** Titanium as a new trace element, *Commun. Soil Sci. Plant Anal.,* 8, 407, 1977.

598. **Palsson, P. A., Georgsson, G., and Petruson, G.,** Fluorosis in farm animals in Iceland related to volcanic eruptions, in *Geomedical Aspects in Present and Future Research,* Låg, J., Ed., Universitesforlaget, Oslo, 1980, 123.

599. **Panov, E. N.,** Behavior of minor elements under the treatment of microcline by solution of organic acids, *Geokhimiya,* 10, 1568, 1980 (Ru).

600. **Parker, R. D. R., Sharma, R. P., and Miller, G. W.,** Vanadium in plants, soil and water in the Rocky Mountain region and its relationship to industrial operations, in *Trace Subst. Environ. Health,* Vol. 12, Hemphill, D. D., Ed., University of Missouri, Columbia, Mo., 1978, 340.

601. **Pauli, W. F.,** Heavy metal humates and their behavior against hydrogen sulfide, *Soil Sci.,* 119, 98, 1975.

602. **Pavlotskaya, F. I.,** On the role of organic matter in the migration of radioactive products of global fallout, in *Contributions to Recent Geochemistry and Analytical Chemistry,* Izd. Nauka, Moscow, 1972, 521 (Ru).

603. **Pavlotskaya, F. I.,** *Migration in Soils of Radioactive Products of Global Nuclear Processes,* Atomizdat, Moscow, 1974, 215 (Ru).

604. **Pavlotskaya, F. I., Goriachenkova, T. A., and Blokhina, M. I.,** Behavior of Sr-90, in the system: ammonium humate-stable strontium-iron (III), *Pochvovedenie,* 11, 33, 1976 (Ru).

605. **Pawlak, L.,** Trace Element Pollution of Soils and Plants in the Vicinity of the Oil Refinery Plant near Plock, Doctoral thesis, Agricultural University, Warsaw, 1980, 165 (Po).

605a. **Peneva, N.,** Effect of organic matter on zinc retention and liability in the soil, *Pochvozn. Agrokhim.,* 11, 14, 1976 (Bu).

606. **Perkins, D. F., Millar, R. O., and Neep, P. E.,** Accumulation of airborne fluoride by lichens in the vicinity of an aluminium reduction plant, *Environ. Pollut.,* 21, 155, 1980.

607. **Perrott, K. W., Smith, B. F. L., and Inkson, R. H. E.,** The reaction of fluoride with soils and soil minerals, *J. Soil Sci.,* 27, 58, 1976.

608. **Pesek, F. and Kolsky, V.,** Studies on trace contaminants of edible parts of crops, *Rostl. Vyroba,* 13, 445, 1967 (Cz).

609. **Peterson, P. J.,** Unusual accumulations of elements by plants and animals, *Sci. Prog.,* 59, 505, 1971.

610. **Peterson, P. J., Burton, M. A. S., Gregson, M., Nye, S. M., and Porter, E. K.,** Tin in plants and surface waters in Malayasian ecosystems, in *Trace Subst. Environ. Health,* Vol. 10, Hemphill, D, D., Ed., University of Missouri, Columbia, Mo., 1976, 123.

611. **Petrov, I. I. and Tsalev, D. L.,** Atomic absorption methods for determination of soil arsenic based on arsine generation, *Pochvozn. Agrokhim.,* 14, 20, 1979 (Bu).

612. **Petrov, I. I., Tsalev, D. L., and Lyotchev, I. S.,** Investigation on arsenic and cadmium content in soils, *Higiena Zdravyeopazv.,* 22, 574, 1979 (Bu).

613. **Petrunina, N. S.,** Geochemical ecology of plants from the provinces of high trace element contents, in *Problems of Geochemical Ecology of Organisms,* Izd. Nauka, Moscow, 1974, 57 (Ru).

614. **Peyve, I. V.,** About biochemical role of microelements in fixation of molecular nitrogen, in *Biologicheskaya Rol Mikroelementov i ikh Primeneniye w Sielskom Khozyastvie i Medicinie,* Izd. Nauka, Moscow, 1974, 3 (Ru).

615. **Pike, J. A., Golden, M. L., and Freedman, J.,** Zinc toxicity in corn as a result of a geochemical anomaly, *Plant Soil,* 50, 151, 1978.

616. **Pilegaard, K.,** Heavy metal uptake from soil in four seed plants, *Bot. Tidsskr.,* 73, 167, 1978.

617. **Pinta, M. and Ollat, C.,** Researches physicochimiques des éléments traces dans les sols tropicaux, Etude de quelques sols du Dahomey, *Geochim. Cosmochim. Acta,* 25, 14, 1961.

618. **Piotrowska, M.,** The mobility of heavy metals in soils contaminated with the copper smelter dusts, and metal uptake by orchard grass, *Materialy IUNG,* 159-R, Pulawy, Poland, 1981, 88 (Po).

618a. **Piotrowska, M.,** Occurrence of selenium in arable soils of Poland, *Rocz. Glebozn.,* in press, (Po).

619. **Piotrowska, M. and Wiacek, K.,** Fluorine content in some Polish soils, *Rocz. Nauk Roln.,* 101a, 93, 1975 (Po).

620. **Piotrowska, M. and Wiacek, K.,** Effect of the long-term phosphorus fertilization on the content of some minor elements in soils and plants, *Rocz. Nauk Roln.,* 103a, 7, 1978 (Po).

621. **Pokatilov, J. G.,** Content of iodine in soils of Barguzine depression in Buriatia, *Agrokhimiya,* 8, 96, 1979 (Ru).

622. **Ponnamperuma, F. N.,** Screening rice for tolerance to mineral stress, paper presented at Workshop on Plant Adaptation to Mineral Stress in Problem Soils, Wright, M. J., Ed., Cornell University, Ithaca, N.Y., 1976, 341.

623. **Popović, Z., Pantovic, M., and Jakovljević, M.,** The content of micronutrients in alfalfa fertilized with monoammonium phosphate, *Agrokhimiya,* 11-12, 497, 1981 (Sh).

624. **Porter, E. K. and Peterson, P. J.,** Arsenic accumulation by plants on mine waste, United Kingdom, *Sci. Total Environ.,* 4, 365, 1975.

625. **Powley, F.,** Denmark's heavy report on cadmium, *Ambio,* 10, 190, 1981.

626. **Prather, R. J.,** Sulfuric acid as an amendment for reclaiming soils high in boron, *Soil Sci. Soc. Am. J.,* 41, 1098, 1977.

627. **Preer, J. R. and Rosen, W. G.,** Lead and cadmium content of urban garden vegetables, in *Trace Subst. Environ. Health,* Vol. 11, Hemphill, D. D., Ed., University of Missouri, Columbia, Mo., 1977, 399.

628. **Preer, J. R., Sekhon, H. S., Weeks, J., and Stephens, B. R.,** Heavy metals in garden soil and vegetables in Washington, D. C., in *Trace Subst. Environ. Health,* Vol. 14, Hemphill, D. D., Ed., University of Missouri, Columbia, Mo., 1980, 516.

629. **Presant, E. W.,** Geochemistry of iron, manganese, lead, copper, zinc, antimony, silver, tin and cadmium in the soils of the Bathurst area, New Brunswick, *Geol. Surv. Can. Bull.,* 174, 1, 1971.

630. **Price, C. A., Clark, H. E., and Funkhouser, E. A.,** Functions of micronutrients in plants, in *Micronutrients in Agriculture,* Mortvedt, J. J., Giordano, P. M., and Lindsay, W. L., Eds., Soil Science Society of America, Madison, Wis., 1972, 231.

631. **Prikhodko, N. N.,** Vanadium, chromium, nickel and lead in soils of Pritissenskaya lowland and piedmonts of Zakarpatie, *Agrokhimiya,* 4, 95, 1977 (Ru).

632. **Prince, N. B. and Calvert, S. E.,** The geochemistry of iodine in oxidised and reduced recent marine sediments, *Geochim. Cosmochim. Acta,* 37, 2149, 1973.

633. **Pulford, I. D.,** Controls on the solubility of trace metals in soils, paper presented at 9th Int. Coll. Plant Nutrition, Coventry, August 22, 1982, 486.

634. **Purves, D.,** The contamination of soil and food crops by toxic elements normally found in municipal wastewaters and their consequences for human health, in *Wastewater Renovation and Reuse,* D'Itri, F. M., Ed., Marcel Dekker, New York, 1977, 257.

635. **Quinche, J. P.,** La pollution mercurielle de diverses espèces de champignons, *Rev. Suisse Agric.,* 8, 143, 1976.

636. **Quinche, J. P.,** L'*Agaricus bitorquis,* un champignon accumulateur de mercure, de sélénium et de cuivre, *Rev. Suisse Vitic. Arboric. Hortic.,* 11, 189, 1979.

637. **Quinche, J. P., Bolay, A., and Dvorak, V.,** La pollution par le mercure des vegetaux et des sols de la suisse romande, *Rev. Suisse Agric.,* 8, 130, 1976.

638. **Quirk, J. P. and Posner, A. M.,** Trace element adsorption by soil minerals, in *Trace Elements in Soil-Plant-Animal Systems,* Nicholas, D. J. D., Ed., Academic Press, New York, 1975, 95.

639. **Rajaratinam, J. A., Lowry, J. B., and Hock, L. I.,** New method for assessing boron status of the oil palm, *Plant Soil,* 40, 417, 1974.

640. **Ranadive, S. J., Naik, M. S., and Das, N. B.,** Copper and zinc status of Naharashtra soils, *J. Ind. Soc. Soil. Sci.,* 12, 243, 1964.

641. **Randhawa, S. J., Kanwar, J. S., and Nijhawan, S. D.,** Distribution of different forms of manganese in the Punjab soils, *Soil Sci.,* 92, 106, 1961.

642. **Rashid, M. A.,** Amino acids associated with marine sediments and humic compounds and their role in solubility and complexing of metals, in Proc. 24th Int. Geol. Congr., Sect. 10, Montreal, 1979, 346.

643. **Rauser, W. E.,** Entry of sucrose into minor veins of bean seedlings exposed to phytotoxic burdens of Co, Ni or Zn, paper presented at Int. Symp. Trace Element Stress in Plants, Los Angeles, November 6, 1979, 33.

644. **Ravikovitch, S., Margolin, M., and Navroth, J.,** Microelements in soils of Israel, *Soil Sci.,* 92, 85, 1961.

645. **Raychaudhuri, S. P. and Biswas, D. N. R.,** Trace element status of Indian soils, *J. Indian Soc. Soil Sci.,* 12, 207, 1964.

646. **Reid, D. A.,** Aluminum and manganese toxicities to the cereal grains, in Proc. of Workshop on Plant Adaptation to Mineral Stress in Problem Soils, Wright, M. J., Ed., Cornell University, Ithaca, N.Y., 1976, 55.

647. **Reilly, A. and Reilly, C.,** Copper-induced chlorosis in *Becium homblei* (De Wild.) Duvig. et Plancke, *Plant Soil,* 38, 671, 1973.

648. **Reisenauer, H. M., Walsh, L. M., and Hoeft, R. G.,** Testing soil for sulphur, boron, molybdenum, and chlorine, in *Soil Testing and Plant Analysis,* Walsh, L. M. and Beaton, J. D., Eds., Soil Science Society of America, Madison, Wis., 1973, 173.

649. **Reuter, D. J.,** The recognition and correction of trace element deficiencies, in *Trace Elements in Soil-Plant-Animal Systems,* Nicholas, D. J. D., and Egan, A. R., Eds., Academic Press, New York, 1975, 291.

650. **Riadney, C.,** Techniques d'Études des Carens en Molybdène Pouvant Affecter les Cultures en Régions Tropicales, Thèse, Conservatoire National des Arts et Métiers, Paris, 1964, 117.

651. **Richards, B.N.,** *Introduction to the Soil Ecosystem,* Logman, Essex, 1974, 266.

652. **Rieder, W. and Schwertmann, U.,** Kupferanreichung in hopfengenutzen Böden der Hallertau, *Landwirtsch. Forsch.,* 25, 170, 1972.

653. **Riffaldi, R., Levi-Minzi, R., and Soldatini, G. E.,** Pb absorption by soils, *Water Air Soil Pollut.,* 6, 119, 1976.

654. **Rinkis, G. J.,** *Optimalization of Mineral Nutrition of Plants,* Zinante, Riga, 1972, 355 (Ru).

655. **Rippel, A., Balažova, G., and Palušova, O.,** Das Kadmium in der Lebensumwelt, in *Proc. Kadmium Symp.* Anke, M. and Schneider, H. J., Eds., Friedrich-Schiller University, Jena, E. Germany, 1979, 238.

656. **Risch, M. A., Atadschanev, P., and Teschabaev, C.,** Arsenhaltige biogeochemische Provinzen Usbekistans in *Proc. Arsen Symp.,* Anke, M., Schneider, H. J., and Brückner, Chr., Eds., Friedrich-Schiller University, Jena, E. Germany, 1980, 91.

657. **Roberts, T. M.,** A review of some biological effects of lead emissions from primary and secondary smelters, paper presented at Int. Conf. on Heavy Metals, Toronto, October 27, 1975, 503.

658. **Roberts, T. M., Gizyn, W., and Hutchinson, T. C.,** Lead contamination of air, soil, vegetation and people in the vicinity of secondary lead smelters, in *Trace Subst. Environ. Health,* Vol. 8, Hemphill, D. D., Ed., University of Missouri, Columbia, Mo., 1974, 155.

659. **Roberts, R. D., Johnson, M. S., and Firth, J. N. M.,** Predator-prey relationship in the food-chain transfer of heavy metals, in *Trace Subst. Environ. Health,* Vol. 12, Hemphill, D. D., Ed., University of Missouri, Columbia, Mo., 1979, 104.

660. **Robson, A. D. and Reuter, D. J.,** Diagnosis of copper deficiency and toxicity, in *Copper in Soils and Plants,* Loneragan, J. F., Robson, A. D., and Graham, R. D., Eds., Academic Press, New York, 1981, 287.

661. **Rolfe, G. L. and Bazzaz, F. A.,** Effect of lead contamination on transpiration and photosynthesis of Loblolly Pine and Autumn Olive, *Forest Sci.,* 21, 33, 1975.

662. **Romney, E. M., Wallace, A., and Alexander, G. V.,** Responses of bush bean and barley to tin applied to soil and to solution culture, *Plant Soil,* 42, 585, 1975.

663. **Roques, A., Kerjean, M., and Auclair, D.,** Effects de la pollution atmospherique par le fluor et le dioxyde de soufre sur l'appareil reproducteur femelle de *Pinus silvestris* en foret de Roumare, *Environ. Pollut.,* 21, 191, 1980.

664. **Roslakov, N. A.,** *Geochemistry of Gold in Hypergene Zone,* Idz. Nauka, Novosibirsk, 1981, 237 (Ru).

665. **Roszyk, E.,** Contents of vanadium, chromium, manganese, cobalt, nickel, and copper in Lower Silesian soils derived from loamy silts and silts, *Rocz. Glebozn.,* 19, 223, 1968 (Po).

666. **Roucoux, P. and Dabin, P.,** The effect of cadmium on the nitrogen fixation, paper presented at Semin. Carbohydrate and Protein Synthesis, Giessen, September 7, 1977, 215.

667. **Ruebenbauer, T. and Stopczyk, K.,** The content and distribution of some microelements in wheat, diploid an tetraploid rye as well as wheat-rye, *Acta Agrar. Silvestria,* 12, 75, 1972 (Po).

668. **Rühling, A. and Tyler, G.,** An ecological approach to the lead problem, *Bot. Not.,* 121, 321, 1968.

669. **Rühling, A. and Tyler, G.,** Ecology of heavy metals: a regional and historical study, *Bot. Not.,* 122, 248, 1969.

670. **Russel, S. and Šwiecicki, C.,** Fluorine effect on biological activity of black earth and pseudopodsolic soil, *Rocz. Nauk Roln.,* 103a, 47, 1978 (Po).

671. **Ruszkowska, M., Ed.,** Dynamics and balance of mineral nutrients in a lysimetric experiment, *Rocz. Nauk Roln.,* 173d, 1, 1979 (Po).

672. **Ruszkowska, M., Lyszcz, S., and Sykut, S.,** The activity of catechol oxidase in sunflower leaves as indicator of copper supply in plants, *Pol. J. Soil Sci.,* 8, 67, 1975.

673. **Ryan, J., Miyamoto, S., and Stroehlein, J. L.,** Solubility of manganese, iron, and zinc as affected by application of sulfuric acid to calcereous soils, *Plant Soil.,* 40, 421, 1974.

674. **Ryan, J., Miyamoto, S., and Stroehlein, J. L.,** Relation of solute and sorbed boron to the boron hazard in irrigation water, *Plant Soil,* 47, 253, 1977.

675. **Saeed, M., and Fox, R. L.,** Influence of phosphate fertilization on Zn adsorption by tropical soils, *Soil Sci. Soc. Am. J.,* 43, 683, 1979.

676. **Salnikov, V. G., Pavlotskaya, F. J., and Moisieyev, J. T.,** Effect of lime and peat on the bonding of Sr-90 with the components of soil organic matter, *Pochvovedenie,* 5, 87, 1976 (Ru).

677. **Sanders, J. R. and Bloomfield, C.,** The influence of pH, ionic strength and reactant concentrations on copper complexing by humified organic matter, *J. Soil Sci.,* 31, 53, 1980.

678. **Sandmann, G. and Boger, O.,** Copper-mediated lipid peroxidation process in photosynthetic membranes, *Plant Physiol.,* 66, 797, 1980.

679. **Sankla, N. and Sankla, D.,** Effects of germanium on growth of higher plants, *Naturwissenschaften,* 54, 621, 1967.

680. **Sanko, P. M. and Anoshko, V. S.,** Zinc in meadow soils and grasses of Byelorussia, *Agrokhimiya,* 7, 109, 1975 (Ru).

681. **Sapek, A.,** The role of the humus substances in podzol soil development, *Stud. Soc. Sci. Torun.,* 7, 1, 1971 (Po).

682. **Sapek, B.,** Copper behavior in reclaimed peat soil of grassland, *Rocz. Nauk Roln.,* 80f, 13, and 65, 1980.

683. **Sapek, B. and Okruszko, H.,** Copper content of hay and organic soils of northeastern Poland, *Z. Probl. Post. Nauk Roln.,* 179, 225, 1976 (Po).

684. **Sapek, A. and Sapek, B.,** Nickel content in the grassland vegetation, in *Proc. Nickel Symp.,* Anke, M., Schneider, H. J., and Brückner, Chr., Eds., Friedrich-Schiller University, Jena, E. Germany, 1980, 215.

685. **Sapek, A. and Sklodowski, P.,** Concentration of Mn, Cu, Pb, Ni, and Co in rendzinas of Poland, *Rocz. Glebozn.,* 27, 137, 1976 (Po).

686. **Sarosiek, J. and Klys, B.,** Studies on tin content of plants and soils from Sudetian, *Acta Soc. Bot. Pol.,* 31, 737, 1962 (Po).

687. **Savic, B.,** The dynamics of molybdenum in some soils of native grass land in Bośnia, *Rad. Poljopr. Fak. Univ. Sarajevo,* 13, 69, 1964 (Sh).

688. **Scharpenseel, H. W., Pietig, F., and Kruse, E.,** Urankonzentration in Böden und ihre mögliche Nutzung als Prospektionshilfe, *Z. Pflanzenernaehr. Bodenkd.,* 2, 131, 1975.

688a. **Scheffer, K., Stach, W., and Vardakis, F.,** Über die Verteilung der Schwermatallen Eisen. Mangan, Kupfer und Zink in Sommergesternpflanzen, *Landwirtsch. Forsch.,* 1, 156, 1978; 2, 326, 1979.

689. **Schlichting, E. and Elgala, A. M.,** Schwermetallverteilung und Tongehalte in Böden, *Z. Pflanzenernaehr. Bodenkd.,* 6, 563, 1975.

690. **Schnitzer, M. and Kerndorff, H.,** Reactions of fulvic acid with metal ions, *Water Air Soil Pollut.,* 15, 97, 1981.

691. **Schnitzer, M. and Khan, S. U.,** *Humic Substances in the Environment,* Marcel Dekker, New York, 1972, 327.

692. **Schnitzer, M. and Khan, S. U.,** *Soil Organic Matter,* Elsevier, Amsterdam, 1978, 319.

693. **Schnug, E. and Finck, A.,** Trace element mobilization by acidifying fertilizers, paper presented at 9th Int. Coll. Plant Nutrition, Conventry, August 22, 1982, 582.

694. **Schönborn, W. and Hartmann, H.,** Enterung von Schwermetallen aus Klärschlämmen durch bakterielle Laugung, *Gwf. Wasser-Abwasser,* 120, 329, 1979.

695. **Schroeder, H. A. and Balassa, J. J.,** Abnormal trace metals in man: germanium, *J. Chron. Dis.,* 20, 211, 1967.

696. **Schroeder, H. A., Balassa, J. J., and Tipton, I. H.,** Abnormal trace metals in man: chromium, *J. Chron. Dis.,* 15, 941, 1962.

697. **Schroeder, H. A., Buckman, J., and Balassa, J. J.,** Abnormal trace elements in man: tellurium, *J. Chron. Dis.,* 20, 147, 1967.

698. **Schulz, R. K. and Babcock, K. L.,** On the availability of carrier-free Ru-106 from soils, *Soil Sci.,* 117, 171, 1974.

699. **Schwertmann, U. and Taylor, R. M.,** Iron oxides, in *Minerals in Soil Environmens,* Dixon, J. B. and Weed, S. B., Eds., Soil Science Society of America, Madison, Wis., 1977, 145.

700. **Selezniev, Y. M. and Tiuriukanov, A. N.,** Some factors of changing iodine compound forms in soils, *Biol. Nauki,* 14, 128, 1971 (Ru).

701. **Senesi, N. and Polemio, M.,** Trace element addition to soil by application of NPK fertilizers, *Fert. Res.,* 2, 289, 1981.

702. **Severne, B. C.,** Nickel accumulation by *Hybanthus floribundus, Nature (London),* 248, 807, 1974.

703. **Shacklette, H. T,** Mercury content of plants, *U.S. Geol. Surv. Prof. Pap.,* 713, 35, 1970.

704. **Shacklette, H. T.,** Cadmium in plants, *U.S. Geol. Surv. Bull.,* 1314g, 28, 1972.

705. **Shacklette, H. T.,** Elements in fruits, and vegetable from areas of commercial production in the Conterminous United States, *U.S. Geol. Surv. Prof. Pap.,* 1178, 149, 1980.

705a. **Shacklette, H. T., Boerngen, J. G., Cahill, J. P., and Rahill, R. L.,** Lithium in surficial materials of the conterminous United States and partial data on cadmium, *U.S. Geol. Surv. Circ.,* 673, 7, 1973.

706. **Shacklette, H. T. and Boerngen, J. G.,** Element concentrations in soils and other surficial materials of the conterminous United States, *U.S. Geol. Surv. Prof. Pap.,* 1270, in press, 1984.

707. **Shacklette, H. T., Boerngen, J. G., and Keith, J. R.,** Selenium, fluorine, and arsenic in surficial materials of the conterminous United States, *U.S. Geol. Surv. Circ.,* 692, 14, 1974.

708. **Shacklette, H. T. and Connor, J. J.,** Airborne chemical elements in Spanish moss, *U.S. Geol. Surv. Prof. Pap.,* 574e, 46, 1973.

709. **Shacklette, H. T. and Cuthbert, M. E.,** Iodine content of plant groups as influenced by variation in rock and soil type, *Geol. Soc. Am. Spec. Pap.,* 90, 31, 1967.

710. **Shacklette, H. T., Erdman, J. A., and Harms, T. F.,** Trace elements in plant foodstuffs, in *Toxicity of Heavy Metals in the Environments, Part I,* Oehme, F. W., Ed., Marcel Dekker, New York, 1978, 25.

711. **Shacklette, H. T., Lakin, H. W., Hubert, A. E., and Curtin, G. C.,** Absorption of gold by plants, *U.S. Geol. Surv. Bull.,* 1314b, 1970.

712. **Shakuri, B. K.,** Trace elements in northern Caucasian chernozems of the Rostov Region, *Izv. Akad. Nauk Az. SSR Ser. Biol.,* 4, 81, 1964 (Ru).

713. **Shakuri, B. K.,** Lithium content in soils of Nakhichevan Azerbaizhan, SSR, *Pochvovedenie,* 8, 130, 1976 (Ru).

714. **Shakuri, B. K.,** Nickel, vanadium, chromium and strontium in soils of Nakhichevan Azerbaizhan SSR, *Pochvovedenie,* 4, 49, 1978 (Ru).

715. **Shaw, G.,** Concentrations of twenty-eight elements in fruiting shrubs downwind of the smelter at Flin Flon, Manitoba, *Environ. Pollut.,* A25, 197, 1981.

715a. **Shcherbakov, V. P., Dvornikov, A. T., and Zakrenichnya, G. L.,** New data on occurrence and forms of mercury in the Donbas coals, *Dopov. ANUR,* 2b, 126, 1970 (Uk).

716. **Shcherbina, V. V.,** Geochemistry of molybdenum and tungsten in oxidic zone, in *Geokhimiya Molybdena i Volframa,* Izd. Nauka, Moscow, 1971, 81 (Ru).

717. **Shklyaev, J. N.,** Effect of top dressing with cobalt on distribution of cobalt-60 and carbon-14 in vetch plants, *Agrokhimiya,* 1, 115, 1978 (Ru).

718. **Shkolnik, M. J.,** *Microelements in Plant Life,* Izd. Nauka, Leningrad, 1974a, 323 (Ru).

719. **Shkolnik, M. J.,** General conception of the physiological role of boron in plants, *Physiol. Rasteniy,* 21, 174, 1974b (Ru).

720. **Sholkovitz, E. R. and Copland, D.,** The coagulation, solubility and adsorption properties of Fe, Mn, Cu, Ni, Cd, Co and humic acids in a river water, *Geochim. Cosmochim. Acta,* 45, 181, 1981.

721. **Shukla, U. C., Mittal, S. B., and Gupta, R. K.,** Zinc adsorption in some soils as affected by exchangeable cations, *Soil Sci.,* 129, 366, 1980.

722. **Shukla, U. C. and Yadav, O. P.,** Effect of phosphorus and zinc on nodulation and nitrogen fixation in chickpea (*Cicer arietinum* L.), in Abstr., 12th Int. Soil Sci. Congr., New Delhi, 1982, 54.

723. **Shupe, J. L. and Sharma, R. P.,** Fluoride distribution in a natural ecosystem and related effects on wild animals, in *Trace Subst. Environ. Health,* Vol. 10, Hemphill, D. D., Ed., University of Missouri, Columbia, Mo., 1976, 137.

724. **Siegel, S. M. and Siegel, B. Z.,** A note on soil and water mercury levels in Israel and the Sinai, *Water Air Soil Pollut.,* 5, 263, 1976.

725. **Sievers, M. L. and Cannon, H. L.,** Disease patterns of Pima Indians of the Gila River Indian Reservation of Arizona in relation to the geochemical environments, in *Trace Subst. Environ. Health,* Vol. 7, Hemphill, D. D., Ed., University of Missouri, Columbia, Mo., 1973, 57.

726. **Sikora, L. J., Chaney, R. L., Frankos, N. H., and Murray, Ch. M.,** Metal uptake by crops grown over entrenched sewage sludge, *J. Agric. Food Chem.,* 28, 1281, 1980.

727. **Sillanpää, M.,** Micronutrients and the Nutrient Status of Soils: A Global Study, Food and Agriculture Organization and United Nations, Rome, 1982, 444.

728. **Simon, E.,** Cadmium tolerance in populations of *Agrostis tenuis* and *Festuca ovina, Nature (London),* 265, 328, 1977.

729. **Singh, M.,** Other trace elements, in Abstr. 12th Int. Soil Sci. Congr., Part I, New Delhi, 1982, 412.

730. **Singh, S. and Singh, B.,** Trace element studies on some alkali and adjoining soils of Uttar-Pradesh, *J. Indian Soc. Soil Sci.,* 14, 19, 1966.

731. **Singh, M., Singh, N., and Bhandari, D. K.,** Interaction of selenium and sulfur on the growth and chemical composition of raya, *Soil Sci.,* 129, 238, 1980.

732. **Singh, B. R. and Steinnes, E.,** Uptake of trace elements by barley in zinc-polluted soils: lead, cadmium, mercury, selenium, arsenic, chromium and vanadium, in barley, *Soil Sci.,* 121, 38, 1976.

733. **Singh, B. B. and Tabatabai, M. A.,** Factors affecting rhodanese activity in soils, *Soil Sci.,* 125, 337, 1978.

734. **Sippola, J.,** Selenium content of soils and timothy in Finland, *Ann. Agric. Fenn.,* 18, 182, 1979.

735. **Skripnichenko, J. J. and Zolotaryeva, B. N.,** Supply of mercury to plants with increasing concentration of pollutants in nutritive medium, *Agrokhimiya,* 9, 110, 1980 (Ru).

736. **Smeltzer, G. G., Langille, W. M., and MacLean, K. S.,** Effects of some trace elements on grass legume production in Nova Scotia, *Can. J. Plant Sci.,* 42, 46, 1962.

737. **Smeyers-Verbeke, J., de Graeve, M., François, M., de Jaegere, R., and Massart, D. L.,** Cd uptake by intact wheat plants, *Plant Cell Environ.,* 1, 291, 1978.

738. **Smilde, K. W., Koukoulakis, P., and van Luit, B.,** Crop response to phosphate and lime on acid sandy soils high in zinc, *Plant Soil,* 41, 445, 1974.

739. **Smith, I. C. and Carson, B. L.,** *Trace Metals in the Environment,* Vol. 2, Ann Arbor Scientific Publications, Ann Arbor, Mich., 1977a, 469.

740. **Smith, I. C. and Carson, B. L.,** *Trace Metals in the Environment,* Vol. 1, Ann Arbor Scientific Publications, Ann Arbor, Mich., 1977b, 394.

741. **Smith, I. C. and Carson, B. L.,** *Trace Metals in the Environment,* Vol. 3, Ann Arbor, Mich., 1978, 405.

742. **Smith, I. C., Carson, B. L., and Hoffmeister, F.,** *Trace Metals in the Environment,* Vol. 5, Ann Arbor Scientific Publications, Ann Arbor, Mich., 1978a, 552.

743. **Smith, I. G., Carson, B. L., and Ferguson, T. L.,** *Trace Metals in the Environment,* Vol. 4, Ann Arbor Scientific Publications, Ann Arbor, Mich., 1978b, 193.

744. **Smith, I. C. and Carson, B. L.,** *Trace Metals in the Environment,* Vol. 6, Ann Arbor Scientific Publications, Ann Arbor, Mich., 1981, 1202.

745. **Soltanpour, P. N. and Workman, S. M.,** Use of NH_4HCO_3 + DTPA soil test to assess availability and toxicity of selenium to alfalfa plants, *Commun. Soil Sci. Plant Anal.,* 11, 1147, 1980.

746. **Somers, E.,** Fungitoxicity of metal ions, *Nature (London),* Suppl., 7, 184, 475, 1959.

747. **Soon, Y. K.,** Solubility and sorption of cadmium in soils amended with sewage sludge, *J. Soil Sci.,* 32, 85, 1981.

748. **Sorterberg, A.,** The effect of some heavy metals on oats in pot experiments with three different soil types, in *Geomedical Aspects in Present and Future Research,* Låg, J., Ed., Universitetsforlaget, Oslo, 1980, 209.

749. **Souty, N., Guennelon, R., and Rode, C.,** Quelgues observations sur l'absorption du potassium, du rubidium-86 et du césium-137 par des plantes cultivées sur solutions nutritives, *Ann. Agron.,* 26, 58, 1975.

750. **Söremark, R.,** Vanadium in some biological specimens, *J. Nutr.,* 92, 183, 1967.

751. **Staerk, H. and Suess, A.,** Bromine content of vegetables and its accumulation after soil fumigation with methyl bromide, using neutron activation analysis, in *Comparative Studies of Food and Environ. Contam.,* IAEA, Vienna, 1974, 417.

752. **Stanchev, L., Gyurav, G., and Mashev, N.,** Cobalt as a trace element in Bulgarian soils *Izv. Centr. Nauch. Inst. Pochvozn. Agrotekh. ''Pushkarov'',* 4, 145, 1962 (Bu).

753. **Stärk, H., Suss, A., and Trojan, K.,** Das Vorkommen von Brom in verschiedenen Gemüsepflanzen, *Landwirtsch. Forsch.,* 24, 193, 1971.

754. **Steinnes, E.,** Regional distribution of arsenic, selenium and antimony in humus layers of Norwegian soils, in *Geomedical Aspects in Present and Future Research,* Låg, J., Ed., Universitetsforlaget, Oslo, 1980, 217.

754a. **Stegnar, P., Kosta, I., Byrne, A. R., and Ravnik, V.,** The accumulation of mercury by, and the occurrence of methyl mercury in, some fungi, *Chemosphere,* 2, 57, 1973.

755. **Stekar, J. and Pen, A.,** Sodium, zinc and manganese contents in feed from grassland, *Agrokhemiya,* 1, 7, 1980 (Sh).

755a. **Stenström, T. and Vahter, M.,** Cadmium and lead in Swedish commercial fertilizers, *Ambio,* 3, 91, 1974.

756. **Stepanova, M. D.,** *Microelements in Organic Matter of Soils,* Izd. Nauka, Novosibirsk, 1976, 105 (Ru).

757. **Stevenson, F. J. and Ardakani, M. S.,** Organic matter reactions involving micronutrients in soils, in *Micronutrients in Agriculture,* Mortvedt, J. J., Giordano, P. M., and Lindsay, W. L., Eds., Soil Science Society of America, Madison, Wis., 1972, 79.

758. **Stevenson, F. J. and Fitch, A.,** Reactions with organic matter, in *Copper in Soils and Plants,* Loneragan, J. F., Robson, A. D., and Graham, R. D., Eds., Academic Press, New York, 1981, 69.

759. **Stevenson, F. J. and Welch, L. F.,** Migration of applied lead in a field soil, *Environ. Sci. Technol.,* 13, 1255, 1979.

760. **Stiepanian, M. S.,** Content of total and water-soluble iodine in soils of Armenia, *Agrokhimiya,* 10, 138, 1976 (Ru).

761. **Street, J. J., Lindsay, W. L., and Sabey, B. R.,** Solubility and plant uptake of cadmium in soils amended with cadmium and sewage sludge, *J. Environ. Qual.,* 6, 72, 1977.

762. **Strzelec, A. and Koths, J. S.,** Effects of Mn, Zn, and Cd on processes of ammonification and nitrification in soils, in *Proc. Conf. No. 2, Effects of Trace Element Pollution and Sulfur on Agric. Environ. Quality,* Kabata-Pendias, A., Ed., IUNG, Pulawy, Poland, 1980, 9.

763. **Stuanes, A.,** Adsorption of Mn^{2+}, Zn^{2+}, Cd^{2+}, and Hg^{2+} from binary solutions by mineral material, *Acta Agric. Scand.,* 26, 243, 1976.

764. **Sunderman, F. W.,** Chelation therapy in nickel poisoning, in *Proc. Nickel Symp.,* Anke, M., Schneider, H. J., and Brückner, Chr., Eds., Friedrich-Schiller University, Jena, 1980, 359.

765. **Susaki, H., Ishida, N., and Kawashima, R.,** Selenium concentrations in Japanese fodder, *Jpn. J. Zootechnol. Sci.,* 51, 806, 1980.

766. **Suttie, J. W.,** Effects of fluoride on livestock, *J. Occup. Med.,* 19, 40, 1977.

767. **Suzuki, Y., Suzuki, Y., Fujii, N., and Mouri, T.,** Environmental contamination around a smelter by arsenic, *Shikoku Igaku Zasshi,* 30, 213, 1974 (Ja).

768. **Szalonek, I. and Warteresiewicz, M.,** Yields of cultivated plants in the vicinity of an aluminum works, *Arch. Ochr. Srodowiska,* 3, 63, 1979 (Po).

769. **Szentmihalyi, S., Regius, A., Anke, M., Grün, M., Groppel, B., Lokay, D., and Pavel, J.,** The nickel supply of ruminants in the GDR, Hungary and Czechoslovakia dependent on the origin of the basic material for the formation of soil, in *Proc. Nickel Symp.,* Anke, M., Schneider, H. J., and Brückner, Chr., Eds., Friedrich-Shiller University, Jena, E. Germany, 1980, 229.

770. **Takahashi, T., Kushizaki, M., and Ogata, T.,** Mineral composition of Japanese grassland under heavy use of fertilizers, in Proc. Int. Sem. SEFMIA, Tokyo, 1977, 118.

771. **Takamatsu, I. and Yoshida, T.,** Determination of stability constants of metal-humic complexes by potentiometric titration and ion-selective electrodes, *Soil Sci.,* 125, 377, 1978.

772. **Takkar, P. N.,** Micronutrients: forms, content, distribution in profile, indices of availability and soil test methods, in Abstr., 12th Int. Soil Sci. Congr., Part 1, New Delhi, 1982, 361.

773. **Taskayev, A. J., Ovchenkov, V. J., Aleksahkin, R. M., and Shuktomova, J. J.,** Forms of Ra-226 in soil horizons with its high concentrations, *Pochvovedenie,* 2, 18, 1978 (Ru).

774. **Tchuldziyan, H. and Khinov, G.,** On the chemistry of copper pollution of certain soils, *Pochvozn. Agrokhim.,* 11, 41, 1976 (Bu).

775. **Temple, P. J. and Bisessar, S.,** Uptake and toxicity of nickel and other metals in crops grown on soil contaminated by a nickel refinery, paper presented at Int. Symp. Trace Element Stress in Plants, Los Angeles, November 6, 1979, 43.

776. **Temple, P. J. and Linzon, S. N.,** Contamination of vegetation, soil, snow and garden crops by atmospheric deposition of mercury from a chlor-alkali plant, in *Trace Subst. Environ. Health,* Vol. 11, Hemphill, D. D., Ed., University of Missouri, Columbia, Mo., 1977, 389.

777. **Temple, P. J., Linzon, S. N., and Chai, B. L.,** Contamination of vegetation and soil by arsenic emissions from secondary lead smelters, *Environ. Pollut.,* 12, 311, 1977.

778. **Terry, N.,** Physiology of trace element toxicity and its relation to iron stress, paper presented at Int. Symp. Trace Element Stress in Plants, Los Angeles, November 6, 1979, 50.

779. **Theng, B. K. G. and Scharpenseel, H. W.,** The adsorption of ^{14}C-labelled humic acid by montmorillonite, in Proc. Int. Clay Conf., Mexico City, 1975, 649.

780. **Thomas, W.,** Monitoring organic and inorganic trace substances by epiphytic mosses — a regional pattern of air pollution, in *Trace Subst. Environ. Health,* Vol. 13, Hemphill, D. D., Ed., University of Missouri, Columbia, Mo., 1979, 285.

781. **Thomas, J., Jr., Glass, H. D., White, W. A., and Trandel, R. M.,** Fluoride content of clay minerals and argillaceous earth materials, *Clays Clay Miner.,* 25, 278, 1977.

782. **Thompson, L. K., Sidhu, S. S., and Roberts, B. A.,** Fluoride accumulation in soil and vegetation in the vicinity of a phosphorus plant, *Environ. Pollut.,* 18, 221, 1979.

783. **Thoresby, P. and Thornton, I.,** Heavy metals and arsenic in soil, pasture herbage and barley in some mineralised areas in Britain, in *Trace Subst. Environ. Health,* Vol. 13, Hemphill, D. D., Ed., University of Missouri, Columbia, Mo., 1979, 93.

784. **Thornton, I.,** Biogeochemical studies on molybdenum in the United Kingdom, in *Molydenum in the Environmental,* Chappell, W. R. and Peterson, K. K., Eds., Marcel Dekker, New York, 1977, 341.

785. **Thornton, I. and Plant, J.,** Rgional geochemical mapping and health in the United Kingdom, *J. Geol. Soc.,* 137, 575, 1980.

786. **Thornton, I. and Webb, J. S.,** Trace elements in soils and surface waters contaminated by past metalliferous mining in parts of England, in *Trace Subst. Environ. Health,* Vol. 9, Hemphill, D. D., Ed., University of Missouri, Columbia, Mo., 1975, 77.

786a. **Thornton, I. and Webb, J. S.,** Aspects of geochemistry and health in the United Kingdom, in *Origin and Distribution of the Elements,* Vol. 11, Ahrens, L. H., Ed., Pergamon Press, 1979, 791.

787. **Tidball, R. R.,** Lead in soils, in *Lead in the Environment,* Lovering, T. G., Ed., *U.S. Geol. Surv. Prof. Pap.,* 957, 43, 1976.

788. **Tiffin, L. O.,** Translocation of micronutrients in plants, in *Micronutrients in Agriculture,* Mortvedt, J. J., Giordano, P. M., and Lindsay, W. L., Eds., Soil Science Society of America, Madison, Wis., 1972, 199.

789. **Tiffin, L. O.,** The form and distribution of metals in plants: an overview, in Proc. Hanford Life Sciences Symp. U.S. Department of Energy, Symposium Series, Washington, D.C., 1977, 315.

790. **Tikhomirov, F. A., Rerikh, V. I., and Zyrin, N. G.,** Accumulation by plants of natural and applied cobalt and zinc, *Agrokhimiya,* 6, 96, 1979 (Ru).

791. **Tikhomirov, F. A., Sandzharova, I. I., and Smirnov, J. G.,** Accumulation of Sr-90 in grasses grown in meadow and forest, *Lesovedenie,* 5, 78, 1976 (Ru).

792. **Tiller, K. G.,** Weathering and soil formation on the dolerite in Tasmania, with particular reference to several trace elements, *Aust. J. Soil Res.,* 1, 74, 1963.

793. **Tiller, K. G.,** The interaction of some heavy metal cations and silicic acid at low concentrations in the presence of clays, in Trans. 9th Int. Soil Sci. Congr., Part 2, Adelaide, 1968, 567.

793a. **Tiller, K. G.,** Environmental pollution of the Port Pirie region, paper presented at 1st Australian Workshop on Environmental Studies, Cowes, Phillip Island, October 24, 1977, 194.

794. **Tiller, K. G.,** The availability of micronutrients in paddy soils and its assessment by soil analysis including radioisotopic techniques, in *Proc. Symp. on Paddy Soil,* Science Press, Beijing and Springer-Verlag, Berlin, 1981, 273.

795. **Tiller, K. G. and Merry, R. H.,** Copper pollution of agricultural soils, in *Copper in Soils and Plants,* Loneragan, J. F., Robson, A. D., and Graham, R. D., Eds., Academic Press, New York, 1981, 119.

796. **Tiller, K. G., Nayyar, V. K., and Clayton, P. M.,** Specific and non-specific sorption of cadmium by soil clays as influenced by zinc and calcium, *Aust. J. Soil Res.,* 17, 17, 1979.

797. **Tiller, K. G., Suwadji, E., and Beckwith, R. S.,** An approach to soil testing with special reference to the zinc requirement of rice paddy soils, *Commun. Soil Sci. Plant Anal.,* 10, 703, 1979.

798. **Tinker, P. B.,** Levels, distribution and chemical forms of trace elements in food plants, *Philos. Trans. R. Soc. London,* 294b, 41, 1981.

799. **Tipping, E.,** The adsorption of aquatic humic substances by iron oxides, *Geochim. Cosmochim. Acta,* 45, 191, 1981.

800. **Tiutina, N. A., Aleskovskiy, V. B., and Vasiliyev, V. B.,** An experiment in biogeochemical prospecting and the method of determination of niobium in plants, *Geokhimiya,* 6, 550, 1959 (Ru).

801. **Tjell, J. Ch. and Hovmand, M. F.,** Metal concentrations in Danish arable soils, *Acta Agric. Scand.,* 28, 81, 1972.

802. **Tjell, J. Ch., Hovmand, M. F., and Mosbaek, H.,** Atmospheric lead pollution of grass grown in a background area in Denmark, *Nature (London),* 280, 425, 1979.

803. **Tölgyesi, G. and Kárpáti, I.,** Some regularities in the nutrient content of the meadow vegetation along the river Zala, *Agrokem. Talajtan,* 26, 63, 1977 (Hu).

804. **Tölgyesi, G. and Kozma, A.,** Investigation on factors affecting boron uptake by grasses, *Agrokem. Talajtan,* 23, 83, 1974 (Hu).

805. **Treshow, M.,** Fluorides as air pollutants affecting plants, *Annu. Rev. Phytopath.,* 9, 21, 1971.

806. **Treyman, A. A.,** Copper and manganese in soils, plants and waters of Salair Mountain-Ridge and Prisalair Plain, in *Copper, Manganese and Boron in Landscapes of Baraba Depression and the Novosibirsk Region,* Ilyin, W. B., Ed., Izd. Nauka, Novosibirsk, 1971, 55 (Ru).

807. **Trolldenier, G.,** Influence of plant nutrition on the microbial activity in the rhizosphere, paper presented at 16th Coll. Agric. Yield Potentials in Continental Climates, Warsaw, June 22, 1981, 127.

808. **Troyer, L., Olson, B. H., Hill, B. C., Thornton, I., and Matthews, H.,** Assessment of metal availability in soil through the evaluation of bacterial metal resistance, in *Trace Subst. Environ. Health,* Vol. 14, Hemphill, D. D., Ed., University of Missouri, Columbia, Mo., 1980, 129.

809. **Trudinger, P. A. and Swaine, D. J., Eds.,** *Biogeochemical Cycling of Mineral-Forming Elements,* Elsevier, Amsterdam, 1979, 612.

810. **Turner, M. A. and Rust, R. H.,** Effect of chromium on growth and mineral nutrition of soybeans, *Soil Sci. Soc. Am. Proc.,* 35, 755, 1971.

811. **Tyler, G.,** Heavy metals in the terrestrial environment, in *Proc. Nordic Symp. Biological Parameters for Measuring Global Pollution,* IBPJ Norden, 1972, 99.

812. **Tyler, G.,** Effect of Heavy Metal Pollution on Decomposition in Forest Soil, SNV/PM, Lund University, Lund, Sweden, 1975, 47.

813. **Tyler, G.,** Heavy metal pollution, phosphatase activity and mineralization of organic phosphorus in forest soil, *Soil Biol. Biochem.,* 8, 327, 1976a.

814. **Tyler, G.,** Influence of vanadium on soil phosphatase activity, *J. Environ. Qual.,* 5, 216, 1976b.

815. **Tyler, G.,** Leaching rates of heavy metal ions in forest soil, *Water Air Soil Pollut.,* 9, 137, 1978.

816. **Tyler, G.,** Leaching of metals from the A-horizon of a spruce forest soil, *Water Air Soil Pollut.,* 15, 353, 1981.

817. **Udo, E. J., Ogunwale, J. A., and Fagbami, A. A.,** The profile distribution of total and extractable copper in selected Nigerian soils, *Commun. Soil Sci. Plant Anal.,* 10, 1385, 1979.

818. **Ure, A. M. and Bacon, J. R.,** Comprehensive analysis of soils and rocks by spark-source mass spectrometry, *Analyst,* 103, 807, 1978.

819. **Ure, A. M., Bacon, J. R., Berrow, M. L., and Watt, J. J.,** The total trace element content of some Scottish soils by spark source mass spectrometry, *Geoderma,* 22, 1, 1979.

820. **Van Dijk, H.,** Cation binding of humic acids, *Geoderma,* 5, 53, 1971.

820a. **Van Goor, B. J.,** Distribution of mineral nutrients in the plant in relation to physiological disorder, paper presented at 19th Int. Horticultural Congr., Warsaw, September 11, 1974, 217.

821. **Van Goor, B. J. and Wiersma, D.,** Redistribution of potassium, calcium, magnesium and manganese in the plant, *Physiol. Plant.,* 31, 163, 1974.

822. **Van Goor, B. J. and Wiersma, D.,** Chemical form of manganese and zinc in phloem exudates, *Physiol. Plant.,* 36, 213, 1976.

823. **Van Hook, R. I., Harris, W. F., and Henderson, G. S.,** Cadmium, lead, and zinc distribution and cycling in a mixed deciduous forest, *Ambio,* 6, 281, 1977.

824. **Vesper, S. J. and Weidensaul, T. C.,** Effects of cadmium, nickel, copper, and zinc on nitrogen fixation by soybeans, *Water, Air Soil Pollut.,* 9, 413, 1978.

825. **Vlasov, N. A. and Mikhaylova, A. I.,** Interaction coal-humic acid fractions with some metallic cations and the effect of coal-humic fertilizers on the distribution of micronutrients in soil and plants, in Trans. 6th Int. Symp. Humus et Planta, Praha, 1975, 131 (Ru).

826. **Vlasyuk, P. A.,** Current research on microelements and the perspective for a further development in USSR and MSSR, in *Microelements in Environment,* Vlasyuk, P. A., Ed., Naukova Dumka, Kiyev, 1980, 5 (Ru).

827. **Vlek, P. L. G. and Lindsay, W. L.,** Thermodynamic stability and solubility of molybdenum minerals in salts, *J. Soil Sci. Soc. Am.,* 41, 42, 1977.

828. **Vochten, R. C. and Geyes, J. G.,** Pyrite and calcite in septarian concretions from the Rupelian clay at Rumst (Belgium) and their geochemical composition, *Chem. Geol.,* 14, 123, 1974.

829. **Vogt, P. and Jaakola, A.,** The effect of mineral elements added to Finnish soils on the mineral contents of cereal, potato, and hay crops, *Acta Agric. Scand. Suppl.,* 20, 69, 1978.

830. **Vought, R. L., Brown, F. A., and London, W. T.,** Iodine in the environment, *Arch. Environ. Health,* 20, 516, 1970.

831. **de Vries, M. P. C. and Tiller, K. G.,** Sewage sludges as a soil amendment, with special reference to Cd, Cu, Mn, Ni, Pb and Zn — comparison of results from experiments conducted inside and outside a glasshouse, *Environ. Pollut.,* 16, 231, 1978.

832. **Vyas, B. N. and Mistry, K. B.,** Influence of clay mineral type and organic matter content on the uptake of Pu-239 and Am-241 by plants, *Plant Soil,* 59, 75, 1981.

833. **Wada, K.,** Allophane and imogolite, in *Minerals in Soil Environments,* Dixon, J. B. and Weed, S. B., Eds., Soil Science Society of America, Madison, Wis., 1977, 603.

834. **Wada, K. and Abd-Elfattah, A.,** Characterization of zinc adsorption sites in two mineral soils, *Soil Sci. Plant Nutr.,* 24, 417, 1978.

835. **Wada, K. and Kakuto, Y.,** Selective adsorption of zinc on halloysite, *Clays Clay Miner.,* 28, 321, 1980.

836. **Wada, K., Seirayosakol, A., Kimura, M., and Takai, Y.,** The process of manganese deposition, *Soil Sci. Plant Nutr.,* 24, 319, 1978.

837. **Wagner, G. H., Konig, R. H., Vogelpohl, S., and Jones, M. D.,** Base metals and other minor elements in the manganese deposits of West-Central Arkanas, *Chem. Geol.,* 27, 309, 1979.

838. **Walczyna, J., Sapek, A., and Kuczyńska, I.,** Content of mineral elements in more important species of grasses and other meadow plants from peat soils, *Z. Probl. Post. Nauk Roln.,* 175, 49, 1975 (Po).

839. **Wallace, A.,** Monovalent-ion carrier effects on transport of Rb-86 and Cs-137 into bush bean plants, *Plant Soil,* 32, 526, 1970.

840. **Wallace, A.,** *Regulation of the Micronutrient Status of Plants by Chelating Agents and Other Factors,* Wallace, A., Ed., Los Angeles, 1971, 309.

841. **Wallace, A., Alexander, G. V., and Chaudhry, F. M.,** Phytotoxicity of cobalt, vanadium, titanium, silver and chromium, *Commun. Soil Sci. Plant Anal.,* 8, 751, 1977a.

842. **Wallace, A. and Romney, E. M.,** Some interactions of Ca, Sr and Ba in plants, *Agron. J.,* 3, 245, 1971.

843. **Wallace, A., Romney, E. M., and Cha, J. W.,** Nickel-iron interaction in bush beans, *Commun. Soil Sci. Plant Anal.,* 8, 787, 1977b.

844. **Wallace, A., Romney, E. M., Cha, J. W., Soufi, S. M., and Chaudhry, F. M.,** Nickel phytotoxicity in relationship to soil pH manipulation and chelating agents, *Commun. Soil Sci. Plant Anal.,* 8, 757, 1977c.

844a. **Wallace, A., Romney, E. M., Cha, J. W., and Chaundhry, F. M.,** Lithium toxicity in plants, *Commun. Soil Sci. Plant Anal.,* 8, 773, 1977d.

845. **Wallace, A., Mueller, R. T., and Wood, R. A.,** Arsenic phytotoxicity and interactions in bush bean plants grown in solution culture, *J. Plant Nutr.,* 2, 111, 1980a.

846. **Wallace, A., Romney, E. M., and Alexander, G. V.,** Zinc-cadmium interactions on the availability of each to bush bean plants grown in solution culture, *J. Plant Nutr.,* 2, 51, 1980b.

847. **Walsh, L. M. and Beaton, J. D., Eds.,** *Soil Testing and Plant Analysis,* Soil Science Society of America, Madison, Wis., 1973, 491.

848. **Warren, H. V.,** Environmental geochemistry in Canada — a challenge, paper presented at Int. Conf. on Heavy Metals in the Environment, Toronto, October 27, 1975, 55.

849. **Warren, H. V.,** Biogeochemical prospecting for lead, in *The Biogeochemistry of Lead in the Environment,* Nriagu, J. O., Ed., Elsevier, Amsterdam, 1978, 395.

850. **Warren, H. V. and Delavault, R. E.,** Mercury content of some British soils, *Oikos,* 20, 537, 1969.

851. **Warren, H. V., Delavault, R. E., and Barakso, J.,** The arsenic content of Douglas Fir as a guide to some gold, silver and base metal deposits, *Can. Mining Metall. Bull.,* 7, 1, 1968.

852. **Warren, H. V., Delavault, R. E., Fletcher, K., and Wilks, E.,** Variation in the copper, zinc, lead and molybdenum content of some British Columbia vegetables, in *Trace Subst. Environ. Health,* Vol. 4, Hemphill, D. D., Ed., University of Missouri, Columbia, Mo., 1970, 94.

853. **Wasserman, R. H., Romney, E. M., Skougstad, M. W., and Siever, R.,** Strontium, in *Geochemistry and the Environment,* Vol. 2, Wasserman, R. H., Ed., N.A.S., Washington, D.C., 1977, 73.

854. **Wazhenin, I. G. and Bolshakov, W. A.,** All-Union Conference of the joint departments on "Methodical Principles of Mapping Soil Contamination with Heavy Metals and Methods of their Determination", *Pochvovedenie,* 2, 151, 1978 (Ru).

855. **Wedepohl, K. H., Ed.,** *Handbook of Geochemistry,* Springer-Verlag, Berlin, 1969-1974 (several volumes).

856. **Weinberg, E. D., Ed.,** *Microorganisms and Minerals,* Marcel Dekker, New York, 1977, 492.

857. **Weinstein, L. H.,** Fluoride and plant life, *J. Occup. Med.,* 19, 49, 1977.

858. **Welch, R. M.,** Vanadium uptake by plants, *Plant Physiol.,* 51, 828, 1973.

859. **Welch, R. M.,** The biological significance of nickel, paper presented at Int. Symp. Trace Element Stress in Plants, Los Angeles, November 6, 1979, 36.

860. **Welch, R. M. and Cary, E. E.,** Concentration of chromium, nickel, and vanadium in plant materials, *J. Agric. Food Chem.,* 23, 479, 1975.

861. **Wells, N.,** Total elements in top soils from igneous rocks: an extension of geochemistry, *J. Soil Sci.,* 11, 409, 1960.

862. **Wells, N.,** Selenium in horizons of soil profiles, *N.Z. J. Sci.,* 10, 142, 1967.

863. **Wells, N.,** Element composition of soils and plants, in *Soils of New Zealand,* Vol. 2, New Zealand Soil Bureau, Wellington, 1968, 115.

864. **Wells, N. and Whitton, J. S.,** A pedochemical survey. I. Lithium, *N.Z. J. Sci.,* 15, 90, 1972.

865. **Wells, N. and Whitton, J. S.,** A pedochemical survey. III. Boron, *N.Z. J. Sci.,* 20, 317, 1977.

866. **Whitby, L. M., Stokes, P. M., Hutchinson, T. C., and Myslik, G.,** Ecological consequence of acidic and heavy metal discharge from the Sudbury smelters, *Can. Miner.,* 14, 47, 1976.

867. **White, C. L., Robson, A. D., and Fisher, H. M.,** Variation in nitrogen, sulfur, selenium, cobalt, manganese, copper and zinc contents of grain from wheat and two lupin species grown in a range of Mediterranean environments, *Aust. J. Agric. Res.,* 32, 47, 1981.

867a. **Whitehead, D. C.,** Nutrient minerals in grass-land herbage, *Pastures Field Crops Publ.,* 1, 83, 1966.

868. **Whitehead, D. C.,** The sorption of iodide by soil components, *J. Sci. Food Agric.,* 25, 73, 1974.

869. **Whitehead, D. C.,** Uptake by perennial ryegrass of iodine, elemental iodine and iodate added to soil as influenced by various amendments, *J. Sci. Food Agric.,* 26, 361, 1975.

870. **Whitton, J. S. and Wells, N.,** A pedochemical survey. II. Zinc, *N.Z. J. Sci.,* 17, 351, 1974.

871. **Widera, S.,** Contamination of the soil and assimilative organs of the pine tree at various distances from the source of emission, *Arch. Ochrony Srodowiska,* 3, 147, 1980 (Po).

872. **Wiersma, D. and Van Goor, B. J.,** Chemical forms of nickel and cobalt in phloem of *Ricinus communis, Physiol. Plant.,* 45, 440, 1979.

873. **Wildung, R. E. and Garland, T. R.,** The relationship of microbial processes to the fate and behavior of transuranic elements in soils and plants, in *The Transuranic Elements in the Environment,* Hanson, W. C., Ed., ERDA Publ. Ser. TIC-22800, NTIS, Springfield, Va., 1980, 300.

874. **Wilkins, C.,** The distribution of lead in soils and herbage of West Pembrokeshire, *Environ. Pollut.,* 15, 33, 1978a.

875. **Wilkins, C.,** The distribution of Br in the soils and herbage of North-West Pembrokeshire, *J. Agric. Sci. Camb.,* 90, 109, 1978b.

876. **Wilkins, C.,** The distribution of Mn, Fe, Cu, and Zn in topsoils and herbage of North-West Pembrokeshire, *J. Agric. Sci. Camb.,* 92, 61, 1979.

877. **Williams, S. T., McNeilly, T., and Wellington, E. M. H.,** The decomposition of vegetation growing on metal mine waste, *Soil Biol. Biochem.,* 9, 271, 1977.

878. **Williams, C. H. and David, D. J.,** The accumulation in soil of cadmium residues from phosphate fertilizers and their effect on the cadmium content of plants, *Soil Sci.,* 121, 86, 1976.

879. **Williams, C. H. and David, D. J.,** The effect of superphosphate on the cadmium content of soils and plants, *Aust. J. Soil Res.,* 11, 43, 1981.

880. **Williams, C. H. and Thornton, I.,** The effect of soil additives on the uptake of molybdenum and selenium from soils from different environments, *Plant Soil,* 36, 395, 1972.

881. **Wilson, M. J. and Berrow, M. L.,** The mineralogy and heavy metal content of some serpentinite soils, *Chem. Erde Bd.,* 37, 181, 1978.

882. **Wilson, D. O. and Cline, J. F.,** Removal of plutonium-239, tungsten-185, and lead-210 from soils, *Nature (London),* 209, 941, 1966.

883. **Wilson, D. O. and Reisenauer, H. M.,** Effects of some heavy metals on the cobalt nutrition of *Rhizobium meliloti, Plant Soil,* 32, 81, 1970.

884. **Woldendorp, J. W.,** Nutrients in the rihizosphere, paper presented at 16th Coll. on Agric. Yield Potentials in Continental Climates, Warsaw, June 22, 1981, 89.

885. **Wood, T. and Bormann, F. H.,** Increases in foliar leaching caused by acidification of an artificial mist, *Ambio,* 4, 169, 1975.

886. **Woolhouse, H. W. and Walker, S.,** The physiological basis of copper toxicity and copper tolerance in higher plants, in *Copper in Soils and Plants,* Loneragan, J. F., Robson, A. D., and Graham, R. D., Eds., Academic Press, New York, 1981, 235.

887. **Woolson, E. A., Axley, J. H., and Kearney, P. C.,** The chemistry and phytotoxicity of arsenic in soils, II. Effects of time and phosphorus, *Soil Sci. Soc. Am. Proc.,* 37, 254, 1973.

888. **Wright, T. L. and Fleischer, M.,** Geochemistry of the platinum metals, *Geol. Surv. Bull.,* 1214a, 1965.

889. **Woytowicz, B.,** Lead effect on the accumulation of nitrates in soil, *Rocz. Glebozn.,* 31, 309, 1980 (Po).

890. **Yamada, Y.,** Occurrence of bromine in plants and soil, *Talanta,* 15, 1135, 1968.

891. **Yamagata, N. and Shigematsu, I.,** Cadmium pollution in perspective, *Bull. Inst. Public Health,* 19, 1, 1970.

892. **Yamasaki, S., Yoshino, A., and Kishita, A.,** The determination of sub-microgram amounts of elements in soil solution by flameless atomic absorption spectrophotometry with a heated graphite atomizer, *Soil Sci. Plant Nutr.,* 21, 63, 1975.

893. **Yeaple, D. S.,** Mercury in *Bryophytes, Nature, (London),* 235, 229, 1972.

894. **Yen, T. F.,** Vanadium chelates in recent and ancient sediments, in *Trace Subst. Environ. Health,* Vol. 6, Hemphill, D. D., Ed., University of Missouri, Columbia, Mo., 1972, 347.

895. **Yläranta, T., Jansson, H., and Sippola, J.,** Seasonal variation in micronutrient contents of wheat, *Ann. Agric. Fenn.,* 18, 218, 1979.
896. **Yudintseva, E. V., Mamontova, L. A., and Levina, E. M.,** Accumulation of radiocesium in the yield of plants depending on the application of lime, peat and peast ashes, *Dokl. Vses. Akad. Skh. Nauk.,* 8, 27, 1979 (Ru).
897. **Yuita, K., Shibuya, M., and Nozaki, T.,** The accumulation of bromine and iodine in Japanese soils, in Abstr. 11th Int. Soil Sci. Congr., Edmonton, Alberta, 1978, 260.
898. **Zajic, J. E.,** *Microbial Biogeochemistry,* Academic Press, New York, 1969, 345.
899. **Zakharov, E. P. and Zakharova, G. R.,** On geochemical ecology and teratology of plants grown in cobalt and cobalt-copper mineralized area of Central Tuva, in *Biologicheskaya Rol Mikroelementov i ikh Primeneniye v Selskom Khozaystviye i Mediciniye,* Izd. Nauka, Moscow, 1974, 140 (Ru).
900. **Zborishchuk, J. N. and Zyrin, N. G.,** Copper and zinc in the ploughed layer of soils of the European USSR, *Pochvovedenie,* 1, 31, 1978 (Ru).
901. **Zelazny, L. W. and Calhoun, F. G.,** Palygorskite (attapulgite), sepiolite, talc, pyrophyllite, and zeolites, in *Minerals in Soil Environments,* Dixon, J. B. and Weed, S. B., Eds., Soil Science Society of America, Madison, Wis., 1977, 435.
902. **Zimdahl, R. L.,** Entry and movement in vegetation of lead derived from air and soil sources, paper presented at 68th Annu. Meeting of the Air Pollution Control Association, Boston, Mass., June 15, 1975, 2.
903. **Zimdahl, R. L. and Hassett, J. J.,** Lead in soil, in *lead in the Environment,* Bogges, W. R. and Wixson, B. G., Eds., Report NSF, National Science Foundation, Washington, D.C., 1977, 93.
904. **Zimdahl, R. L. and Koeppe, D. E.,** Uptake by plants, in *Lead in the Environment,* Boggess, W. R. and Wixson, B. G., Eds., Report NSF, National Science Foundation, Washington, D.C., 1977, 99.
905. **Żmijewska, W. and Minczewski, J.,** A study on the reproducibility of obtaining soil solution, *Rocz. Nauk Roln.,* 95a, 239, 1969.
906. **Zook, E. G., Greene, F. E., and Morris, E. R.,** Nutrient composition of selected wheats and wheat products, *Cereal Chem.,* 47, 72, 1970.
907. **Zöttl, H. W., Stahr, K., and Hädrich, F.,** Umsatz von Spurenelementen in der Bärhalde und ihren Okosystemen, *Mitt. Dtsch. Bodenkundl. Ges.,* 29, 569, 1979.
908. **Zunino, H., Aquilera, M., Caiozzi, M., Peirano, P., Borie, F., and Martin, J. P.,** Metal binding organic macromolecules in soil, *Soil Sci.,* 128, 257, 1979.
909. **Zýka, V.,** Thallium in plants from Alsar, *Sb. Geol. Ved. Technol. Geochem.,* 10, 91, 1972.
910. **Zyrin, N. G. and Malinina, M. S.,** Mobile molybdenum in soils of West Georgia and molybdenum content of tangerine tree and tea plant, *Agrokhimiya,* 3, 113, 1978 (Ru).
911. **Zyrin, N. G., Rerich, W. J., and Tikhomirov, F. A.,** Forms of zinc compounds in soils and its supply to plants, *Agrokhimiya,* 5, 124, 1976 (Ru).
912. **Zyrin, N. G. and Zborishchuk, J. N.,** Boron in the ploughed layer of soils of the European part of the USSR, *Pochvovedenie,* 5, 44, 1975a (Ru).
913. **Zyrin, N. G. and Zborishchuk, J. N.,** Content of iodine in the ploughed layer of the USSR soils, *Pochvovedenie,* 9, 49, 1975b (Ru).
914. **Zyrin, N. G. and Zborishchuk, J. N.,** Molybdenum in the ploughed layer of soils of the European part of the USSR., *Agrokhimiya,* 8, 105, 1976 (Ru).
915. Authors' unpublished data.
916. **Curtin, G. C. and King, H. D.,** unpublished data.
917. **Vandeputte, M.,** personal communication.

INDEX

H